T0134953

Engineering Cyber-Physical Systems and Critical Infrastructures

Volume 4

Series Editor

Fatos Xhafa⬛, Departament de Ciències de la Computació, Technical University of Catalonia, Barcelona, Spain

The aim of this book series is to present state of the art studies, research and best engineering practices, real-world applications and real-world case studies for the risks, security, and reliability of critical infrastructure systems and Cyber-Physical Systems. Volumes of this book series will cover modelling, analysis, frameworks, digital twin simulations of risks, failures and vulnerabilities of cyber critical infrastructures as well as will provide ICT approaches to ensure protection and avoid disruption of vital fields such as economy, utility supplies networks, telecommunications, transports, etc. in the everyday life of citizens. The intertwine of cyber and real nature of critical infrastructures will be analyzed and challenges of risks, security, and reliability of critical infrastructure systems will be revealed. Computational intelligence provided by sensing and processing through the whole spectrum of Cloud-to-thing continuum technologies will be the basis for real-time detection of risks, threats, anomalies, etc. in cyber critical infrastructures and will prompt for human and automated protection actions. Finally, studies and recommendations to policy makers, managers, local and governmental administrations and global international organizations will be sought.

Ahmed A. Abd El-Latif · Yassine Maleh ·
Wojciech Mazurczyk · Mohammed ELAffendi ·
Mohamed I. Alkanhal
Editors

Advances in Cybersecurity, Cybercrimes, and Smart Emerging Technologies

 Springer

Editors
Ahmed A. Abd El-Latif ⓘD
EIAS Data Science Lab
College of Computer and Information
Sciences
Prince Sultan University
Riyadh, Saudi Arabia

Faculty of Science, Department
of Mathematics and Computer Science
Menoufia University
Shebin El-Koom, Egypt

Wojciech Mazurczyk
Institute of Computer Science
Warsaw University of Technology
Warsaw, Poland

Mohamed I. Alkanhal
Prince Sultan University
Riyadh, Saudi Arabia

Yassine Maleh
Sultan Moulay Slimane University
Khouribga, Morocco

Mohammed ELAffendi
EIAS Data Science Lab, College
of Computer and Information Sciences
Department of Computer Science
Prince Sultan University
Riyadh, Saudi Arabia

ISSN 2731-5002   ISSN 2731-5010 (electronic)
Engineering Cyber-Physical Systems and Critical Infrastructures
ISBN 978-3-031-21185-0   ISBN 978-3-031-21101-0 (eBook)
https://doi.org/10.1007/978-3-031-21101-0

This Springer imprint is published by the registered company Springer Nature Switzerland AG
The registered company address is: Gewerbestrasse 11, 6330 Cham, Switzerland

Committes

General Chairs

Prof. Ahmed A. Abd El-Latif, Prince Sultan University, Saudi Arabia
Prof. Yassine Maleh, SMIEEE, Sultan Moulay Slimane University, Khouribga, Morocco
Prof. Mohammed ELAffendi, Prince Sultan University, Saudi Arabia.

Honorary Committee

Prof. Ahmed S. Yamani, President of Prince Sultan University, Saudi Arabia
Prof. Mohamed I. Alkanhal, Prince Sultan University, Saudi Arabia
Prof. Dusit (Tao) Niyato, IEEE Fellow, Nanyang Technological University, Singapore
Prof. Wojciech Mazurczyk, Warsaw University of Technology (WUT), Poland.

Steering Committee

Prof. Vincenzo Piuri, University of Milan, Italy
Prof. Eleni (Helen) Karatza, Aristotle University of Thessaloniki, Greece.

Organizing Chair(s)

Prof. Ahmed A. Abd El-Latif, Prince Sultan University, Saudi Arabia
Prof. Yassine Maleh, SMIEEE, Sultan Moulay Slimane University, Khouribga, Morocco

Prof. Mohammed ELAffendi, Prince Sultan University, Saudi Arabia
Prof. Yasir Javed, Prince Sultan University, Saudi Arabia.

Program Chairs

Dr. Bassem Abd-El-Atty, Luxor University, Egypt
Dr. Suliman Mohamed Fati, Prince Sultan University
Dr. Peng jialiang, Heilongjiang University, China
Dr. Jiawen Kang, Nanyang Technological University, Singapore
Dr. Zehui Xiong, Singapore University of Technology and Design, Singapore
Dr. Loai A Tawalbeh, Department of Computing and Cyber Security at Texas A&M University-San Antonio, USA.

Publicity Chairs

Dr. Bassem Abd-El-Atty, Luxor University, Egypt
Dr. Mohamed Hammad, Menoufia University, Egypt
Dr. Loai A Tawalbeh, Department of Computing and Cyber Security at Texas A&M University-San Antonio, USA
Dr. Yassine Maleh, SMIEEE, National School of Applied Sciences, Khouribga, Morocco.

Publication Chairs

Prof. Ahmed A. Abd El-Latif, Prince Sultan University, Saudi Arabia
Prof. Yassine Maleh, SMIEEE, National School of Applied Sciences, Khouribga, Morocco.

Technical Program Committee

Xuehu Yan, National University of Defense Technology, China
Matthieu Perrin, Nantes University, France
Wei-Chiang Hong, Asia Eastern University of Science and Technology, Taiwan
Yakubu Makeri Ajiji, Kampala International University, Uganda
Dr. Mai Moodley Maiendra, Moodley and Associates, India
Benabdellah Mohammed, Mohammed First University, Oujda, Morocco

Patel Ankit R., University of Minho, Guimaraes, Portugal
Marium Malik The Superior University
Yasmine Harbi Universite Ferhat Abbas Setif1, Algeria
Abdultaofeek Abayomi, Durban University of Technology, South Africa
Giuseppe Ciaburro, Università della Campania Luigi Vanvitelli, Italy
Mani Zarei, Islamic Azad University, Tehran, Iran
Zakaria Sabir, Ibn Tofail University, Morocco
Mohammad Samadi Gharajeh, Polytechnic Institute of Porto, Portugal
Ghizlane Orhanou, Mohammed V University in Rabat, Morocco
Della Krachai Mohamed, University of Science and technologies of Oran, Algeria
Pedro Antonio Martin Cervantes, Universidad de Almeria, Spain
Yilun Shang Northumbria University, UK
Mounia Zaydi, University Hassan 1st, Morocco
Dr. Narina Thakur, Bhagwan Parshuram Institute of Technology, Delhi, India
Dalibor Dobrilovic, Technical Faculty "Mihajlo Pupin" Zrenjanin, Serbia
Khalid El Makkaoui, Mohammed First University, Nador, Morocco
Yassine Sadqi, University Sultan Moulay Slimane, Beni Mellal, Morocco
Ramgopal Kashyap, Amity University Chhattisgarh, India
Rathin Shit, International Institute of Information Technology, Indioa
Khalid El Gholami, University Sultan Moulay Slimane, Beni Mellal, Morocco
Pankaj Pal, RCC Institute of Information Technology, India
Badr Bentalha, ENCG Fez, Moroocco
Rhoulami Khadija, Faculty of science Rabat, Morocco
Tekouabou Koumetio Cédric Stéphane, Faculty of Sciences, El Jadida, Morocco
Sheikh Shah Mohammad Motiur Rahman, Daffodil International University, Bangladesh
Dilbag Singh, School of Computing and information technology, Manipal University Jaipur, India.
Ahmed Sedik, Kafrelsheikh University, Egypt
Said Fathy El-Zoghdy, Menoufia University, Egypt
Edmond S. L. Ho, Northumbria University, UK
Ibrahim A. Elgendy, Harbin Institute of Technology, China
Ahmed Ghoneim, King Saud University, Saudi Arabia
Samir Elmougy, Mansoura University, Egypt
Praveenkumar Padmapriya, SASTRA University, India
Manoranjan Mohanty, University of Technology, Sydney
Paweł Pławiak, Cracow University of Technology, Poland
Ali Asghar Heidari, School of Computing, National University of Singapore
Laith Mohammad Abualigah, Amman Arab University
Robertas Damasevicius, Silesian University of Technology
Kashif Hussain, Bahria University, Karachi, Pakistan
Brahim Lejdel, Eloued University
Reza Moghdani, Persian Gulf University
Ibrahim A. Elgendy, Computer Science and Technology, Harbin Institute of Technology, China

Ammar Muthanna, The Bonch-Bruevich Saint-Petersburg State University of Telecommunications, Russia.
Muhammad Ibrahim, Virtual University of Pakistan.
Reem I. Alkanhel, Princess Nourah bint Abdulrahman University, Riyadh, Saudi Arabia.
Faisal Jamil, Jeju National University, Jeju, South Korea
Amir Chaaf, Chongqing University of Posts and Telecommunications, Chongqing, China
Mohammed Saleh Ali Muthanna, Chongqing University of Posts and Telecommunications, China
Soha Alhelaly, Saudi Electronic University, Saudi Arabia
Khizar Abbas, Jeju National University, Korea
Ahsan Rafiq, Chongqing university of posts and telecommunications, China
Mehdhar Al-gaashani, Chongqing University of Posts and Telecommunications, China
Fengjun Shang, Chongqing University of Posts and Telecommunications, China
Mashael Khayyat, College of Computer Science and Engineering, University of Jeddah, Jeddah, Saudi Arabia
Rhoulami Khadija, Mohammed V University, Morocco
Roose Philippe, University of Pau, France
Sadqi Yassine, FP—Beni Mellal, Morocco
Samadi Gharajeh Mohammad, Polytechnic Institute of Porto, Portugal
Sarea Adel, Ahlia University, Bahrain
Sea Alex Denioux, Africa Fintech Network, Ivory Coast
Shaker Noha, Africa Fintech Network, Egypt
Shang Yilun, Northumbria University, UK
Shariar Houssain, Kennesaw State University, USA
Sheraz Anwar, Xiamen University, China
Sheta Alaa, Electronics Research Institute, Egypt
Shiu Hung-Jr, Tunghai University Taichung, Taiwan
Shojafar Mohammad, SMIEEE University of Surrey, UK
Siarry Patrick, University of Paris 12, France
Soule-Dupuy Chantal, U. Paul Sabatier, France
Souri Alireza, Halic University—Istanbul, Turkey
Su Chao-Ton, National Tsing Hua University, Taiwan
Tarbalouti Said, Cadi Ayyad University-Marrakech, Morocco
Tardif Pierre-Martin, UdeS, Canada
Tawalbeh Lo'ai A., SMIEEE—Texas A&M University—San Antonio, USA
Tekouabou Cédric Stéphane, Mohammed VI Polytechnic University, Morocco
Thakur Narina, Bharati Vidyapeeth College of Engineering New Delhi, India
Thaseen Sumaiya, VIT University, India
Tsai Sang-Bing, University of Electronic Science and Technology of China, China

Hong Wei-Chiang, Jiangsu Normal University, China
Weizhi Meng, Technical University of Denmark, Denmark
Yuan Xiaohong, North Carolina A&T State University, USA
Zahrane Tarik, Cadi Ayyad University Marrakech, Morocco
Zarei Mani, IAU of Shahr-e-Qods—Tehran, Iran
Zohdy A. Mohamed, Oakland University, USA

Preface

Industry 4.0 is changing the way manufacturers think about data security. In the past, the IT and operational environments were completely separate. Factories were largely autonomous, with only controlled communication between systems. This is no longer the case. Today, huge amounts of data are generated and shared in operations through Industry 4.0 technologies such as IIoT, cloud, digital wires, and real-time data analytics. There is an increased need for cybersecurity today.

In addition, wireless sensors, networks, and connected devices, such as smartphones, tablets, and other wearable technologies are emerging in the workplace. Modern Industrial Control Systems (ICS) allow engineers to deploy fully automated and virtually unmanned sites. Suppliers of Supervisory Control and Data Acquisition (SCADA) systems, distributed control systems, and Manufacturing Execution Systems (MES) offer human-machine interfaces and wireless communication devices. These allow operators and engineers to take control of equipment from locations both inside and outside the plant. In addition, distributed system controllers now have built-in servers that allow them to access the web. The devices that do the most critical and difficult work in our societies, such as controlling power generation and distribution, water purification and distribution, and chemical production and refining, are the most vulnerable in an industrial network. As industrial control systems are increasingly connected to the Internet, the threat of security breaches and potential damage to facilities and processes have become very real. The spectre of cyberattacks on industrial networks and systems is growing exponentially, making cybersecurity in the manufacturing sector a more important issue than ever.

This book volume constitutes the refereed proceedings of the International Conference on Cybersecurity, Cybercrimes, and Smart Emerging Technologies, held in Riyadh, Saudi Arabia, during May 10–11, 2022. CCSET2022 was organized by Data Science & Blockchain LAB EIAS, College of Computer & Information Sciences (CCIS), Prince Sultan University, Saudi Arabia.

The 27 full papers were carefully reviewed and selected from 89 submissions. The papers presented in the volume are organized in topical sections on synergies between Machine Intelligence and information retrieval; Smart Systems and Networks; online learning; Smart Healthcare; Cybersecurity, and Information Assurance. All contributions were subject to a double-blind review. The review process was highly competitive. We had to review near to 89 submissions. A team of over 80 program committee members and reviewers did this terrific job. Our special thanks go to all of them.

CCSET2022 offered again an exciting technical program as well as networking opportunities. Outstanding scientists and industry leaders accepted the invitation for keynote speeches:

CCSET2022 offered again an exciting technical program as well as networking opportunities. Distinguished scientists accepted the invitation for keynote speeches:

- Prof. Helen Karatza
  Aristotle University of Thessaloniki, Greece
- Prof. Amir Hussain
  Edinburgh Napier University, UK
- Prof. Mohammed M. Alani
  Seneca College of Applied Arts and Technology, Toronto, Canada
- Prof. Mansaf Alam
  Department of Computer Science, Jamia Millia Islamia, India.

We want to take this opportunity and express our thanks to the contributing authors, the members of the Program Committee, the reviewers, the organizers and sponsors (Prince Sultan University, College of Computer & Information Sciences (CCIS), and Data Science & Blockchain LAB EIAS), and the Organizing Committee for their hard and precious work. Thanks for your help—CCSET2022 would not exist without your contribution.

Shebin El-Koom, Egypt   Ahmed A. Abd El-Latif
Khouribga, Morocco   Yassine Maleh
Riyadh, Saudi Arabia   Mohammed ELAffendi
Warsaw, Poland   Mohamed I. Alkanhal
Riyadh, Saudi Arabia   Wojciech Mazurczyk

Contents

Advances in Cybersecurity and Cybercrimes

Malware Detection Using RGB Images and CNN Model Subclassing . . .   3
Ikram Ben Abdel Ouahab, Yasser Alluhaidan, Lotfi Elaachak,
and Mohammed Bouhorma

Ensemble Feature Selection for Android SMS Malware Detection   15
Syed F. Ibrahim, Md. Sakir Hossain, Md. Moontasirul Islam,
and Md. Golam Mostofa

**Analyzing Malware From API Call Sequences Using Support
Vector Machines**   27
Qasem Abu Al-Haija and Moez Krichen

**Android Ransomware Attacks Detection with Optimized Ensemble
Learning**   41
Shaharia Sifat, Md. Sakir Hossain, Sadia Afrin Tonny,
Bejoy Majumder, Riftana Mahajabin, and Hossain Md. Shakhawat

**A Dual Attack Tree Approach to Assist Command and Control
Server Analysis of the Red Teaming Activity**   55
Atul Rana, Sachin Gupta, and Bhoomi Gupta

**A Propagation Model of Malicious Objects via Removable Devices
and Sensitivity Analysis of the Parameters**   69
Apeksha Prajapati

**Improving Deep Learning Model Robustness Against Adversarial
Attack by Increasing the Network Capacity**   85
Marco Marchetti and Edmond S. L. Ho

**Intelligent Detection System for Spoofing and Jamming Attacks
in UAVs**   97
Khadeeja Sabah Jasim, Khattab M. Ali Alheeti,
and Abdul Kareem A. Najem Alaloosy

AST-Based LSTM Neural Network for Predicting Input Validation
Vulnerabilities 111
Abdalla Wasef Marashdih, Zarul Fitri Zaaba, and Khaled Suwais

IoT Trust Management as an SIoT Enabler Overcoming Security
Issues 123
Assiya Akli and Khalid Chougdali

An Overview of the Security Improvements of Artificial
Intelligence in MANET 135
Hafida Khalfaoui, Abderrazak Farchane, and Said Safi

Cloud Virtualization Attacks and Mitigation Techniques 147
Syed Ahmed Ali, Shahzad Memon, and Nisar Memon

Advances in Smart Emerging Technologies

Short Survey on Using Blockchain Technology in Modern Wireless
Networks, IoT and Smart Grids 163
Moez Krichen, Meryem Ammi, Alaeddine Mihoub,
and Qasem Abu Al-Haija

Predicting Sleeping Quality Using Convolutional Neural Networks 175
Vidya Rohini Konanur Sathish, Wai Lok Woo, and Edmond S. L. Ho

A Dynamic Routing for External Communication in Self-driving
Vehicles 185
Khattab M. Ali Alheeti and Duaa Al Dosary

Analysis of the Air Inlet and Outlet Location Effect on Human
Comfort Inside Typical Mosque Using CFD 199
Mohammed O. Alhazmi and Abdulaziz S. Alaboodi

Hybrid Feature-Based Multi-label Text
Classification—A Framework 211
Nancy Agarwal, Mudasir Ahmad Wani, and Mohammed ELAffendi

Voice Recognition and User Profiling 223
Bahaa Eddine Elbaghazaoui, Mohamed Amnai, and Youssef Fakhri

Compression-Based Data Augmentation for CNN Generalization 235
Tajeddine Benbarrad, Salaheddine Kably, Mounir Arioua,
and Nabih Alaoui

A Study of Scheduling Techniques in Ad Hoc Cloud Using Cloud
Computing 245
Rajdip Das and Umesh Pal

Performance Improvement of SAC-OCDMA Network Utilizing an Identity Column Shifting Matrix (ICSM) Code 263
Mohanad Alayedi, Abdelhamid Cherifi, Abdelhak Ferhat Hamida,
Rima Matem, and Somia A. Abd El-Mottaleb

Localization of Pashto Text in the Video Frames Using Deep Learning 279
Syeda Freiha Tanveer, Sajid Shah, Ahmad Khan,
Mohammed ELAffendi, and Gauhar Ali

Exploration of Epidemic Outbreaks Using Machine and Deep Learning Techniques 289
Farah Jabeen, Fiaz Gul Khan, Sajid Shah, Bilal Ahmad,
and Saima Jabeen

Propaganda Identification on Twitter Platform During COVID-19 Pandemic Using LSTM 303
Akib Mohi Ud Din Khanday, Qamar Rayees Khan,
Syed Tanzeel Rabani, Mudasir Ahmad Wani, and Mohammed ELAffendi

Model of the Internet of Things Access Network Based on a Lattice Structure 315
A. Paramonov, S. Bushelenkov, Alexey Tselykh, Ammar Muthanna,
and Andrey Koucheryavy

Study of Methods for Remote Interception of Traffic in Computer Networks 323
Maxim Kovtsur, Ammar Muthanna, Victoria Konovalova,
Abramenko Georgii, and Shterenberg Olga

Advances in Cybersecurity and Cybercrimes

Malware Detection Using RGB Images and CNN Model Subclassing

Ikram Ben Abdel Ouahabⓘ**, Yasser Alluhaidan, Lotfi Elaachak**ⓘ**, and Mohammed Bouhorma**ⓘ

Abstract Malware authors often upload their software to third-party application repositories to allow hackers to take control of a device by stealing passwords or providing access to contacts. Therefore, the development of an efficient malware detection tool is urgently needed. Malware detection researchers usually begins by extracting characteristics from specific sections of malware files, and this technique failed in case of zero-day malware. Unfortunately, malware classification remains a challenge, even if current state-of-the-art classifiers generally achieve excellent results, especially in image processing. To support efficient and powerful malware classification, we propose a CNN model subclassing architecture using RGB images. Malicious and benign files samples are converts to RGB images then the proposed classifier is able to recognize either it is malicious or not. We build our model using high-level API, then we examined out many optimizers. Finally, we got a Malware Detection Model as efficient as fast using RGB images, CNN from scratch with subclassing and Nadam optimizer. The end result we were given is 96% precision on a small database.

Keywords Malware detection · Cybersecurity · Deep learning · Visualization technique

I. B. A. Ouahab (✉) · L. Elaachak · M. Bouhorma
Computer Science, Systems and Telecommunication Laboratory (LIST), FSTT, University Abdelmalek Essaadi, Tangier, Morocco
e-mail: ibenabdelouahab@uae.ac.ma

L. Elaachak
e-mail: lelaachak@uae.ac.ma

M. Bouhorma
e-mail: mbouhorma@uae.ac.ma

Y. Alluhaidan
Saudi Data and Artificial Intelligence Authority, Riyadh, Saudi Arabia
e-mail: ylhaidan@nic.gov.sa

© The Author(s), under exclusive license to Springer Nature Switzerland AG 2023
A. A. Abd El-Latif et al. (eds.), *Advances in Cybersecurity, Cybercrimes, and Smart Emerging Technologies*, Engineering Cyber-Physical Systems and Critical Infrastructures 4,
https://doi.org/10.1007/978-3-031-21101-0_1

1   Introduction

Some of the most impactful incidents of 2020 to date include: SolarWinds hack, Twitter, Marriott fraud, MGM Resorts, Zoom, Magellan Health, Finastra, Greek Banking System and others. The SolarWinds hack compromised multiple systems of governments and companies worldwide, where on the server providing access to updates and patches for SolarWinds Orion tools was compromised. This allows attackers to inject code into the software updates and infect multiple clients at once very fast. This code allowed data modification and exfiltration as well as remote access to devices that had the software installed. Companies and high-level entities such as Microsoft and the US Department of Defense were affected by this hack. In addition, two significant travel-related incidents were detected in 2020 targeting: Marriott and MGM Resorts. Marriott is one of the largest hotels brands; it was a victim of data breach of their guest information. About 5.2 million hotel guests were fraudulently accessed in 2020. Similarly, MGM Resorts is a global entertainment company with luxury resorts and casinos, was hacked and more than 10.6 million guests' personal data was shared on a hacking forum.

Why these attacks were such successful? Many answers are available: Social Engineering, data security, Ransomware and patch management are popular methods of attack. According to the ISACA's State of Cybersecurity 2020 report [1], social engineering is behind 15% of compromised respondents as a vehicle of entry to huge companies. The success of social engineering attacks is thanks to one vulnerable person even in presence of a strong cybersecurity staff. Also, data security is very important. Some systems are infected due to the leak present at this stage. So, companies should ensure robust encryption policies to reduce the harmful impact if the data is stolen or exfiltrated. Moreover, the increase of ransomware in health care has been the most significant trend in cybersecurity in the past year. Ransomware are resulting from phishing exploits under healthcare organizations.

A malware is any malicious software that aims to harm a user computer, a company network or IoT devices. Types of malwares can include computer viruses, worms, Trojan horses, ransomware, spyware and others. Also, researchers classify malware variants into families based on their malicious behavior. The malicious behavior of malware is not limited to stealing, encrypting or deleting sensitive data, altering or hijacking core computing functions and monitoring users' computer activity. A simple malware can lead to catastrophic cyberattacks and damages could be as physical as financial. So, detecting the presence of malware in any systems early is a challenging task nowadays.

Our contribution aims to use advanced deep learning methods to detect malicious software. We build a classifier able to make difference between malicious and benign files with a precision of 96%. To get this result, we used visualization technique of malware which transforms a malware to an RGB image. So, instead of dealing directly with malicious software, we have an image classification problematic. It's a binary classification problem that we handle using CNN model subclassing. After that we use various optimizers. The remaining of this paper is organized as follow. Section 2 concerns related works about malware detection and classification. After that, in Sect. 3, we present the malware visualization technique. In Sect. 4, we detailed our proposed CNN model using subclassing. Then we discuss the experimental results in Sect. 5, before concluding and giving some of our future perspectives.

2 Related Works

Recently, many researchers across the globe are working with the malware visualization approach. This approach was firstly initialized by Nataraj [2] in 2011. In his research work, he introduced the conversion of any PE file to executable. Then with time he proves the efficiency of this technique using the power of AI algorithms. Afterwards, Microsoft and Intel have recently collaborated on a new research project that has explored a new approach to detect and classify malware.

In [3], They proposed the STAMINA stand for STAtic Malware-as-Image Network Analysis. In their approach they used malware visualization technique with transfer learning and deep learning model, they got a 99% accurate model.

In [4], authors used API calls of malicious and benign files, with 6434 benign and 8634 malware samples. They build a supervised machine learning classifier which produced 99.1% accuracy using ensemble algorithms. However, this feature failed to detect recent and zero-day malwares.

Also, authors in [5] proposed the MSIC (Malware Spectrogram Image Classification) for malicious and benign classification which scored 96% F-measure and accuracy. It uses spectrogram images in conjunction with a convolutional neural network to classify a malware file into its appropriate family and distinguish it from a benign file.

In addition, here [6] researchers focus on ransomware detection. So, the paper proposed a method to distinguish between ransomware and malware. They used Windows Native API invocation sequences as features, and CF-NCF (Class Frequency–Non-Class Frequency), and machine learning algorithms to build the classifier. Experimental results shows that the ransomware recognition accuracy is up to 98.65%.

In a recent work [7], we perform a malicious-benign classifier using machine learning algorithms. In order to distinguish between malicious and benign files we extracted HOG and DAISY features from grayscale malware images. Then we combined these features and passed them as input to machine learning algorithms. Finally, we conclude that Random Forest Algorithms was efficient with these features and database with an accuracy of 97%.

In this paper, we would like to work with RGB images for malware-benign classification, to compare with the previous work. Further, we implemented the CNN model using TensorFlow and Keras subclassing API, which give us more flexibility and possibilities.

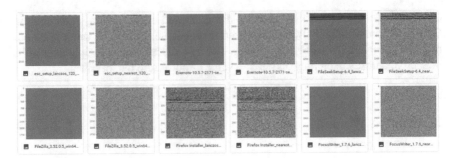

Fig. 1 Benign samples from RGB database

3 Malware Visualization Technique

'Malware as images' database was generated using Mallook [8] technique. Mallook (Malware look) is a Python script for generating image representations of portable executables (PEs). This script is designed to iterate byte-by-byte through an executable and generate a 3-channel RGB image with square dimensions as output for use in training image recognition models. In particular, the intended datasets are meant to be image representations of malware and benign software.

'Malware as images' database [9] is available on Kaggle. It contains image representations of Portable Executables of malware and ordinary software (benign). Each folder contains 4 DPI versions of the images (120, 300, 600, and 1200), using two different methods of interpolation: Nearest and Lanczos [10]. Interpolation is the process of finding a value between two points on a line or a curve. In our experimentation, we'll be using only 120 DPI versions of images using both interpolation methods: *nearest* and *lanczos*. The total number of images in our database is 282 where 122 present benign images and 160 of malicious samples.

Benign samples. Benign samples was collected from Best Free Software of 2020 Magazine[11], and downloaded from official sources (Fig. 1).

Malicious samples. Malicious Portable executable files were collected from the Zoo [12] (Fig. 2).

4 CNN Model Subclassing for Malware Detection

TensorFlow and Keras deliver us 3 methods to do the implementation of Neural Network architectures: Sequential API, Functional API, and Model Subclassing. They are mentioned based on the developer's expertise from beginners to experts. It's true that model subclassing is much harder to use that Sequential API and Functional API, however, the fact that it's fully-customizable and enables us the implementation of our own custom forward-pass of the model was our biggest motivation to do so.

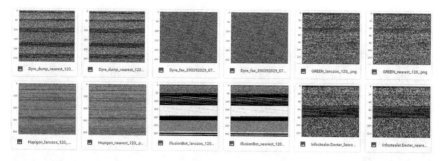

Fig. 2 Malicious samples from RGB database

Model subclassing provide a full control over every nuance of the network and the training process.

4.1 The Proposed Architecture

In this paper, we use a high-level API style to build our CNN model. Using a pure Oriented Object Programming concepts of Python using TensorFlow and Keras to use model subclassing. This method allows us as experts to create new layers, have hands-on many variables and manage various details that was inaccessible using classical sequential or function API. Inside of Keras the 'Model' class is the root class used to define a model architecture. Since Keras utilizes object-oriented programming, we can actually subclass the Model class and then insert a customized architecture.

Class diagram of our architecture is presented in Fig. 3. Then we give a detailed architecture layer per layer in Fig. 4, where we defined 2 blocks. First, the CNNBlock is composed of a single conv2D layer with relu activation function. Second, the ResBlock is made of 3 CNNBlocks and MaxPooling2D layer. Also, we used one identity connection. Regarding the Malware Detection Model, we made it up using 2 ResBlocks and others classical layers, like MaxPooling2D, Flatten, Dense and for activation functions we used both relu and softmax.

4.2 Model Optimizers

An optimizer updates the weight parameters to minimize the loss function of a deep learning algorithm. The loss function serves as a guide to the terrain and tells the optimizer whether it is moving in the right direction to reach the bottom of the valley, the global minimum. Types of optimizers are: Momentum, Nesterov accelerated gradient (NAG), Adaptive Gradient Algorithm (Adagrad), Adadelta, Root Mean

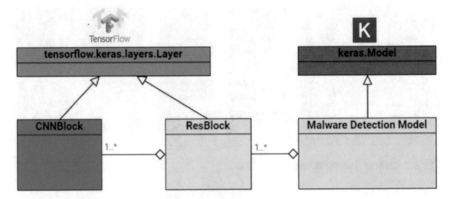

Fig. 3 Class diagram of malware detection model

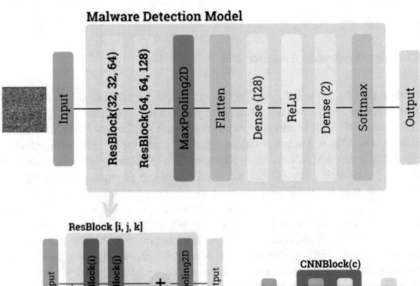

Fig. 4 The proposed malware detection model architecture

Square Propagation (RMSProp), Adaptive Moment Estimation (Adam), Nesterov-accelerated Adaptive Moment Estimation (Nadam), and others. In our experimentation, we'll be using Adam, Nadam and Adamax to see which one is the most efficient in our case.

Adam. Adam optimization is a stochastic gradient descent method that is based on adaptive estimation of first-order and second-order moments [13]. This is the wide used optimizer in most of cases. It's proved to be computationally efficient, requires little memory, is invariant to diagonal scaling of gradients, and is like minded for issues that are massive in terms of data/parameters [14].

Nadam. Nadam is Adam with Nesterov momentum [15]. That said, Nadam combines Nesterov accelerated gradient and Adam. It is utilized for noisy gradients or for gradients with high curvatures. The learning process is accelerated by adding the exponential decay of the moving averages for the previous and current gradient [16].

Adamax. Adamax is another variant of Adam which is based on the norm of infinity. Adamax is sometimes superior to Adam, especially on models with embeddings [17]. When Adam does a change weight reciprocally proportional to the scaled L2 norm (squared) of past gradients, Adamax extends this to the alleged infinite norm (max) of past gradients [14].

5 Experimentation and Result

In this section, we present and discuss experimental results base on the proposed subclassing model and RGB malware images.

The Malware Detection Model was implemented based on the proposed architecture (Fig. 5), using TensorFlow and Keras API. As hardware equipment, we use a NVIDIA Quadro T1000 with Max-Q Design GPU [18], which have a computation capacity of 7.5. It contains an Intel(R) Core(TM) i7-10750H CPU processor. In order to set up TensorFlow and Keras environment in our device, we should do the manual installation to avoid any type of issues. And based in our experience, we suggest to do manual installation for TensorFlow and Keras in GPU devices, it may take longer but it's ensure having the rights versions of required python libraries and avoid the conflicts of versions. We take notes of the manual installation in our GitHub reposity with full references and details [19].

Moving to the implementation, you find below in Fig. 5 the summary of layers. As detailed in the previous section. At the beginning, we have 2 ResBlocks then MaxPooling, Flatten and Dense Layers. The total of trainable is 1.263.138. The training process don't take a long time in our device (about 2–3 min).

The malware detection is a binary classification problem. So, for a given malicious file, we convert it into a RGB image then this image is passed as an input to our trained model, after that, the output of our model is the classification results to shown either the file is malicious or benign.

```
Layer (type)                    Output Shape             Param #
=================================================================
res_block (ResBlock)            multiple                 29536

res_block_1 (ResBlock)          multiple                 184640

max_pooling2d_2 (MaxPooling     multiple                 0
2D)

flatten (Flatten)               multiple                 0

dense (Dense)                   multiple                 1048704

dense_1 (Dense)                 multiple                 258

=================================================================
Total params: 1,263,138
Trainable params: 1,263,138
Non-trainable params: 0
```

Fig. 5 Summary of malware detection model

In our experimentation, we fix the epochs number at 50, input size images as 64 × 64 and we keep varying the optimization function. So that, we use 3 optimizers: Adam, Adamax and Nadam. Then, we got the results presented below. Results are evaluated using 4 evaluation metrics: Loss, Accuracy, Precision and Recall for each case (Table 1). Also, we present the plot of training/validation loss and training/validation accuracy with each optimizer separately (Table 3).

Using Adam optimizer, we got a loss of 0.17 and an accuracy of 94.11%. And using Adamax the loss is slightly better with a value of 0.16 keeping same accuracy value. Then using Nadam optimizer, all metrics improves. We got a loss of 0.15, and the accuracy is higher than the previous results with a value of 95.58%.

The classification report (Table 2) gives a clear vision on the model performance for each class. In addition to the previous metrics, here we can see how accurate is our classification model for each class (malicious/benign). Without doubt, Nadam

Table 1 Results of CNN subclassing model using various optimizers and evaluation metrics

Optimizer	Evaluation metrics			
	Loss	Accuracy	Precision	Recall
Adam	0.170847	0.941176	0.941176	0.941176
Adamax	0.167235	0.941176	0.941176	0.941176
Nadam	0.154963	0.955882	0.955882	0.955882

Table 2 Classification report for each optimizer

Using Adam

	Precision	Recall	F1-score	Support
0	0.94	0.94	0.94	32
1	0.94	0.94	0.94	36
Accuracy			0.94	68
Macro avg	0.94	0.94	0.94	68
Weighted avg	0.94	0.94	0.94	68

Using Adamax

	Precision	Recall	F1-score	Support
0	0.91	0.91	0.94	32
1	0.97	0.97	0.94	36
Accuracy			0.94	68
Macro avg	0.94	0.94	0.94	68
Weighted avg	0.94	0.94	0.94	68

Using Nadam

	Precision	Recall	F1-score	Support
0	0.94	0.97	0.95	32
1	0.97	0.94	0.96	36
Accuracy			0.96	68
Macro avg	0.96	0.96	0.96	68
Weighted avg	0.96	0.96	0.96	68

optimizer was the most accurate one for all classes using evaluation data. Moreover, the obtained results are good as compared to other found on literature (Table 3).

6   Conclusion

Malware evolved, added new features, and raised the bar in terms of infection rates. Despite these advances, malware nevertheless remained very plenty a self-replicating message in a bottle. The strength of malware was largely predetermined at the time of writing. In this paper, we proposed a Malware detection model to make the difference between malicious and benign files. We take advantage of malware visualization technique, so we used RGB malware images. Then we build a CNN model from scratch using subclassing. After that, we perform a comparison of various optimizers to see which one fit most in our case. Finally, we reach an accuracy of 96%.

In the future, we are looking forward to collect more data and work with it to improve our model performances. Also, we would like to use different visualization

Table 3 Visualization of training and validation, loss and accuracy for each used optimizer

Using Adam optimizer

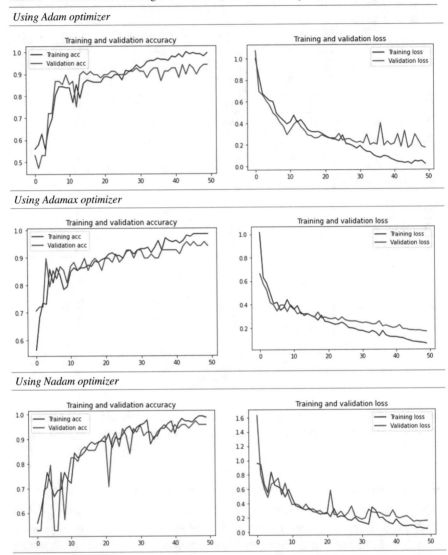

Using Adamax optimizer

Using Nadam optimizer

techniques of colored images. Otherwise, we are working on fine tuning pretrained models to be compared with the actual findings.

Acknowledgements Authors acknowledge financial support for this research from the "Centre National pour la Recherche Scientifique et Technique", CNRST, Rabat, Morocco.

References

1. State of Cybersecurity (2020). https://www.isaca.org/go/state-of-cybersecurity-2020. Accessed 27 Jun 2021
2. Nataraj L, Karthikeyan S, Jacob G, Manjunath BS (2011) Malware images: visualization and automatic classification. In: Proceedings of the 8th international symposium on visualization for cyber security. Association for Computing Machinery, New York, NY, USA, pp 1–7. https://doi.org/10.1145/2016904.2016908
3. Chen L, Sahita R, Parikh J, Marino M (2020) STAMINA: scalable deep learning approach for malware classification 11.
4. Singh J, Singh J (2020) Assessment of supervised machine learning algorithms using dynamic API calls for malware detection. Int J Comput Appl 0:1–8. https://doi.org/10.1080/1206212X.2020.1732641
5. Azab A, Khasawneh M (2020) MSIC: Malware spectrogram image classification. IEEE Access. 8:102007–102021. https://doi.org/10.1109/ACCESS.2020.2999320
6. Bae SI, Lee GB, Im EG (2020) Ransomware detection using machine learning algorithms. Concurr Comput Pract Exp 32:e5422. https://doi.org/10.1002/cpe.5422
7. Ben Abdel Ouahab I, Elaachak L, Bouhorma M (2022) Classification of malicious and benign binaries using visualization technique and machine learning algorithms. In: Baddi Y, Gahi Y, Maleh Y, Alazab M, Tawalbeh L (eds) Big data intelligence for smart applications. Springer International Publishing, Cham, pp 297–315. https://doi.org/10.1007/978-3-030-87954-9_14
8. Fields M (2021) Mallook
9. Malware as Images. https://kaggle.com/matthewfields/malware-as-images. Accessed 01 Aug 2021
10. Interpolations for imshow—Matplotlib 3.4.2 documentation. https://matplotlib.org/stable/gallery/images_contours_and_fields/interpolation_methods.html. Accessed 04 Aug 2021
11. The Best Free Software of 2020. https://www.pcmag.com/news/best-free-software. Accessed 01 Aug 2021
12. Nativ Y (2019) A repository of LIVE malwares for your own joy and pleasure. The Zoo is a project created to make the possibility of malware analysis open and available to the public ytisf/theZoo
13. Team K (2022) Keras documentation: Adam. https://keras.io/api/optimizers/adam/. Accessed 27 Mar 2022
14. Kingma DP, Ba J (2017) Adam: a method for stochastic optimization. arXiv:1412.6980 [cs]
15. Dozat T (2016) Incorporating Nesterov Momentum into Adam. 6
16. Team K (2022) Keras documentation: Nadam. https://keras.io/api/optimizers/Nadam/. Accessed 27 Mar 2022
17. Team K (2022) Keras documentation: Adamax. https://keras.io/api/optimizers/adamax/. Accessed 27 Mar 2022
18. CUDA GPUs. https://developer.nvidia.com/cuda-gpus. Accessed 27 Mar 2022
19. Ouahab IBA (2022) This is a step by step guide to install latest version of TensorFlow on GPU

Ensemble Feature Selection for Android SMS Malware Detection

Syed F. Ibrahim, Md. Sakir Hossain⑩, Md. Moontasirul Islam, and Md. Golam Mostofa

Abstract As Android is the most popular operating system of mobile devices, the mobile devices with the Android operating system are a common target of the attackers. To protect the devices, various solutions are presented to date. Of them, the machine learning assisted solutions are considered more effective. However, such solutions often suffer from a limited detection accuracy and high computational complexity. To this end, we propose an ensemble feature selection method for detecting Android SMS malware. We exploit six feature selection algorithms to find the most important 12 features from the traffic of various Android SMS malware. Then, various machine learning algorithms are trained to predict whether an incoming traffic is benign or SMS malware. Through extensive experiments, the gradient boost-based classifiers are found to be the most effective. The gradient boost classifiers can detects the SMS malware with the highest accuracy (93.34%) compared to the other classifiers. The proposed method achieves 3.17% performance improvement with respect to the state-of-the-art SMS malware methods. The number of features selected by the proposed method is 33% less compared to the existing methods. In addition, up to 3-times more true positive rate can be obtained by the proposed metho.

Keywords SMS malware · Android · Ensemble feature selection · CICAndMal2017 · Machine learning

S. F. Ibrahim · Md. S. Hossain (✉) · Md. M. Islam · Md. G. Mostofa
American International University-Bangladesh (AIUB), 408/1, Kuratoli Road, Khilkhet, Dhaka 1229, Bangladesh
e-mail: sakir.hossain@aiub.edu

S. F. Ibrahim
e-mail: 17-34347-1@student.aiub.edu

Md. M. Islam
e-mail: 17-34336-1@student.aiub.edu

Md. G. Mostofa
e-mail: 17-34611-2@student.aiub.edu

1 Introduction

Smartphones are no longer just for making phone calls, as they traditionally were. They are currently used to store personally identifiable information, health care information, financial details, and other e services. As a result, the number of smartphone users in the world has increased. Android is the most widely used smartphone operating system in the world. Due to its popularity, it has become a target of cyber attackers. Malware dangers are increasing, as are measures to manage such risks. Security researchers use two ways to identify malware in this regard: the static-based method, which aims to discover malware without executing it, and the dynamic-based method, which analyzes malware behavior inside an isolated environment (i.e., monitoring malware generated traffic). To make the Android devices secure, it is imperative to develop a method to detect Android malware.

In last two decades are variety of methods are proposed. In this section, we outline the latest development in devising Android malware detection methods.

In [2], a long short-term memory model is proposed for Android ransomware detection. The CI-CAndMal2017 dataset is used. Eight different feature selection algorithms are used in the feature selection process. The common features suggested by all the eight algorithms are selected. In this way, a total of 19 features are selected. Binary classification is considered with the class names: benign (normal) and ransomware (attack). The detection accuracy of the model is 97.08%. In the experiment, a highly imbalanced dataset is used. This leads to a biased machine learning model. For this reason, the detection accuracy that is found in [2] will not be found in real-world networks. Another investigation is carried out in [12] on the CICAndMal2017 dataset. Nine features are selected. Instead of considering the whole dataset, a subset of 0.6 million instances is considered for training and testing various machine learning models. The detection accuracy of 94% for 11 families of adware is obtained in all the five classifiers: random forest, J48 decision tree, random tree, K-nearest neighbor (KNN), and logistic regression.

Instead of considering the traffic of a specific kind of malware, three malwares, such as adware, scareware, and ransomware, are considered in [3]. From the CICAndMal2017 dataset, the traffic of 10 randomly chosen malware families is used. A total of 15 time-related and packet-related features are selected for training three classifiers named random forest, decision tree, and KNN. Two types of classification are performed in the experiment. In binary classification, three machine learning models are trained to detect malware traffic in the presence of benign traffic. In the multiclass classification, the models are trained against four kinds of traffic: benign, adware, scareware, and ransomware. The detection accuracy of 95% and 86% are achieved in the binary and multiclass classification, respectively.

In [8], the authors propose a method for detecting malicious files on a mobile device and classifying them as particular malware. Information Gain, Cfs Subset, and SVM algorithms are used in the feature selection process. Nine common features are selected. To identify malware families, classification techniques are used for packet-based and flow-based features. Random forest, decision tree (J48), random tree, k-nearest neighbors, and regression are five frequent classifiers that are used. The proposed technique achieved average accuracy, precision, and false-positive rates of 91.41%, 91.24%, and (0.085) respectively.

In [9], the authors describe a framework for making android malware datasets using actual smartphones. Network traffic features is used to detect Android malware. Network traffic analysis is used to extract more than 80 features to detect and categorize malicious families of malware by CICFlowMeter. For feature selection, the authors utilize two features selection techniques: Information gain with the Ranker search method and CfsSubsetEval with the best first search method. A total nine common features among the features selected from the feature selection algorithms are selected. As classifiers, the random forest, KNN, and decision tree are used. The accuracy is 88% for the binary classification (two classes) and the recall is 85%, while the accuracy and recall for category classification (4 classes) is 27% and 25%, respectively. Finally, the accuracy and recall for family categorization (43-Classes) are 27% and 25%, respectively.

In [14], the authors use the second portion of the CICAndMal2017 dataset, which uses API calls as dynamic features, and intents and permissions as static characteristics. Malware classification algorithms include random forest, decision tree, and (KNN). The author proposes a two-layer analysis system: static binary classification and dynamic malware classification. The model results 95.3% precision in malware binary classification. Some 83.3% and 59.7% precision are found in dynamic-based malware category classification and dynamic-based malware family categorization.

In [5], the authors purpose a model to analyze the most effective factors for detecting malware in smartphones, particularly those in Android. The CICAndMal2017 Dataset is used with 600 samples from the benign class and 400 samples from the malware class. In order to reduce the number of features, the features selection algorithms such as wrapper method, forward selection, backward elimination, and optimized selection are used. Five sets of features are selected with 75, 45, 25, 15, and 9 features. To identify each malware category (Adware, Scareware, Ransomware, and SMS Malware), these features are utilized. Machine learning methods such as Decision tree, Random forest, Naive Bays, Linear Regression, K-NN, and Support vector machine (SVM) are used for implementation. The best results is from the Naïve bias with an F1 of 74.83% for detecting malware families. The model focuses on the F1 score.

The information gain and Chi-square tests are used to prioritize network traffic of Android malware in [1]. Following that, network traffic features are reduced using a developed technique to improve detection accuracy and shorten the training and testing stages. To rank features, statistical analysis approaches are applied. The proposed method indicates that 9 out of 22 features are sufficient to get the increased detection accuracy. Similarly, the study findings reveal that it can minimize the time required for the training and testing stages by 50% and 30%, respectively.

CNN are used in [10] to detect Android malware through static analysis. The APK format Android applications are first converted into grayscale photos, and then Machine learning models are trained with the dataset as well as the CICAndMal2017 dataset. This system gets a malware detection capability of 84.9

In [13], the author explained a framework for examining android API calls performed by an application. The authors design a CNN to detect malware from the sequence of API-calls done by an application. A pseudo-random analyzer of the programs is developed to find the sequence of API-calls a program makes in its run. The sequence is used to train CNN. Multiple layers of CNN are implemented in order to reduce the number of features. The model is tested on a dataset of 1016 APK records and found to obtain 99.4% accuracy. The CNN model outperforms the LSTM-based approach.

A framework, named MalDozer, is proposed in [7] for effectively detecting malware based on API-calls of applications. It exploits deep CNN for automatically identifying malware and malware families. The API-calls' raw sequence is given as inputs to the deep CNN model. It attains 96% to 99% of F1-score with 0.06% false positive rate.

In [6], a two-layer technique is proposed for recognizing Android malware depending on the network traffic elements of Applications, and it is a combination CNN and CACNN strategy engineering. The CICAndMal2017 dataset is used. Static analysis method is used to extract static features, then mobile traffic analysis method is used to extract traffic features. Two standard deep learning classifiers, namely CNN and the combination of CNN and cascaded CNN based on the three cases: binary malware classification, category classification, and family classification. The binary, category, and family classifications accuracies are 99.3%, 98.2%, and 71.48%, respectively. Since this technique uses CNN, it requires considerable training time in the training phase. The CICAndMal2017 dataset is used to detect malware in Android apps using gated recurrent units (GRU), which is the architecture of recurrent neural network (RNN) in [4]. The dataset consists of 347 benign samples and 365 malicious samples of Android applications. Deep learning techniques (GRU) as well as machine learning techniques decision tree, random forest, svm, naive bayes algorithms are used in order to select the most efficient model that delivers the best results in identifying Android malware. The deep learning approach (GRU) achieves best among all machine learning approaches, with a 98.2% level of accuracy and 98% F-measure score. However, the dataset used in the model is biased and may favor benign traffic in real world testing. Using the same dataset, the detection accuracy of various Android SMS malware using various machine learning algorithms is investigated in [11]. As a feature selection algorithm, the ANOVA-F is used. A total 18 features are selected. The best accuracy of 90.17% is obtained.

From the above discussion, it is evident that machine learning assisted method of Android malware detection is more effective compared to the traditional methods. Furthermore, researchers are more interested in developing solutions using traffic of malwares. Although a significant improvement has been achieved to date in developing machine learning assisted Android SMS malware detection, the detection accuracy is still limited in [11]. In this paper, we want to develop a machine learning method for detecting various SMS malware attacks which will provide a higher detection accuracy with a lower computational overhead. In this paper, an ensemble feature selection method is proposed to select the most influential set of features. Thereafter, various machine learning classifiers are trained using the selected features. Finally, we evaluate the proposed models for detecting families of various Android SMS malware. Furthermore, we also investigate the reason behind getting low detection accuracy of some malware. We carry out a comparison of the performance of the proposed model with the state-of-the-art solutions.

2 Proposed Method

In this section, we will provide a high-level description of the proposed methodology. Then, a detail description of various steps will be provided.

2.1 System Model

The architecture of this model is shown in Fig. 1, and the phases are briefly discussed below.

1. The first phase is data preprocessing, which involves removing missing values from columns and features with low variance values. Standardization scaling method is used in which the values are scaled with a unit standard deviation and centered around the mean. The formula for standardized scaling is given below:

$$X' = \frac{X - \mu}{\sigma} \tag{1}$$

   where X is the original data point, μ and σ represent mean and variance of the data. Unlike the normalization method, this method is not affected by the data outliers. For this reason, we select this method of scaling. It's important to note that the dataset utilized in this study has no missing values.
2. The second phase is feature selection. A dataset may contain some irrelevant features. These irrelevant features cause two problems: escalate system complexity and reduce classification accuracy. To overcome these two problems, it is a good practice to select the most relevant features. There are various feature selection algorithms. Of them, we use six feature selection methods.Details of this stage will be provided later.
3. Then we perform the majority voting method to select the most common features among the features suggested by the given feature selection methods. The majority voting is understood as a technique where the various feature selection methods select various sets of features as the best set of features, and the common among the produced set of features are selected for the classification algorithms.
4. Once the most relevant features are selected, we use them to train various machine learning classification algorithms. In this paper, we use the most effective five classifiers, named decision tree, KNN, random forest, gradient boost, and extreme gradient boost. Of them them, the last three classifiers are ensemble classifiers.

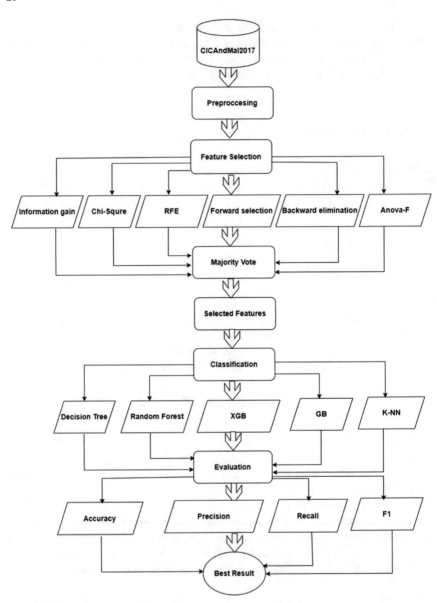

Fig. 1 Proposed system architecture

The dataset utilized in this study was split into two parts: 70% of the dataset is used to train the algorithms, while the rest is used to test them.

5. Finally, the efficacy of each classifier is evaluated with respect to four classifiers, named accuracy, precision, recall, and F1-score.

2.2 Dataset

In this paper we use android malware dataset CICAndMAL2017 which is published by Canadian Institute for Cybersecurity (CIC) [8]. The dataset consists of 10854 apps (6500 benign and 4354 malware). Some 426 malware and 5065 benign apps are installed on real Android phones in order to perform a real-world situation. The dataset traffic is classified into 5 categories (benign, ransomware, adware, SMS malware, and scareware). Malware categories include ransomware, adware, SMS malware, and scareware.

2.3 Data Preprocessing

The dataset contains a total of 84 features. Of them, there is a duplication of features. The feature Fwd Header Length appears twice in the dataset. One of them is removed. There are some features which have zero values for all samples. In addition, some features have zero or a very low variance. When a dataset has features that have zero values or very little variance or the same value for all rows in a column, these features do not provide any important information to the model. These features unnecessarily increase the computational complexity in the training phase. For this reason, these features should be removed. We use variance threshold method to select the feature which should be removed. It eliminates all features whose variance comes below a certain level and removes all zero-variance features by default. A total 12 features are removed after applying the variance threshold methods, and the remaining 71 features are selected for further processing.

Some of the features (e.g., Flow ID, source IP address, destination IP address, and Protocol, Timestamps) have string type values on which numerical computation cannot be done. To overcome this problem, we use a label encoder. Label encoding is the process of changing labels into a numeric format so that various numeric calculations can be carried out.

We apply the standardization technique to improve the performance of the machine learning models. Standardization is a scaling method in which the values are scaled with a unit standard deviation and centered around the mean. This means that the data points' mean will be 0 and the standard deviation will be 1.

2.4   Feature Selection

Features selection is a dimensionality reduction technique to select a limited subset of features by deleting irrelevant, noisy features that are not important for the model construction. It is one of the most important assets in machine learning. It could largely lead to improved learning performance, higher learning accuracy, reduced computing cost, reduced training time, decrees overfitting, and improved model interpretability.

The proposed feature selection technique is shown in Fig. 1. In the proposed feature selection method, we use multiple feature selection algorithms for selecting features Then, we perform a majority voting to select those features which are common at the output of all feature selection algorithms. The following feature selection algorithms are applied independently: Anova-F, Information gain, RFE, Chi-Square, Forward selection, and Backward eliminations algorithms. The feature selection experiment is performed in Python. After applying the above-mentioned six feature selection algorithms, a total of 12 features are chosen by majority voting. The selected features after the majority voting include Flow Id, Total Length of Bwd Packets, Timestamp, Fwd Packet Length Max, Source IP, Flow IAT Std, Source Port, Flow IAT Min, Destination IP, Fwd IAT Min, Destination Port, and Bwd Packest/s.

2.5   Performance Metrics

Some evaluation metrics are used to measure the performance of the trained classifiers. When a trained classifier is evaluated against test data, these metrics measure and describe its quality. There are some key metrics to consider while evaluating classifications. The metrics are given in Table 1. In the Table, TP means true positive, while TN means true negative. TN and FN indicate the true negative and false negative, respectively. FP represents the false positive.

Table 1 Performance metrics

Parameter	Equation
Accuracy	$\frac{TP+TN}{TP+FP+TN+FN}$
Percision	$\frac{TP}{TP+FP}$
Recall	$\frac{TP}{TP+TN}$
F1-score	$\frac{2*Precision*Recall}{Precision+Recall}$

3 Experimental Results

We have used a Python development platform for the experiment. In training and testing phase, we split the dataset into 70% and 30% for training and testing, respectively. The widely used classifiers such as decision tree (DT), random forest (RF), K-NN, XGBoost (XGB), and gradient boost (GB) are used in the experiments. In this section, we evaluate the performance of the proposed ensemble feature selection method. The evaluation will be done in two phases: (1) thorough investigation in each category of malware (we call it general investigation), (2) performance comparison with the existing method

We carried out an experiment to detect SmsMalware on Android smartphones. For this experiment, we applied the dataset of Benign traffic and 11 types of SMS malware families such as Beanbot, Biige, Fakeinst, Fakemart, Fakenotify, Jifake, Mazarbot, Nandrobox, Plankton, SMSsniffer, and Zsone. The SMS malware detection consists of a total of 464272 samples and 12 features. The performance of various classifiers in detecting the SMS malware is given in Fig. 2. In this case, the gradient Boost (GB) achieves the highest with an accuracy of 93.34%, precision of 93%, recall of 93%, and F1-score of 93%. In this case, the detection accuracy is almost 7% less compared to the the detection accuracy of the ransomware. With an accuracy of 83.26%, precision of 83%, recall of 83%, and F1- score of 83%, the K-NN performs the worst. Next, we investigate the precision, recall and F1-score of various types of SMS malware (see Fig. 3). While the benign, biige, mazarbot, nandrobox, and Zsone can be detected with 100% precision, recall and F1-score, the parameters go below 60% in case of the beanbot, fakeinst, and fakenotify. This means that the beanbot, fakenotify, and jifake are difficult to detect. We will investigate the reason behind such low performance criteria in detecting these SMS malware are difficult to detect. To this end, we use the confusion matrix.

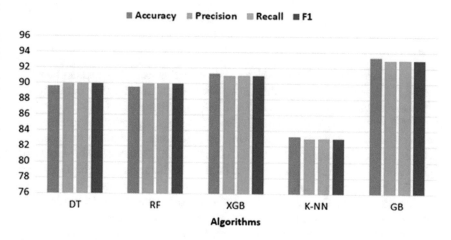

Fig. 2 Performance comparison of various classifiers in detection SMS malware families

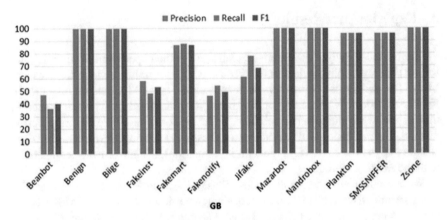

Fig. 3 Precision, recall, and f1-score of Benign and SmsMalware families

Table 2 Confusion matrix for detection of SMS Malware families [%]

	BN	BB	BG	FI	FM	FN	JF	MB	NB	PT	SS	ZS
BN	100	0	0	0	0	0	0	0	0	0	0	0
BB	0	35.51	0	0.079	0	45.98	18.41	0	0	0	0	0
BG	0	0	100	0	0	0	0	0	0	0	0	0
FI	0	0.13	0	48.24	0	51.62	0	0	0	0	0	0
FM	0	0.22	0	0	88.23	0	11.1	0	0	0.44	0	0
FN	0	21.31	0	24.98	0	53.64	0.04	0	0	0	0	0
JF	0	6.84	0	0.15	14	0	73.56	0	0	0	0	0
MB	0	0	0	0	0	0	0	100	0	0	0	0
NB	0	0	0	0	0	0	0	0	100	0	0	0
PT	0	0	0	0	0.01	0	0	0	0	96.37	3.61	0
SS	0	0	0	0	0	0	0	0	0.01	4.21	95.76	0
ZS	0	0	0	0	0	0	0	0	0	0	0	100

The confusion matrix of the SMS malware detection using the gradient boosting classifier is shown in Table 2. In this table, BN: Benign, BB: Beanbot, BI: Biige, FI: FakeInst, FM: Fakemart, FN: Fakenotify, JF: Jifake, MB: Mazarbot, ND: Nandrobox, PT: Plankton, SS: SMSsniffer, and ZS: ZsoneThe bigge, mazarbot, nandrobox, and zsone can be classified with 100% accuracy. On the other hand, the detection accuracy for the beanbot, fakeinst, and fakenotify are less than 50%. Beanbot and fakenotify are highly confusing. Similary, fakenotify and fakeinst families are also highly confusing.

In SMS malware detection, the benchmark is Mim [11] where decision tree is used as a classifier and ANOVA-F feature selection is used for selecting 18 features. The performance comparison is shown in Table 3 In our proposed method, the gradient boost provides 3.17% higher accuracy compared to Mim [30]. Our proposed model Gradient Boost (GB) achieves the highest with an accuracy of 93.34%, precision of

Table 3 Performance comparison between the proposed method and Mim [11] [%]

Method	Accuracy	Precision	Recall	F1-score
Proposed	93.34	93	93	93
Mim [11]	90.17	90	91.2	90.6

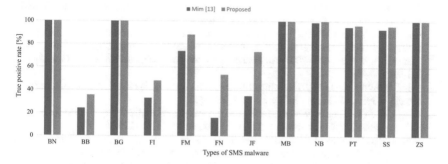

Fig. 4 Performance comparison with the benchmark in terms of true positive rate

93%, recall of 93%, and F1-score of 93% (see Table 3). On the other hand, Mim [30] achieved with an accuracy of 90.17%. Figure 4 shows the comparison of the true positive rate (TPR) of the two methods. We see that the proposed method clearly outperformed the benchmark. Up to 3 times improvement is achieved. For example, the TPRs of the proposed method and benchmark are 53.3% and 16%, respectively. Higher improvement in TPR is obtained for those types of the SMS malware (such as beanbot, fakenotify, jifake, and fakeInst) which are difficult to detect.

4 Conclusions

In this paper, we proposed an ensemble feature selection method for Android SMS malware detection. We used six feature selection methods such as information gain, chi-square, RFE, ANOVA-F, forward selection, and backward elimination to select features. Then, the most common 12 features among all the feature selection methods are selected. Five classifiers, named decision tree, random forest, gradient boosting, XGB, and KNN, are trained against the selected 12 features. It was found that the gradient boosting based classifiers outperformed the other classifiers. The gradient boosting method outperformed all other methods. The proposed method outperformed the existing methods in terms of accuracy, precision, recall, and F1-score. As a future work, the investigation of the detection of other Android malware by the proposed ensemble feature selection method can be carried in future.

References

1. Arora A, Peddoju SK (2017) Minimizing network traffic features for android mobile malware detection. In: Proceedings of the 18th international conference on distributed computing and networking, pp 1–10
2. Bibi I, Akhunzada A, Malik J, Ahmed G, Raza M (2019) An effective android ransomware detection through multi-factor feature filtration and recurrent neural network. In: 2019 UK/China Emerging Technologies (UCET). IEEE, pp 1–4
3. Chen R, Li Y, Fang W (2019) Android malware identification based on traffic analysis. In: International conference on artificial intelligence and security. Springer, pp 293–303
4. Elayan ON, Mustafa AM (2021) Android malware detection using deep learning. Proc Comput Sci 184:847–852
5. Fallah S, Bidgoly AJ (2019) Benchmarking machine learning algorithms for android malware detection. Jordan J Comput Inf Technol (JJCIT) 5(03)
6. Feng J, Shen L, Chen Z, Wang Y, Li H (2020) A two-layer deep learning method for android malware detection using network traffic. IEEE Access 8:125786–125796
7. Karbab EB, Debbabi M, Derhab A, Mouheb D (2018) Maldozer: automatic framework for android malware detection using deep learning. Digit Investig 24:S48–S59
8. Lashkari AH, Kadir AFA, Gonzalez H, Mbah KF, Ghorbani AA (2017) Towards a network-based framework for android malware detection and characterization. In: 2017 15th annual conference on privacy, security and trust (PST). IEEE, pp 233–23309
9. Lashkari AH, Kadir AFA, Taheri L, Ghorbani AA (2018) Toward developing a systematic approach to generate benchmark android malware datasets and classification. In: 2018 international carnahan conference on security technology (ICCST). IEEE, pp 1–7
10. Lekssays A, Falah B, Abufardeh S (2020) A novel approach for android malware detection and classification using convolutional neural networks. In: ICSOFT, pp. 606–614
11. Mim KR, Hossain MS, Tisha SA, Kalpo KR, Bakul MH, Hossain MS (2021) Traffic analysis based android sms malware detection using machine learning. In: Proceedings of the international conference on 4th industrial revolution and beyond. Springer
12. Murtaz M, Azwar H, Ali SB, Rehman S (2018) A framework for android malware detection and classification. In: 2018 IEEE 5th international conference on engineering technologies and applied sciences (ICETAS). IEEE, pp 1–5
13. Nix R, Zhang J (2017) Classification of android apps and malware using deep neural networks. In: 2017 international joint conference on neural networks (IJCNN). IEEE, pp 1871–1878
14. Taheri L, Kadir AFA, Lashkari AH (2019) Extensible android malware detection and family classification using network-flows and api-calls. In: 2019 international carnahan conference on security technology (ICCST). IEEE, pp 1–8

Analyzing Malware From API Call Sequences Using Support Vector Machines

Qasem Abu Al-Haija🆔 and Moez Krichen🆔

Abstract Malware (malicious software) are available as software or program that is deliberately developed to cause disturbance in the computation systems such as computers, servers, or networks. Typically, malware aims to drip private data/information, gain unlawful access to system resources (hardware, software, and information/data), deny authorized users from accessing system resources, or even destroy or corrupt system resources. While the level of impact for malware might range from limited to severe, it is essential to detect malware in the system at earlier stages to enable the proper defense to be activated in response to malware. In this paper, we propose a machine learning-based model for identifying malware from goodware by analyzing the API call sequences over the operating system (Windows OS) using support vector machines (SVM). The experimental results show that our model can analyze API call sequences to malware provide identification with an accuracy rate of 98.7% in 13.5 μs only. Besides, the comparison with other state-of-the-art models exhibits the advantage of our model in terms of detectability at high inferencing rates.

Keywords Malware · Goodware · Machine learning · Classification · Support vector machines (SVM) · API calls

1 Introduction

Malware [1], short for "malicious software", refers to any intrusive program created by cybercriminals (commonly referred to as "hackers") in order to steal data, damage, or destroy computers and computer systems. Viruses, worms, Trojan horses,

Q. A. Al-Haija (✉)
Department of Computer Science/Cybersecurity, Princess Sumaya University for Technology (PSUT), Amman 11941, Jordan
e-mail: q.abualhaija@psut.edu.jo

M. Krichen
FCSIT, Al-Baha University, Al-Baha, Saudi Arabia
e-mail: mkreishan@bu.edu.sa; moez.krichen@redcad.org

ReDCAD Laboratory, University of Sfax, Sfax, Tunisia

spyware, adware, and ransomware are examples of common malware [2]. Every day, according to AV-Test, an independent research institute for IT security, 350,000 new malware samples are detected. According to this institute, there are currently over 972 million malware specimens swarming the internet. As a result, the cybersecurity industrial sector spends billions of dollars seeking to detect and prevent malware. Cybercriminals, on the other hand, are constantly seeking to beat the game by altering their code and strategies.

According to [3], the methods for detecting malware are separated into two categories: behavior-based and signature-based methods. In addition, these methods may be either static or dynamic. Nowadays, Machine Learning (ML) and Artificial Intelligence (AI) approaches [4–6] play a significant role in automatic malware detection and classification. A variety of ML and AI methods, both supervised and unsupervised, have been investigated in order to detect malware and categorize it into classes [7]. More precisely, machine Learning in cybersecurity allows to learn which elements are malicious and which are benign based on patterns discovered by examining massive databases of known good and known bad elements.

In this paper, an ML-based solution is proposed to model an automated smart detection system for distinguishing malware from benign software (Goodware). Support vector machines (SVMs) are used in our model to evaluate API call sequences [8, 9] in the Windows operating system environment and enable early auto-detection for malware applications. The proposed model has been evaluated on a recent dataset using standard machine learning evaluation factors including identification accuracy, positive predictive value, negative predictive value, identification sensitivity, identification specificity, identification F1-sore, and identification time (μs) [10]. The main reason to employ SVM learning technique since SVM is known as one of the most robust and accurate algorithm among the other classification algorithms [11]. SVMs can efficiently perform a non-linear classification using what is called the kernel trick, implicitly mapping their inputs into high-dimensional feature spaces.

1.1 Our Contributions

The major contributions of this paper are briefed as follows:

- We propose an automated smart detection system to discriminate malicious software (Malware) from good software (Goodware) which is modeled as a machine learning problem.
- Our model is implemented using optimizable support vector machines (SVMs) which analyze API call sequences over Windows operating system environment to provide early auto-detection for malware applications.
- The experimental results show that our model can analyze API call sequences to malware provide identification with an accuracy rate of 98.7% in 13.5 μs only.

1.2 Paper Organization

The remaining part of the paper is outlined as follows. Section 2 reviews related works on similar approaches. In Section 3, the adopted malware detection model is presented in details. In Section 4, the main obtained results are presented. Finally, Section 5 concludes the paper.

2 Related Work

Recently, wide range of cyber-attacks have been developed and launched against various levels of communication and computation architectures of IoT and cyber-physical systems. This includes several attacks vectors including the application layer attacks (such as API calls malware attacks), the networking/communication layer attacks (such as sniffing attacks) and even at the physical layer attacks (such as DDoS attacks) [12]. The rate of launched cyber-attacks is usually undeterministic and is model as a stochastic random approach [13]. In this section, we propose a short overview on 20 recently published works which explored API call sequences for the analysis and detection of Malwares. First, the authors of [14] proposed a deep learning based system called ModCGAN for detecting malware in computing and network systems. For this, visual and text representations of the malware's dynamic activity were generated using log files containing API call sequences. In the same direction, the authors of [15] developed a multi-layer convolutional neural network to classify distinct types of malwares for IoT systems based on API call analysis. Similarly the study presented in [16] reported on a dynamic ransomware detector called DRDT. The latter is an enhancement of Text CNN (convolutional neural network for text), which was trained on API call sequences from ransomware and benign applications.

A malware classifier based on a graph convolutional network was presented in [17]. The suggested method entails extracting the API call sequence from malware code, creating a directed cycle graph, extracting the graph's feature map, and finally designing a graph convolutional network-based classifier. Another malware detection system based on deep learning models (Bi-LSTM) was presented in [18]. The proposed system considered special features called intrinsic features of the API sequences. Using a series of API calls, the study presented in [19] achieved malware detection and classification using a deep learning model. For comparison, the learning model was constructed using two alternative RNN architectures, GRU and LSTM. These same two techniques were also used in [20] for classifying different types of malwares by leveraging long sequences of API calls. The authors of [21] suggested HyMalD, a malware detection framework designed specifically for IoT systems. The proposed framework is mainly based on the Bi-LSTM and SPP-Net models.

The research presented in [22] is based on the API call functions that are extracted using the Cuckoo sandbox. Moreover, the authors of this paper used LightGBM for training the model and for the detection and classification of various kinds of ransomware. In [23], the feature-selection technique called TF-(IDF&ICF) was used for improving the performance of the used machine learning models. The authors of [24] employed NLP approaches and LDA (Latent Dirichlet Allocation) for identifying key features and for improving the performance of the malicious API call detection. Generic behavioral graph models were built in the work presented in [25] to represent both malicious and non-malicious processes. The behavioral differences between malicious and non-malicious API calling sequences were demonstrated by the models.

The authors of [26] proposed a purification and optimization step for the API call sequence. They also employed the fastText and BERT algorithms for the detection and classification of malwares. In the work presented in [27], word embedding was used for understanding contextual relationships existing between API functions in malware call sequences. Moreover, the authors proposed a new technique for detecting and predicting malwares using Markov chains. A similar work based on word bending and Markov chains was presented in [28]. In addition, the authors of this work proposed a heuristic technique for identifying malware's mimicry activities. In [29], an incremental malware detection model for meta-feature API and system call sequence was proposed. For creating sequential system calls, the authors used "NITR System call Tracer". For sequential API calls, they generated a list of anomaly scores for every API call sequence using "Numenta Hierarchical Temporal Memory".

Another interesting framework called "Malware Detection using Complex Network" (MDCN) work was presented in [30]. The authors of this paper considered an "API Call Transition Matrix" for generating complex network topology and extracting an appropriate feature set for distinguishing malware and benign applications. The generated feature set is then sent to several machine learning classifiers. The research [31] proposed using FastText as a classifier and word representation to dynamically extract the API call sequence patterns of various types of malwares. The suggested approach was tested on two publicly available malware datasets. The work presented in [32] proposed a joint framework based on local and global features for detecting malwares for solving the problem of network security of smart cities. The proposed framework is called LGMal. It is a combination of "Stacked Convolutional Neural Network" and "Graph Convolutional Networks". Finally, the authors of [33] considered two methodologies of dynamic analysis by making use of the API call features for detecting malwares, namely: "API Call Frequency Mining" and "API Call Transition Matrix Mining". For Malware Detection, the authors used several supervised algorithms.

Unlike the aforementioned models, we present a new predictive model that analyses API Call sequences launched through Windows OS environment to identify malware calls form goodware calls. The proposed system makes use of optimizable support vector machines (O-SVMs) to ensure high performant identification system. O-SVM scans through the search space of hyperparameters to select the optimal configuration for the SVM hyperparameters that can be employed to address the identification problem stated in this research. Accordingly, our optimized SVM showed an outspending performance in detecting malware of API callswith short inference period. Therefore, due to this lightweight model, it can be employed to detect malware into IoT Network Traffic [34].

3 System Modeling

Like several intelligent intrusion/cyber-attacks detection systems [35], In this paper, we model the problem of discriminating malicious software (Malware) from good software (Goodware) based on the API call sequences over the Windows operating system environment as a machine learning task. To do so, we have firstly collected an up-to-date and representative dataset from the Kaggle repository [36]. The dataset comprises over 43,000 samples of API call sequences (~42,000 for malware API call sequences and 1,100 for goodware API call sequences). Also, the dataset is organized into 101 features and one column for class labeling: the first column is the "Hash" which represents an MD5 hash value (32 bytes string) for every API call. Then, the next 100-columns are numerical integer values (0–306), named t0…t99, to represent the API call. The last column is the target class labels which contain either malware (1) or goodware(0). The collected dataset has been undergone a number of preprocessing operations as illustrated in Fig. 1.

Once the dataset is preprocessed, then it can be fed for processing at the support vector machine (SVM) learning model illustrated in Fig. 2. SVM is a supervised non-parametric learning method that has been widely used to address prediction and classification tasks [11] for several machine learning-based applications. Based on Fig. 2, initially several SVMs (1, 2, …N) are created to be allocated actively by an automated classifier to several training subsets obtained from the original training dataset repeatedly. All created SVMs are trained independently each with its subset

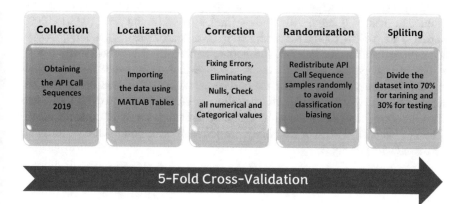

Fig. 1 Preprocessing operations were applied to the collected dataset in sequence

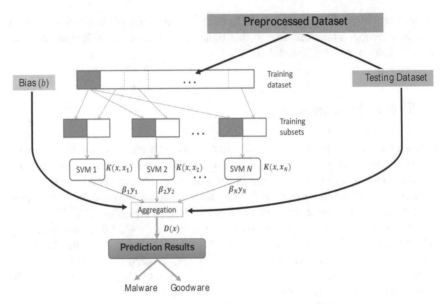

Fig. 2 Illustration of the support vector machine (SVM) learning model

of the training dataset and when the training procedure is accomplished, All results from trained SVMs are accumulated using an appropriate aggregation function to generate the final results and provide the output class.

Besides, the complete system configuration and specification for parameters and hyperparameters of the SVM model are summarized in Table 1. Please note that we are reporting the best Parameters/Hyperparameters configuration of our optimizable SVM model. To obtain the optimizable SVM, the proposed model makes use of several options for the Hyperparameter in order to pick up the optimal SVM model for analyzing the API sequence calls. The hyperparameter search range includes the Multiclass method (One-vs-All, One-vs-One), Box constraint level (0.001–1000), Kernel scale (0.001–1000), Kernel function (Gaussian, Linear, Quadratic, Cubic), and with two options for standardization (i.e., standardize data: true, false).

Finally, the system has been evaluated via several performance factors including identification accuracy (ACC), the positive predictive value/precision (PPV), the negative predictive value (NPV), the identification sensitivity/recall (SEN), the identification specificity (SPC), the harmonic score identification (HSC), and the identification time (IDT) [37]. For simplicity, we summarize the computational formulas for these factors in Fig. 3.

Table 1 Parameters/hyperparameters configuration of our optimizable SVM model

Parameters/hyperparameters	Value
Preset/kernel scale:	Optimizable SVM/scale-1
Kernel function:	Cubic
Box constraint level:	0.0010067
Multiclass method:	One-vs-One
Optimizer:	Bayesian optimization
Acquisition function:	Expected improvement per second plus
Iterations/Training time limit:	30/no limit
Feature Selection	all features used except "hash"
Total Validation cost:	551 samples
Prediction speed:	~74000 sample/s
Training time:	12177 s

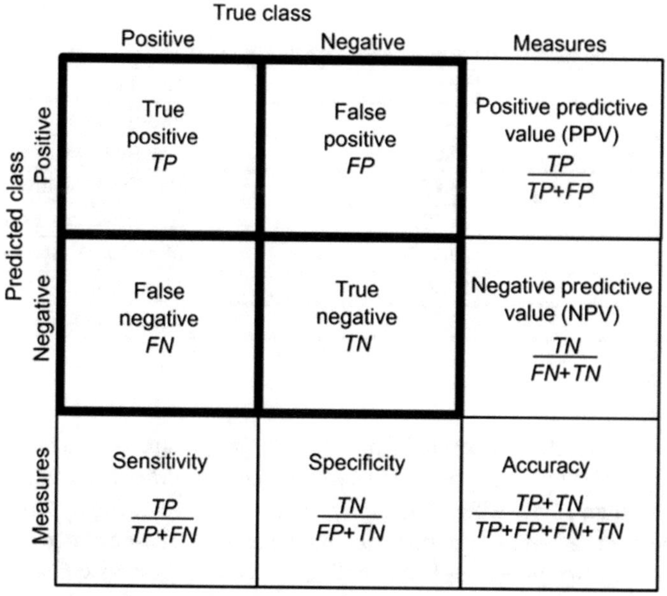

Fig. 3 Summary of performance system of measurements

4 Results and Analysis

In this section, we describe the results of the system evaluation in terms of different key ovulation metrics. In the beginning, Fig. 4 traces the course of the minimum classification error for the proposed SVM model with respect to the learning iterations. According to the figure, the model has been trained and validated for 30-iterations.

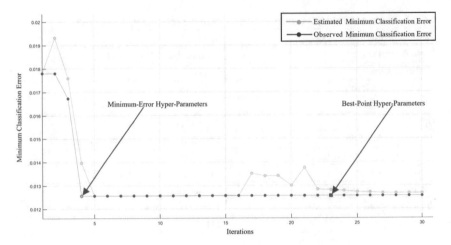

Fig. 4 Analyzing the performance of the proposed model: MCE versus iterations

Table 2 Performance evaluation indicators for the proposed model

Metric	Value	Meaning
ACC	98.74%	Identification Accuracy
PPV	76.90%	Positive Predictive Value
NPV	98.85%	Negative Predictive Value
SEN	95.20%	Identification Sensitivity
SPC	99.87%	Identification Specificity
HSC	85.10%	Identification F1-Sore
IDT	13.51μs	Identification time (μs)

However, the model has achieved the best point hyper-parameters only after iteration number 23. The best point hyper-parameters have been recorded at the minimum error which is less than 1.25% which is the minimum error for selecting hyper-parameters that can be used to build up the detection model at the highest accuracy rates.

Accordingly, the model has been configured with the best specification that can score the optimal performance metrics as reported in Table 2. Table 2 reports the best achievable results from the proposed optimized SVM model to recognize malware from API call sequences. According to the table, the model exhibits a remarkable performance notching high identification accuracy rate of 98.74%, a high NPV rate of 98.85%, a high identification sensitivity rate of 99.87%, a high identification specificity of 99.87%, and a high identification speed with only 13.51μs required to provide the identification for every single sample. However, the system performed tolerably in terms of PPV and HSC achieving 76.90 and 85.10% respectively.

Besides, Fig. 5 demonstrates the plots for the area under the curve (AUC) that correlate the true positive rate (TPR) with the true negative rate (TNR) at different

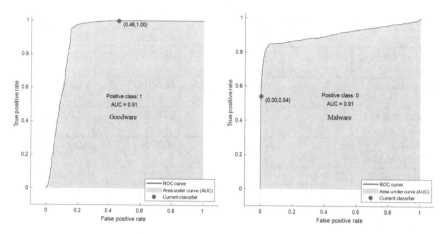

Fig. 5 Analyzing area under the curve for each class

threshold values (0.0, 0.2, 0.4, 0.6, 0.8, and 1.0). The figure shows the plots for every individual class in the system including the AUC value for the positive class (Goodware class) attained an average of 91% after the threshold (0.46, 1.0) and the AUC value for the positive class (Malware class) attained an average of 91% after the threshold (0.00, 0.54). On average, the overall AUC value for the system is 91%.

Lastly, to benchmark the proposed system and show its superiority, we have compared the performance accuracy of the proposed model with several state of art models that concerns with malware detection based on the API Call Sequences. For instance, in [36], the authors utilized three supervised learning models to detect malware of API Call Sequences, namely, one layer-Deep Graph Convolutional Neural Networks (1-layer DGCNN), two layers-Deep Graph Convolutional Neural Networks (2-layer DGCNN), and Long short-term memory (LSTM). Their model-based 2-layer DGCNN has recorded its best accuracy ratio of 92.44%. However, our proposed model surpassed this accuracy by 6.3%. Also, in [38], the researchers have employed a Multi-Dimensional Naïve Bayes Classification (MDNBS) for malware-based API sequence detection. Their model scored a 93.20% of accuracy which is even lower than our best accuracy by 5.5%. Furthermore, contributors of [39] have characterized the performance for several similarity algorithms. Their best accuracy belonged to the model-based longest common subsequence (LCS) recording 94.89% which is lower than our accuracy rate by 4%. In addition, in [40], the authors have examined eight different machine learning models including Multilayer neural Networks (MNN), Convolutional Neural Network (CNN), Recurrent Neural Network (RNN), Gated recurrent unit (GRU), LSTM, Random Forest, Decision Tree, and XGBoost. Their experimental outcomes showed that CNN based model is the best model among others with 94.66% accuracy. Similarly, authors in [41] have employed LSTM for malware detection based on the API Call Sequences and their model recorded a high classification accuracy of 95%. Finally, in [42], the authors have cascaded the CNN with Term Frequency, Inverse Document Frequency (TF-

Table 3 Comparison with state of art models

Paper	Year	Model	Accuracy	Prediction tine
Schranko et al. [36]	2018	1-layer DGCNN	91.05%	>1.0 ms
Schranko et al. [36]	2018	2-layer DGCNN	92.44%	>1.0 ms
Schranko et al. [36]	2018	LSTM	84.88%	~1.0 ms
Jerlin and Marimuthu [38]	2018	MDNBS	93.20%	0.11s
Kim et al. [39]	2019	similarity-LCS	94.89	<1.0 ms
Fadadu et al. [40]	2019	MNN	94.65%	<1.0 ms
Fadadu et al. [40]	2019	CNN	94.66%	<1.0 ms
Fadadu et al. [40]	2019	RNN	92.93%	<1.0 ms
Fadadu et al. [40]	2019	GRU	93.56%	<1.0 ms
Fadadu et al. [40]	2019	LSTM	94.27%	~1.0 ms
Fadadu et al. [40]	2019	Random Forest	92.77%	<1.0 ms
Fadadu et al. [40]	2019	Decision Tree	88.77%	<1.0 ms
Fadadu et al. [40]	2019	XGBoost	90.26%	<1.0 ms
Catak et al. [41]	2020	LSTM	95.00%	~1.0 ms
Schofield et al. [42]	2021	1-D CNN	91.00%	~1.0 ms
Proposed Model	2022	Optimizable SVM	98.74%	13.51μs

IDF) vectors. Their overall system accuracy is 91% which is less than our obtained accuracy of 6.74%. Table 3 summarizes the previously discussed findings.

5 Conclusion

In this paper, an automated smart detection system to discriminate malicious software (Malware) from good software (Goodware) is modeled as a machine learning problem. The proposed model is implemented using support vector machines (SVMs) that analyze API call sequences over Windows operating system environment to provide early auto-detection for malware applications. The proposed model has been evaluated on a recent dataset using standard machine learning evaluation factors including identification accuracy, positive predictive value, negative predictive value, identification sensitivity, identification specificity, identification F1-sore, and identification time (μs). The experimental results show that our model can analyze API call sequences to malware provide identification with an accuracy rate of 98.7% in 13.5 μs only. In future, we intended to evaluate our model using other datasets comprising API call sequences of other operating systems such as LINUX OS and MAC OS to provide early auto-detection for malware applications. Also, we will consider the implementation of several other supervised learning techniques to gain more insights into the system model and solution approach.

References

1. Waghmare JM, Chitmogrekar MM (2022) A review on malware detection methods. SAMRID-DHI: J Phys Sci Eng Technol 14(01):38–43
2. Abu Al-Haija Q, Alsulami A (2021) High performance classification model to identify ransomware payments for heterogeneous bitcoin networks. Electronics 10:2113
3. Souri A, Hosseini R (2018) A state-of-the-art survey of malware detection approaches using data mining techniques. Hum-Centric Comput Inf Sci 8(1):1–22
4. Mihoub A, Fredj OB, Cheikhrouhou O, Derhab A, Krichen M (2022) Denial of service attack detection and mitigation for internet of things using looking-back-enabled machine learning techniques. Comput Electric Eng 98:107716
5. Srinivasan S, Ravi V, Sowmya V, Krichen M, Noureddine DB, Anivilla S, Soman K (2020) Deep convolutional neural network based image spam classification. In: 2020 6th conference on data science and machine learning applications (CDMA). IEEE, pp 112–117
6. Abu Al-Haija Q, Al-Dala'ien M (2022) Elba-iot: An ensemble learning model for botnet attack detection in iot networks. J Sens Actuator Netw 11:18
7. Rani N, Dhavale SV, Singh A, Mehra A (2022) A survey on machine learning-based ransomware detection. In: Proceedings of the seventh international conference on mathematics and computing. Springer, pp 171–186
8. Chen X, Hao Z, Li L, Cui L, Zhu Y, Ding Z, Liu Y (2022) Cruparamer: Learning on parameter-augmented api sequences for malware detection. IEEE Transactions on Information Forensics and Security 17:788–803
9. Cannarile A, Dentamaro V, Galantucci S, Iannacone A, Impedovo D, Pirlo G (2022) Comparing deep learning and shallow learning techniques for api calls malware prediction: A study. Applied Sciences 12(3):1645
10. Abu Al-Haija Q, Ishaiwi A (2022) Multiclass classification of firewall log files using shallow neural network for network security applications. In: Soft computing for security applications. Advances in intelligent systems and computing, vol 1397, pp 27–41
11. Abu Al-Haija Q, Smadi AA, Allehyani MF (2021) Meticulously intelligent identification system for smart grid network stability to optimize risk management. Energies 14(21):6935
12. Abu Al-Haija Q (2022) Top-down machine learning-based architecture for cyberattacks identification and classification in iot communication networks. Front Big Data 4:782902
13. Abu Al-Haija Q (2020) On the security of cyber-physical systems against stochastic cyber-attacks models. In: 2021 IEEE International IOT, electronics and mechatronics conference (IEMTRONICS)
14. Bera P et al (2021) Modcgan: a multimodal approach to detect new malware. In: 2021 international conference on cyber situational awareness, data analytics and assessment (CyberSA). IEEE, pp 1–2
15. Lin Q, Li N, Qi Q, Hu J (2021) Using api call sequences for iot malware classification based on convolutional neural networks. Int J Softw Eng Knowl Eng 31(04):587–612
16. Qin B, Wang Y, Ma C (2020) Api call based ransomware dynamic detection approach using textcnn. In: 2020 international conference on big data, artificial intelligence and internet of things engineering. IEEE, pp 162–166
17. Li S, Zhou Q, Zhou R, Lv Q (2022) Intelligent malware detection based on graph convolutional network. J Supercomput 78(3):4182–4198

18. Li C, Lv Q, Li N, Wang Y, Sun D, Qiao Y (2022) A novel deep framework for dynamic malware detection based on api sequence intrinsic features. Comput Secur 102686
19. Aditya WR, Hadiprakoso RB, Waluyo A et al (2021) Deep learning for malware classification platform using windows api call sequence. In: 2021 international conference on informatics, multimedia, cyber and information system (ICIMCIS). IEEE, pp 25–29
20. Li C, Zheng J (2021) Api call-based malware classification using recurrent neural networks. J Cyber Secur Mobil 617–640
21. Jeon J, Jeong B, Baek S, Jeong Y-S (2021) Hybrid malware detection based on bi-lstm and spp-net for smart iot. IEEE Trans Ind Inf
22. Nguyen DT, Lee S (2021) Lightgbm-based ransomware detection using api call sequences. Int J Adv Comput Sci Appl 12(10)
23. Qin B, Zhang J, Chen H (2021) Malware detection based on tf-(idf&icf) method. J Phys Conf Ser 2024:012030. (IOP Publishing)
24. Voronin V, Morozov A (2021) Analyzing api sequences for malware monitoring using machine learning. In: 2021 3rd international conference on control systems, mathematical modeling, automation and energy efficiency (SUMMA). IEEE, pp 519–522
25. Amer E, Zelinka I, El-Sappagh S (2021) A multi-perspective malware detection approach through behavioral fusion of api call sequence. Comput Secur 110:102449
26. Yesir S, Soğukpinar İ (2021) Malware detection and classification using fasttext and bert. In: 2021 9th international symposium on digital forensics and security (ISDFS). IEEE, pp 1–6
27. Amer E, Zelinka I (2020) A dynamic windows malware detection and prediction method based on contextual understanding of api call sequence. Comput Secur 92:101760
28. Amer E, El-Sappagh S, Hu JW (2020) Contextual identification of windows malware through semantic interpretation of api call sequence. Appl Sci 10(21):7673
29. Kishore P, Barisal SK, Mohapatra DP (2020) An incremental malware detection model for meta-feature api and system call sequence. In: 2020 15th conference on computer science and information systems (FedCSIS). IEEE, pp 629–638
30. Mohanasruthi V, Chakraborty A, Thanudas B, Sreelal S, Manoj B (2020) An efficient malware detection technique using complex network-based approach. In: 2020 national conference on communications (NCC). IEEE, pp 1–6
31. Feng L, Cui Y, Hu J (2020) Detection and classification of malware based on fasttext. In: 2020 IEEE international conference on artificial intelligence and information systems (ICAIIS). IEEE, pp 126–130
32. Chai Y, Qiu J, Su S, Zhu C, Yin L, Tian Z (2020) Lgmal: a joint framework based on local and global features for malware detection. In: 2020 international wireless communications and mobile computing. IEEE, pp 463–468
33. Thanudas B, Sreelal S, Raj VC, Maji S (2020) An efficient approach for detecting malware using api call mining. Int J Adv Sci Technol 29:2254–2274
34. Abu Al-Haija Q, Al-Badawi A (2022) Attack-aware iot network traffic routing leveraging ensemble learning. Sensors 22:241
35. Abu Al-Haija Q, Krichen M, Abu Elhaija W (2022) Machine-learning-based darknet traffic detection system for iot applications. Electronics 11(4)
36. Schranko de Oliveira A, Sassi RJ (2019) Behavioral malware detection using deep graph convolutional neural networks, vol 10043099, p v1. https://doi.org/10.36227/techrxiv
37. Abu Al-Haija Q, Al-Badawi A, Reddy Bojja G (2022) Boost-defence for resilient iot networks: a head-to-toe approach. Expert Syst 39:e12934
38. Jerlin MA, Marimuthu K (2018) A new malware detection system using machine learning techniques for api call sequences. J Appl Secur Res 13(1):45–62
39. Kim H, Kim J, Kim Y, Kim I, Kim KJ, Kim H (2019) Improvement of malware detection and classification using api call sequence alignment and visualization. Cluster Comput 22(1):921–929
40. Fadadu F, Handa A, Kumar N, Shukla SK (2019) Evading api call sequence based malware classifiers. In: International conference on information and communications security. Springer, pp 18–33

41. Catak FO, Yazı AF, Elezaj O, Ahmed J (2020) Deep learning based sequential model for malware analysis using windows exe api calls. PeerJ Comput Sci 6:e285
42. Schofield M, Alicioglu G, Binaco R, Turner P, Thatcher C, Lam A, Sun B (2021) Convolutional neural network for malware classification based on api call sequence. In: 8th international conference on AI and applications

Android Ransomware Attacks Detection with Optimized Ensemble Learning

Shaharia Sifat, Md. Sakir Hossain📵, Sadia Afrin Tonny, Bejoy Majumder, Riftana Mahajabin, and Hossain Md. Shakhawat📵

Abstract Android is the most widely used operating system for mobile devices and is the most common target of various cyber attacks. Ransomware is one of the most common and dangerous malware attacks among of different types of attacks. In the last few years, a sharp escalation of the ransomware attack is observed. Traditional ransomware detection techniques fail to rein in the rise of the ransomware attack. Recently, machine learning is increasingly used in detecting cyberattacks. However, the detection of various types of ransomware is largely overlooked. Even the Android ransomware detection accuracy of the existing solutions are highly limited. In this paper, we propose a machine learning technique for detecting various types of Android ransomware from traffic analysis. The objective is to attain a higher detection rate. To this end, we exploit an ensemble machine learning technique with optimized hyperparameters. The hyperparameters of the Bagging ensemble learning are optimized using the grid search. Through simulations, the proposed machine learning models are found to achieve up to 11% higher detection accuracy than the state-of-the-art solution. The pletor ransomware can be detected with the highest accuracy (95.29%), while the average detection accuracy is the lowest (74.09%) for koler among all the types of Android ransomware.

S. Sifat · Md. S. Hossain (✉) · S. A. Tonny · B. Majumder · R. Mahajabin · H. Md. Shakhawat
American International University-Bangladesh, 408/1, Kuratoli Road, Khilkhet, Dhaka 1229, Bangladesh
e-mail: sakir.hossain@aiub.edu

S. Sifat
e-mail: 17-35333-2@student.aiub.edu

S. A. Tonny
e-mail: 17-34123-1@student.aiub.edu

B. Majumder
e-mail: 17-34959-2@student.aiub.edu

R. Mahajabin
e-mail: 17-35004-2@student.aiub.edu

H. Md. Shakhawat
e-mail: shakhawat@aiub.edu

© The Author(s), under exclusive license to Springer Nature Switzerland AG 2023
A. A. Abd El-Latif et al. (eds.), *Advances in Cybersecurity, Cybercrimes, and Smart Emerging Technologies*, Engineering Cyber-Physical Systems and Critical Infrastructures 4,
https://doi.org/10.1007/978-3-031-21101-0_4

Keywords Ransomware · Android · Ensemble learning · CICAndMal2017 ·
Hyperparameter tuning

1 Introduction

Information technology plays an imperative role in our lives because it allows us
to cope with the ever-changing nature of our daily lives. In our everyday lives, we
are using various online platforms through information technology. For example,
the whole world has witnessed stagnation due to the Covid 19 pandemic. But using
information technology, we can do most of our activities from home. We can continue
our academic activities using online platforms and order any product staying home.
We can make an appointment with the doctor online. We can keep the economy
afloat by using information technology. Otherwise, the pandemic situation could
have become more devastating. The use of hand-held mobile devices makes our lives
more comfortable due to the ubiquitous network facilities. However, our reliance on
information technology poses a big risk due to various cyberattacks. As more than
60% of the mobile devices used around the world use Android operating systems,
Android-based systems are common targets of the cyberattacks.

Some common cyberattacks on Android devices include adware, scareware, SMS
malware [14], and ransomware. Out of the damaging cyberattacks, the ransomware
has become a common attack nowadays. It is a type of malware that restricts access to
the infected Android systems [12]. This form of technological blackmailing exploits
software and hardware vulnerabilities. It encrypts the files of the victim's computer
or mobile devices with a private key. The attacker then demands a ransom from the
victim to rebuild access to the data upon payment. For this reason, we need some
solutions to the problem to drastically reduce its negative impact, prevent the attacks
or if possible, eliminate it from existence.

In literature, a considerable effort is devoted to finding various techniques for
detecting the Android malware attacks. For example, a hybrid technique is proposed
in [10] where a convolution neural network (CNN) is used for selecting important
features. Afterward, a light-weight classifier, Jaccard algorithm, is used for classification purposes. This approach attains a higher malware detection accuracy with
lower resource consumption. In most papers, balanced datasets are used. However,
the proportion of the Android malware among all malware is a little more than 10%
in real-world. To mimic this scenario, an imbalanced dataset with 10.9% malware
is used in [20]. Various feature selection and classifiers are exploited to find the
highest accuracy. The combination of deep learning with two hidden layers and the
significant set of metrics feature selection method provides the highest accuracy of
89.36%. In [19], the object-oriented software metrics are extracted from the Android
apps using reverse engineering. Then, the class of an app is determined using Virus-
Total platform. Many classifiers with various settings of hyperparameters are used
to identify malware. A maximum of 100% accuracy is achieved. Four static features, named permissions, permission rate, monitory system events, and API calls,

are considered for feature selection in [2]. To this end, a new dataset is created combining three public datasets. The frequency counting (for monitory system events) and cosine similarity (for permissions and API calls) are used for feature selection. Five machine learning models and a deep learning model, named recurrent neural networks, are tested on the dataset. The deep learning approach provides the highest accuracy of 98.58%.

A dedicated software-defined networking (SDN)-based method is presented in [6] for detecting and mitigating ransomware attacks, with a focus on HTTP traffic analysis. The ransomware detection rate is up to 98%, with a 1–5% false positives. However, this approach is solely dependent on the scale of the data inserted by the victim into the HTTP POST messages' outgoing sequences. In [18], both static and dynamic host-based analysis techniques are used to examine the wannacry ransomware in the win32 environment. A prototype of software is also demonstrated, which can stop the wannacry from spreading its payload. However, this study does not focus on wannacry's network behaviors in order to derive the behavioral characteristics, but instead provides a minimal detail about the virus's network.

New ransomware uses a variety of strategies to evade anti-virus detection, so the existing technologies are insufficient to handle such kind of ransomware. An API-based ransomware detection system (API-RDS) is proposed in [3] for static analysis of Android apps to detect ransomware applications. The API-RDS service can be delivered via a mobile app, a website app, or an integration with other static or dynamic detection systems [3].

A preventive countermeasure ransomware detection method, called GreatEatlon, is proposed in [21]. The GreatEatlon is a variant of HelDroid [4], but focuses on crypto-ransomware. The authors enhance the text-threatening detector's ability to search images in addition to plain text. This method has the same flaw as HelDroid in that it relies on the availability of text/images and is unsuccessful in languages that lack step structures. The GreatEatlon scans AndroidManifest.xml and then the meta-data file for unsafe policies to detect encryption. It then checks the source code, which is computationally intensive. Additionally, the authors conduct both forward and backward analyses to see the malicious reflection. The GreatEatlon outperforms the HelDroid. In [7], two methods are used for detecting Android malware: the first is based on the hidden Markov model, and the second is based on structural entropy. It is shown that these methods can detect Android malware with high precision.

In [14], various machine learning algorithms are used to detect various types of SMS malware. The ANOVA-F feature selection algorithm is employed. The CICAndMal2017 dataset is used. The decision tree classifier is found to provide the highest detection accuracy. A comprehensive investigation is carried out in [8], where various Android malware detection is carried out. The types of various malware such as adware, scareware, ransomware and SMS malware are detected using machine learning. However, the accuracy of the detection of each type of Android ransomware is not investigated. In [1], some investigations are done to identify Android malware in three different scenarios: detection of malware from the benign traffic, detection of specific kind of malware out of the adware, scareware, ransomware, and SMS malware, and the detection of family of each kind of malware. However, there is no investigation about the types of ransomware. The question "what is the detection accuracy of each kind of Android ransomware?" is not answered. However, this aspect is investigated in [15], where various semi-supervised machine learning algorithms are used for detecting each kind of ransomware. The wrapper-based random forest with One-R feature selection algorithm is used. However, the detection accuracy is very limited for most of the ransomware.

From the above discussion, it is evident that a very few research is carried out to detect various kinds of Android ransomware. Even if such investigation is done in [15], the detection accuracy is very limited. In this paper, our objective is to develop machine learning models to detect various types of ransomware attacks with increased detection accuracy. To this end, we use bagging ensemble learning with optimized hyperparameters obtained through grid research. It is found that the proposed model outperforms the models presented in [15].

2 Android Ransomware

Ransomware is one of the malware that encodes a victim's documents or denies the owner of a device from accessing his device. The attacker then demands a ransom from the victim to rebuild access to the data upon payment. There are two main types of ransomware: crypto-ransomware and locker ransomware. The crypto-ransomware encrypts valuable files of a device so that the user cannot access them. Cyberthieves who conduct crypto-ransomware attacks make money by demanding that the victims pay a ransom to get their files back. Locker ransomware does not encrypt files. Instead, it locks the victim out of their device, preventing them from using it. Once he is locked out, cybercriminals demand ransom to unlock the device. Next, we describe the types of ransomware which are commonly used for attacking Android devices:

1. **Charger**: It is a type of ransomware which copies all data from a device. This malware attacks when users install the power saver app from the play store. It is the most costly ransomware for victims [9].
2. **Pletor**: This malware is a kind of cryptoLocker, which can lock all files and data. It demands ransom to pay. If users do not pay the ransom, the criminals never give back the lost data and files [9].
3. **Jisut**: Jisut first appeared in 2014 in Chinese market. Thereafter, it spread around the world [13]. It first locks the screen of a device with a permanent screen. Then, it asks ransom. Sometimes, the screen is not unlocked even after paying the ransom. There are many variants of this ransomware.
4. **Porndroid**: The users are lured to download a porn. When the victim clicks the file to watch the video, a message is shown that he has watched an illegal porn. For this reason, a security intelligence such as Federal Bureau of investigation (FBI) will hunt for him. To avoid being caught, the victim is asked to pay ransom.

5. **Koler**: Koler is a type of ransomware that is designed as a fake adult theme app. This application is mainly for the United States. This malware locks the phone, and asks to pay a ransom. Koler can scan the victim's device to know his location. Then, it generates a customized lock screen that is based on the victim's location.

6. **RansomBO**: Ransombo is a bot malware that enables the attacker to take control of the infected devices. This malware spreads virus through the Internet, email and other medium like DVD, CD, USB drive etc. The infected device runs slowly and the system often crashes for RansomBO attack.

7. **Lockerpin**: Lockerpin is a type of Locker ransomware that can lock smart phone. Firstly, users install it to their device. Then it downloads a file which is known as a lockerpin. But this fake app resets the screen lock pin. Some victims cannot get back their files unless they pay ransom [9].

8. **Simplocker**: Simplocker is a type of mobile ransomware and uses the AES cypher type encryption. This malware destroys all device's functionalities. For this reason, the device cannot be used. Sometimes, the users install some popular app, and it is a host of Android malware. This malware can scan SD card, then encrypt by using AES cypher.

9. **Svpeng**: Svpeng is mainly a Trojan that affects Android devices. This malware was created for stealing the credit card information of Russian bank customers. Svpeng can check banking information, and they demand a ransom [9].

10. **Wannalocker**: WannaLocker is a type of crypto virus. It is a fake security application. Sometimes this malware shows the fake banking login page, then they collect the banking information. This malware adds a lock screen to user's phone so that victims can not unlock the device [9].

3 Proposed System Design

In this section, we present the proposed framework for detecting various types of Android ransomware. However, before discussing the proposed framework, the background of the ensemble learning will be presented.

3.1 Ensemble Learning

Ensemble learning is a technique of using multiple classifiers on a dataset to obtain a higher classification performance compared to that obtained from an individual classifier. Depending on how multiple classifiers interact with themselves, there are several types of ensemble learning. Of them, the bagging, boosting, and stacking are most popular. In the bagging approach, each classifier works on a subset of the dataset independently. On the other hand, the classifiers are built sequentially in the boosting technique with the objective of a classifier is to rectify the prediction error caused by its previous classifier. The working principle of the stacking ensemble learning is somewhat different from the previous two ensemble learning algorithms. The classifiers are divided into two groups: level 0 which contains multiple classifiers, and level 1 which contains only one classifier. The level 0 classifiers are trained first. Then, the level 1 classifier is used to combine the results produced by the classifiers of the level 0 [5]. In this paper, we consider the bagging ensemble learning as it is efficient in allowing multiple base classifiers to work independently. The details of the bagging classifier is discussed next.

3.2   Bagging Classifier

In bagging ensemble learning, various machine learning algorithms are trained against the respective subset of a dataset. The objective is to train multiple classifiers against the dataset with lower variance. This lower variance in the dataset produces a stable model. To get the low variance dataset, the dataset is randomly divided into multiple subsets. In order to avoid the overfitting problem, bootstrapping can be applied. In bootstrapping, when a sample of the dataset is selected for a subset, a copy of the sample is kept in the dataset so that the original dataset remains unchanged. It makes each subset independent of other subsets. The bagging ensemble learning works best when various base classifiers are trained independent of one another [22].

Figure 1 shows the architecture of the ensemble learning. First, multiple subsets of the dataset are created. Then, a machine learning model is trained against each of the subsets. The classifiers can be of the same type or different types. For example, we can use decision tree as a base classifier, and a decision tree classifier is independently trained against a subset of data. This kind of ensemble learning is called homogeneous ensemble learning. In the later case, we can use various types of classifiers with each classifier is trained against a subset of data. This technique is known as the heterogeneous ensemble learning [22]. Each of the classifiers can either use the same hyperparameters or different hyperparameters. Furthermore, the classifiers may

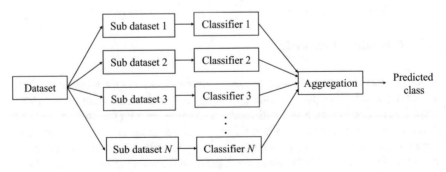

Fig. 1 Architecture of bagging ensemble learning

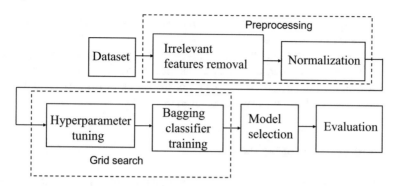

Fig. 2 Proposed Android ransomware detection framework

also work with varying number of features. Finally, the prediction results of the classifiers are aggregated to produce the final prediction. There are various methods of aggregating the results such as averaging, weighted averaging, voting and so on [17]. The higher independence of the classifier leads to a better performance [22].

3.3 Proposed Ransomware Detection Framework

The proposed framework for the Android ransomware detection is shown in Fig. 2. Each of the components of the block diagram is discussed below.

3.3.1 Dataset

In this paper, we use the CICAndMal2017 dataset [11]. This dataset contains traffic information about four different kinds of attacks: adware, scareware, ransomware, and SMS malware. We consider those data which are related to the ransomware attack and benign traffic. Traffic about 10 types of ransomwares is captured. The types of the ransomware include charger, jisut, koler, lockerpin, pletor, porndroid, ransomBO, simplocker, svpeng, and wannalocker. The svpeng and pletor ransomware have the highest (54276) and lowest (4715) number of samples, respectively. There are more than one million of benign traffic samples. Due to the variation of the number of samples per classes, the dataset is highly imbalanced. We make a total of 10 datasets with each dataset consists of samples of a specific type of ransomware and benign traffic. To make each dataset balanced, we make the number of samples of the benign traffic equal to the number of samples available for the particular ransomware type. For example, for svpeng ransomware, the newly created dataset consists of 54276 benign and svpeng ransomware traffic samples individually as the CICAndMal2017 dataset consists of 54276 samples of the svpeng ransomware. For this reason, all the newly created datasets are balanced and contain binary types of traffic (i.e., two classes). In rest of the paper, we will call each dataset according to the ransomware type. For example, the dataset that consists of the traffic of the svpeng and benign data transfer, we will call it as svpeng dataset.

3.3.2 Preprocessing of the Dataset

The preprocessing of a dataset is required to make the dataset suitable for building a classification model. Sometimes, some preprocessing improves the classification accuracy. Next, we will discuss about various preprocessing techniques of the datasets.

The CICAndMal2017 dataset consists of 84 features in each sample. However, some of the features include information about a specific flow of traffic. Of them, the FlowID, SenderIP, DestinationIP, and Timestamp features do not represent any property of traffic of a real world network. The FlowID is a sequence number. The IP addresses used in the dataset creation will not be relevant to the real world attackers' IP addresses. The Timestamp feature represents the date and time of the attack. Since the attacks were artificial, and done in the lab environment, the time of the data collection of the attack is also irrelevant to the real-world attack. For this reason, we remove these four features from the dataset similar to [15]. Furthermore, there are twelve features whose values are zero in all samples. The features numbers are 37, 38, 39, 51, 55, 56, 62, 63, 64, 65, 66, and 67. Since all values of these features are zero, these features have no effect on the type of the traffic. For this reason, we remove these features from the newly created datasets.

The dataset contains a wide range of values of different features. Some features have very large values, while the others have very low values. In any mathematical operation, the large difference between the two or more variables leads to the dominance of the large value on the result of the operation. To make each feature to have an equal effect on a calculation, a technique named normalization is carried out on each feature separately so the feature values are confined between 0 and a small values (usually 1). In this paper, we use the min-max normalization defined by the following formula [16]:

$$x_{new} = \frac{x - x_{min}}{x_{max} - x_{min}} \tag{1}$$

where x is a feature value, x_{new} represents the new value of a feature, and x_{max} and x_{min} represent the largest and smallest values of the feature, respectively. The min-max normalization scales the values of a feature between 0 and 1.

3.3.3   Grid Search

The performance of a classifier is considerably dependent on its hyperparameters where the hyperparameters are those parameters of a classifier which are not optimized during the training process, and are set explicitly before the training phase. For example, the number of neurons of a layer of an artificial neural network is a hyperparameter as this value has to be set before training the model. The number of neurons in a layer has a significant impact on the performance of a neural network. Similarly, there are several parameters of the bagging classifier such as the base estimator, the number of estimators, the number of features used to train the base estimator, the number of samples used to train the classifier, whether bootstrapping is used or not to create the subsets of the dataset, and so on. Each of the hyperparameters has an impact on the performance of a classifier. However, there is no value of a hyperparameter that provides the optimum result in all problems. For this reason, we use the grid search to find the optimal values of some of the hyperparameters of the bagging classifier for each experiment.

In the grid searching, a grid of hyperparameters is created first, then a classifier is trained setting a particular combination of the hyperparameters. Let us assume that we want to optimize the two hyperparameters: base estimator and the number of estimators, of a bagging classifier. We have to first specify the values of the hyperparameters. Suppose that base_estimators = [GaussianNB(), DecisionTreeClassifier()], and num_estimators = [20, 30]. Then, the combinations [GaussianNB(), 20], [GaussianNB(), 30], [DecisionTreeClassifier(), 20], [DecisionTreeClassifier(), 30] will be created. Four bagging classifier models will be created with one using each combination of the hyperparameters. Each model is trained against its respective dataset. The model providing the highest accuracy is selected. Although the grid search is computationally expensive, it can significantly improve the performance of a machine learning algorithm.

4   Experimental Results

In this paper, we use bagging ensemble learning to predict the Android ransomware from the traffic of Android devices. We use CICAndMal2017 dataset. The decision tree classifier is used as the base estimator of the bagging ensemble classifier. The reason behind choosing the decision tree is that it is an unstable classifier because a small change in the dataset leads to a significant change in the output of a classifier [22]. As is mentioned before, the bagging classifier performs better if the base estimators work independently. This independence increases diversity. The diversity increases when the unstable estimators work on each subset of the dataset. We do not use any specific aggregation method as we select the best method through the grid search.

The following hyperparameters of the bagging classifier are optimized: the number of estimators, the number of features used to train the base estimator

($max\_features$), the number of samples used to train the estimator ($max\_samples$), whether the samples are picked up with replacement for a subset, whether the features are picked up with replacement for training the classifier. For the $max\_features$ and $max\_samples$ hyperparameters, we use float type values. The number of features and samples used for training a classifier can be determined as follows for the float type values:

$$No. of\ features = max\_features \times total\_number\_of\_features \quad (2)$$

$$No. of\ samples = max\_samples \times total\_number\_of\_samples \quad (3)$$

Furthermore, we use 10-fold cross validation in the training phase to produce a stable model. We implemented the proposed framework in Python.

To evaluate the performance of the proposed ransomware detection framework, we consider the semi-supervised learning based ransomware detection framework proposed in [15]. In [15], four different semi-supervised machine learning algorithms with feature selection are used to build models individually to detect each type of Android ransomware through binary classification. The best semi-supervised classifier found in [15] is wrapper random forest (RF) with One-R feature selection. Thus, we consider this model as the benchmark. Similar to [15], we will use accuracy of the trained model as the performance criterion.

The evaluation of the proposed solution starts with the searching of the best hyperparameters. Table 1 shows the values of each hyperparameter obtained via experiments for each type of ransomware. The performance comparison between the proposed Android ransomware detection framework and that proposed in [15] is shown in Fig. 3. The proposed detection framework outperforms the wrapper RF-OneR model in detecting all types of the ransomware. The proposed models attain up to 11% higher detection accuracy. In most cases, the detection accuracies of the

Table 1 Optimized hyperparameters of the Bagging classifier

Ransomware	Bootstrap	Bootstrap features	Max features	Max samples	No. of estimators
Charger	False	False	0.5	0.6	50
Pletor	False	False	0.5	0.6	50
Jisut	False	False	0.7	0.6	50
Porndroid	False	False	0.7	0.6	100
Koler	False	True	0.9	0.6	120
RansomBO	True	False	0.7	0.9	100
Lokerpin	True	False	0.7	0.6	150
Simplocker	True	False	0.7	0.9	100
Svpeng	True	False	0.7	0.9	100
Wannalocker	True	False	0.7	0.9	100

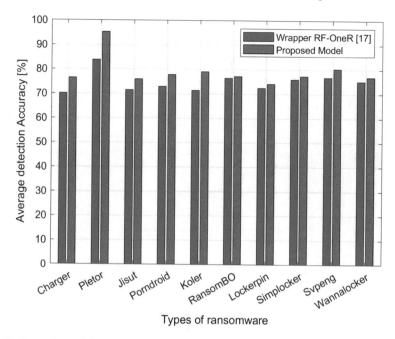

Fig. 3 Comparison of the proposed ransomware detection framework with the semi-supervised machine learning algorithm [15]

proposed models are more than 75%, while the accuracies are a little more than 70% in [15]. The variation of the detection accuracy of the proposed models for various ransomware are shown in Fig. 4. We see that the pletor ransomware has the lowest variation in the accuracy while the wannalocker ransomware has the highest variation. Except charger, jisut and lockerpin, the medians of other ransomware are more than 77.

5   Conclusions

In this paper, we proposed a machine learning based solution for detecting Android ransomware from traffic analysis. The objective was to detect various types of Android ransomware from the traffic analysis. We used grid search for finding the best hyperparameters of the bagging ensemble learning algorithm. The proposed machine learning models outperformed the state-of-the-art solutions of the problem by up to 11%. It was found that the same set of hyperparameters was not equally effective for detecting various kinds of Android ransomware. In addition, not all types of ransomware could be detected with the same accuracy. Rather, the pletor ransomware can be detected with the highest accuracy (95.29%), while the koler ransomware detection accuracy is the minimum (74.09%). As the ransomware detection

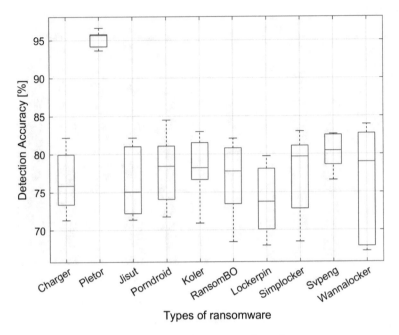

Fig. 4 Illustration of the variance of the detection accuracy for various ransomwares

using machine learning technique provides a very good accuracy, people should use machine learning-based anti-malware software.

In this paper, we used the grid search for hyperparameters tuning. However, this method can search the best combinations of the hyperparameters from the given alternatives. For this reason, it cannot find the actual best set of hyperparameters when these values are not available in the given set. To overcome this issue, we will apply a metaheuristic algorithm for the hyperparameter tuning in future.

References

1. Abuthawabeh M, Mahmoud K (2020) Enhanced android malware detection and family classification, using conversation-level network traffic features. Int Arab J Inf Technol 17(4A):607–614
2. Almahmoud M, Alzu'bi D, Yaseen Q (2021) Redroiddet: Android malware detection based on recurrent neural network. Proc Comput Sci 184:841–846
3. Alsoghyer S, Almomani I (2019) Ransomware detection system for android applications. Electronics 8(8):868
4. Andronio N, Zanero S, Maggi F (2015) Heldroid: dissecting and detecting mobile ransomware. In: Proceedings of the international symposium on recent advances in intrusion detection. Springer, pp 382–404
5. Brownlee J (2021) Ensemble learning algorithms with Python. Machine Learning Mastery
6. Cabaj K, Gregorczyk M, Mazurczyk W (2018) Software-defined networking-based crypto ransomware detection using http traffic characteristics. Comput Electric Eng 66:353–368

7. Canfora G, Mercaldo F, Visaggio CA (2016) An hmm and structural entropy based detector for android malware: an empirical study. Comput Secur 61:1–18
8. Fallah S, Bidgoly AJ (2019) Benchmarking machine learning algorithms for android malware detection. Jordanian J Comput Inf Technol (JJCIT) 5(03)
9. Imaji AO (2019) Ransomware attacks: critical analysis, threats, and prevention methods. Fort Hays State University, Kansas, Hays
10. Kim J, Ban Y, Ko E, Cho H, Yi JH (2022) Mapas: a practical deep learning-based android malware detection system. Int J Inf Secur 1–14
11. Lashkari AH, Kadir AFA, Taheri L, Ghorbani AA (2018) Toward developing a systematic approach to generate benchmark android malware datasets and classification. In: Proceedings of 2018 international Carnahan conference on security technology (ICCST). IEEE, pp 1–7
12. Maigida AM, Abdulhamid SM, Olalere M, Alhassan JK, Chiroma H, Dada EG (2019) Systematic literature review and metadata analysis of ransomware attacks and detection mechanisms. J Reliab Intell Environ 5(2):67–89
13. Martin A, Hernandez-Castro J, Camacho D (2018) An in-depth study of the jisut family of android ransomware. IEEE Access 6:57205–57218
14. Mim KR, Hossain MS, Tisha SA, Kalpo KR, Bakul MH, Hossain MS (2021) Traffic analysis based android sms malware detection using machine learning. In: Proceedings of the international conference on 4th industrial revolution and beyond. Springer
15. Noorbehbahani F, Saberi M (2020) Ransomware detection with semi-supervised learning. In: Proceedings of the 2020 10th international conference on computer and knowledge engineering (ICCKE). IEEE, pp 024–029
16. Ozdemir S, Susarla D (2018) Feature engineering made easy: identify unique features from your dataset in order to build powerful machine learning systems. Packt Publishing Ltd
17. Seni G, Elder JF (2010) Ensemble methods in data mining: improving accuracy through combining predictions. Synth Lect Data Mining Knowl Discov 2(1):1–126
18. Shashidhar NK (2017) Ransomware analysis and defense-wannacry and the win32 environment. Int J Inf Secur Sci 6(4):57–69
19. Tirkey A, Kumar Mohapatra R, Kumar L (2022) Discerning android malwares using extreme learning machine. In: Edge analytics. Springer, pp 17–32
20. Tirkey A, Mohapatra RK, Kumar L (2022) Sniffing android malware using deep learning. In: Edge analytics. Springer, pp 489–505
21. Zheng C, Dellarocca N, Andronio N, Zanero S, Maggi F (2016) Greateatlon: fast, static detection of mobile ransomware. In: Proceedings of the international conference on security and privacy in communication systems. Springer, pp 617–636
22. Zhou ZH (2019) Ensemble methods: foundations and algorithms. Chapman and Hall/CRC

A Dual Attack Tree Approach to Assist Command and Control Server Analysis of the Red Teaming Activity

Atul Rana, Sachin Gupta⬤, and Bhoomi Gupta⬤

Abstract The organizational communication networks are growing at a rapid pace led by the developments in cloud over the last decade. There is a paradigm shift happening from perimeter based networks to perimeter less networks being set up on the cloud and this is adding complexity to both the organizational network structure and the security provisioning at the same time. Increasing instances of ransomware attacks, distributed denial of service and targeted attacks on the crown jewels of organizational network infrastructure has necessitated for the security teams to step up their network defense and test strategies. A comprehensive security coverage decision making plan for an organization can be visualized using the manually drafted attack tree diagram of the security blue team. This paper presents a dual attack tree assisted command and control server activity to ensure enhanced path coverage and test coverage by the red team during security validation and penetration testing.

Keywords Attack tree · Red team · Command and control server · Network security · Penetration testing

A. Rana
MVN University, Haryana, India
e-mail: 20cs9002@mvn.edu.in

S. Gupta (✉)
SoET, MVN University, Haryana, India
e-mail: sachin.gupta@mvn.edu.in

B. Gupta
MAIT, GGSIP University, New Delhi, India
e-mail: bhoomigupta@mait.ac.in

1   Introduction

There exists a multitude of expensive and yet sometimes ineffective approaches to securing the crown jewels of an organization but the literature review suggests that the attack tree approach [1], despite its cost effectiveness and intuitive simplicity has not been utilized to its full potential. Most of the references to attack trees occur in the context of Blue Team [2] activities of the organizational defense, but the versatility of the attack tree can be applied to the domain of a red team [3] offensive strategy for vulnerability discovery and reporting.

While the blue teams design effective security measures based on their knowledge of the network configuration and comparative resource security prioritization, their security measures are usually validated only after a red team exercise which may either be a blind or a double blind process. The red team is responsible for planning attacks to validate or test the security provisions put in place by the blue team. Effective attack planning and pretexting involves preparation of the operation specific to the target, taking into full account Intel gathered from the reconnaissance stages [4]. This commonly includes: creating an initial plan of attack, identification of pretexts, outlining potential alternative plans, crafting custom malicious file payloads, configuring Trojans etc. and several alternative attack paths exist in the planning. For simple networks, and small team sizes, the red teaming activity is fairly straightforward and can be handled without difficulty. However, in complex organizational networks with huge perimeter based and perimeter-less network infrastructure, the red team may take advantage of an attack tree assisted approach for the command and control server activity to ensure both exhaustive path coverage and test coverage. This paper proposes a complimentary dual attack tree approach including a blue team decision attack tree and a blind red team offense attack tree to assist the command and control server analysis of the red teaming activity.

The contributions of this research include the complete setup of an Azure cloud based network as a virtual organizational environment for security testing and a proposed dual attack tree approach for complementing the blue and red teaming exercises. The remainder of this paper has been organized as follows: the present section builds upon the basics and presents the test environment used in this study. Section 2 gives a complete description of the virtual network as a laboratory setup on Azure cloud for this research while Sect. 3 explains the methodology adopted by the attack tree assisted command and control server exercise of the red team. Section 4 briefly presents the preliminary results obtained from the research and Sect. 5 lists the conclusions and future research directions envisioned based on the results obtained.

Our infrastructure environment is deployed on Azure cloud. This will require a subscription to Azure environment and one resource group within the subscription to place resources in one resource group. We also are using Microsoft security tools to create a Security operation center in this test environment. We have a resource group dedicated for this deployment named as annuire-rg as shown in Fig. 1 that consists of all the virtual machines, network interfaces, storages, Data collection rule, disks, log analytics, sentinel etc.

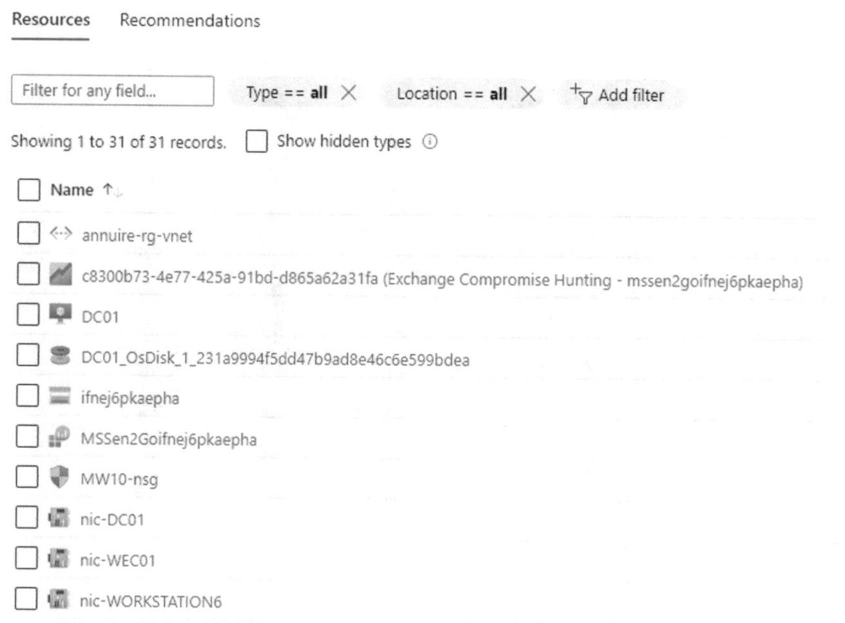

Resources   Recommendations

| Filter for any field... | Type == **all** ✕   Location == **all** ✕   ⁺▽ Add filter |

Showing 1 to 31 of 31 records.   ☐ Show hidden types ⓘ

☐ Name ↑

☐ ⟨··⟩ annuire-rg-vnet

☐ ◢ c8300b73-4e77-425a-91bd-d865a62a31fa (Exchange Compromise Hunting - mssen2goifnej6pkaepha)

☐ 🖥 DC01

☐ 🛢 DC01_OsDisk_1_231a9994f5dd47b9ad8e46c6e599bdea

☐ ▤ ifnej6pkaepha

☐ 🖧 MSSen2Goifnej6pkaepha

☐ 🛡 MW10-nsg

☐ 🖳 nic-DC01

☐ 🖳 nic-WEC01

☐ 🖳 nic-WORKSTATION6

Fig. 1 Resource Group for deployment on Azure

Log collection is done on all the machines installed on the virtual machine using an OMS agent that is installed on every end point and reports it to the log analytics workplace named as "MSSen2Goifnej6pkaepha". Figure 2 shows the screenshot of the logs collected from every endpoint installed. We have also installed Sentinel as an application above log analytics for use-case creation and dash boarding for a better overview and enabled logs to reach Sentinel.

2 Literature Review

The attack tree was used in the cyber security terminology for the very first time by Bruce Schneier [5] in the year 1999 with an intent to understand the adversary and resource structures for a threat perception visualization. The concept was modeled on the lines of an inverted tree data structure with a root node at the top and leaf nodes as resources along with paths being represented by the decision structures. In his dissertation work at Carnegie Mellon University, Sheyner presented scenario based attack graph structures [6] with possible impact. The work along these lines was furthered by Steven Noel [7] who presented a low complexity visualization mechanism to explain the what-if scenarios in a cyber-attack along with the work

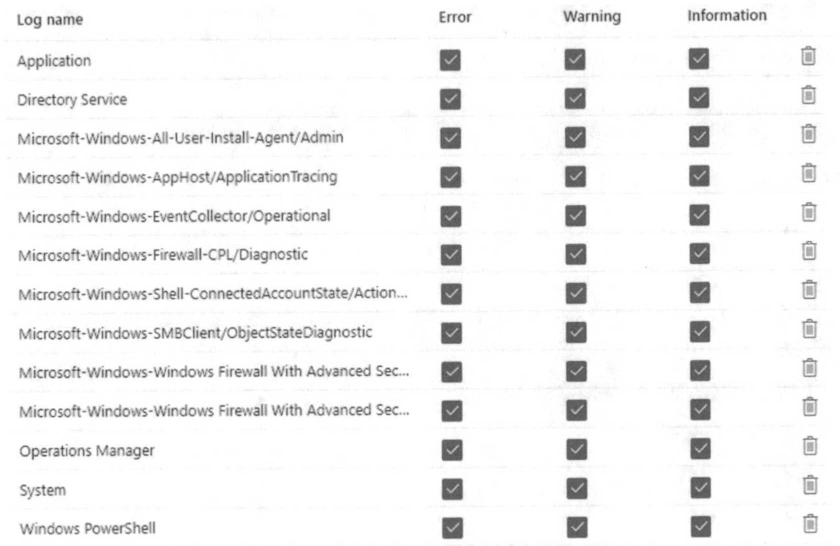

Log name	Error	Warning	Information	
Application	☑	☑	☑	🗑
Directory Service	☑	☑	☑	🗑
Microsoft-Windows-All-User-Install-Agent/Admin	☑	☑	☑	🗑
Microsoft-Windows-AppHost/ApplicationTracing	☑	☑	☑	🗑
Microsoft-Windows-EventCollector/Operational	☑	☑	☑	🗑
Microsoft-Windows-Firewall-CPL/Diagnostic	☑	☑	☑	🗑
Microsoft-Windows-Shell-ConnectedAccountState/Action...	☑	☑	☑	🗑
Microsoft-Windows-SMBClient/ObjectStateDiagnostic	☑	☑	☑	🗑
Microsoft-Windows-Windows Firewall With Advanced Sec...	☑	☑	☑	🗑
Microsoft-Windows-Windows Firewall With Advanced Sec...	☑	☑	☑	🗑
Operations Manager	☑	☑	☑	🗑
System	☑	☑	☑	🗑
Windows PowerShell	☑	☑	☑	🗑

Fig. 2 Agent Configuration for log based analytics

by Mauw and Oosdijk [8] who enhanced the representation of the attack trees with projection functions and additional attributes.

The insider malicious actors were also covered by attack tree visualization in [9] where the authors proposed a framework to distinguish between malicious and benevolent authorized insiders. A formal projection of attack tree semantics [8] was published in 2006 with algebraic conditions for verification. The paper built upon the foundations of the shortcomings found in [10] which was based on petrinets which lacked semantic formalization. The paper also formed the base for another research work [11] on multi modal internet attacks as compared to the singular trees proposed by Schneier.

The research on modifications of the attack trees has been slow but continuous across the last two decades with researchers proposing attack-defense trees in [12] and extended SAND attack trees using sequential operators [13]. There has been adequate ongoing coverage of the attack trees from the perspective of a defense team, with the knowledge of network infrastructure but using attack trees from the perspective of an offensive red team perspective is lacking in the literature. We have represented the blue team attack decision tree for our network resource group hosted on azure in Fig. 3. The arrows represent the direction of flow beginning with the information gathering stage.

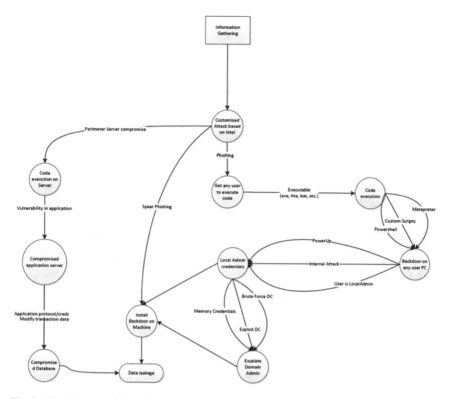

Fig. 3 Blue Team attack decision Tree for our test network

3   Virtual Lab Setup

For the red team engagement we have used another azure environment to create our attacking setup and we are using Visual studio enterprise subscription on Azure for this environment. We have a resource group dedicated for this deployment named as red-team as seen in Fig. 4 that consists of all the virtual machines, network interfaces, storages Data collection rule, disks etc.

We have created two virtual machines, one for deploying Covenant [14] and another for testing commands before using it on the client machine. Covenant is a .NET command and control (C&C) framework that aims to highlight the attack surface of .NET, make the use of offensive .NET tradecraft easier, and serve as a collaborative command and control platform for red teamers. Covenant is an ASP .NET Core, cross-platform application that includes a web-based interface that allows for multi-user collaboration. Covenant is installed on Linux operating system Ubuntu version 20.04 with Standard D2s v3 size and 8 GiB of Ram and 2 vCPU. The machine is also assigned a dynamic public IP to be easily accessed via a Web Interface from Base machine. We also have allowed a few ports for communication with our C&C

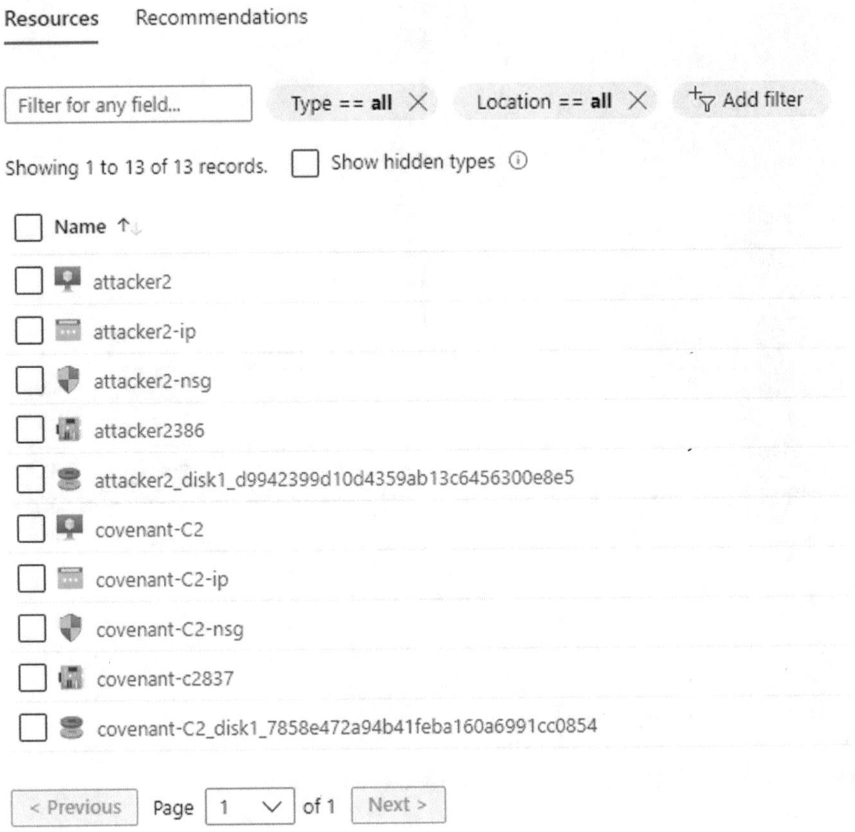

Resources   Recommendations

| Filter for any field... | Type == **all** ✕ | Location == **all** ✕ | ⁺ᵧ Add filter |

Showing 1 to 13 of 13 records. ☐ Show hidden types ⓘ

☐ Name ↑↓

☐ 🖥 attacker2

☐ 🖳 attacker2-ip

☐ 🛡 attacker2-nsg

☐ 🖧 attacker2386

☐ 💿 attacker2_disk1_d9942399d10d4359ab13c6456300e8e5

☐ 🖥 covenant-C2

☐ 🖳 covenant-C2-ip

☐ 🛡 covenant-C2-nsg

☐ 🖧 covenant-c2837

☐ 💿 covenant-C2_disk1_7858e472a94b41feba160a6991cc0854

| < Previous | Page | 1 ⌄ | of 1 | Next > |

Fig. 4 Red team resource group on Azure

server (7443) and also ssh and http port on port number 22 and 80 respectively as shown in Fig. 5.

We need to clone Covenant recursively to initialize the git submodules: using the command—*git clone–recurse-submodules* https://github.com/cobbr/Covenant. In normal and real-world scenarios, we will deploy a C&C server like covenant

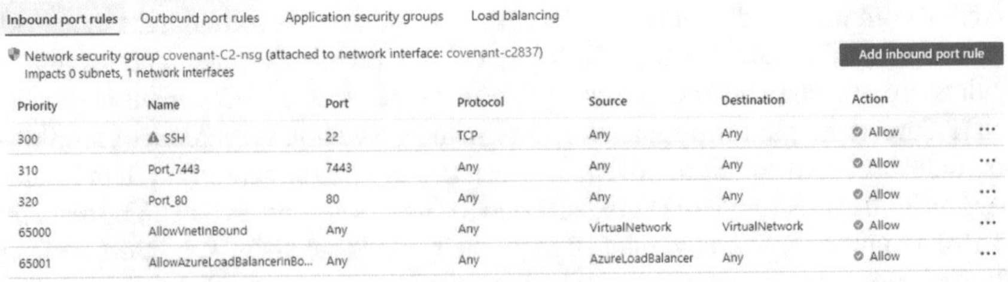

Inbound port rules   Outbound port rules   Application security groups   Load balancing

🛡 Network security group covenant-C2-nsg (attached to network interface: covenant-c2837)
Impacts 0 subnets, 1 network interfaces

Add inbound port rule

Priority	Name	Port	Protocol	Source	Destination	Action	
300	⚠ SSH	22	TCP	Any	Any	⊘ Allow	⋯
310	Port_7443	7443	Any	Any	Any	⊘ Allow	⋯
320	Port_80	80	Any	Any	Any	⊘ Allow	⋯
65000	AllowVnetInBound	Any	Any	VirtualNetwork	VirtualNetwork	⊘ Allow	⋯
65001	AllowAzureLoadBalancerInBo...	Any	Any	AzureLoadBalancer	Any	⊘ Allow	⋯

Fig. 5 Port rules for communication with C&C Server

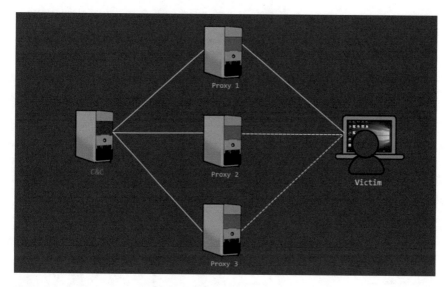

Fig. 6 C&C deployment configuration

behind a few proxy servers so that the real attacker machine IP address is never disclosed and if an IP is blocked we need not change the complete deployment. The configuration has been shown in Fig. 6.

We have also created a Listener that has been created on the IP address of Covenant, to receive all incoming connections on this IP address as a part of preparation for this exercise.

4 Methodology

During this red team exercise, we worked on the below mentioned technique which is a modified version of the cyber kill chain as per the scope of this exercise to get into the organization and achieve the planned target that was within the scope of the exercise. The complete flow of events for the red teaming activity has been shown in Fig. 7.

We spent time researching techniques to be used during the phishing campaign in order to acquire code execution on internal machines. Post getting access to the internal network, the plan of the Red Team is to enumerate the Active Directory environment looking for potential misconfigurations and known vectors that could elevate their privileges during the assignment. The goal of the exercise will be to achieve administration rights over the domain in scope and the team will move closer to it by exploiting weaknesses within the environment and by laterally spreading in order to collect more and more credential material to increase the privileges. After having got administrative access to the target domains, the team will look for users

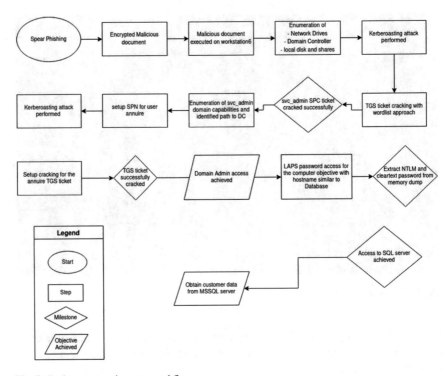

Fig. 7 Red team exercise proposed flow

active on the target applications in order to compromise their machines and acquire the data needed to connect to the service, achieving the ultimate goal of this research. The steps for this exercise are elaborated in the sub-sections ahead.

4.1 Preparation

In order to prepare working and undetected payloads suitable for the phishing campaigns, our Red Team spent research and development time by testing different techniques to obtain code execution via malicious documents. During the preparation phase, the team was mainly focused on two tasks:

- prepare phishing payloads, documents, fake user profiles and emails for multiple campaigns;
- prepare a Covenant instance properly set up for the target environment to remotely control compromised machines.

During the preparation phase, the team sent a phishing document containing a powershell script with a hidden window command attached to it so that the window will be hidden in background after a successful execution of the script. Our phishing payload files were based on XLSM and DOCM documents. We were able to embed malicious code that acts as a downloader for our malware, moreover for some of the scenario's part of this exercise the malware was already embedded in the document either Base64 encoded or as an OLE object.

4.2 Reconnaissance

The first phase in a red team operation is focused on collecting as much information as possible about the target. Reconnaissance is one of the most critical steps. This can be done through the use of tools, such as Dig, Nslookup, DNSRecon etc. or public tools like LinkedIn, Google, Twitter, Facebook etc. As a result, it is usually possible to learn a great deal about the target's people, technology, surroundings, and environment. Reconnaissance consists of techniques that involve adversaries actively or passively gathering information that can be used to support targeting. Such information may include details of the victim organization, infrastructure, or staff/personnel. This step also involves building or acquiring specific tools for the red team test. Reconnaissance is split up into 2 categories based on the type of interaction with the target:

- Active reconnaissance—Actively engaging/interacting with the target network, hosts, employees etc. (Port scanning, vulnerability scans, web app scanning)
- Passive reconnaissance—Utilizing publicly available information using search engines and personal information

During this engagement we used passive reconnaissance and used the publicly available information and used email of the user to send malicious encoded powershell to invade the defense.

4.3 Intrusion

In this phase the adversary is trying to get into the target network. Intrusion consists of techniques that use various entry vectors to gain their initial foothold within a network. Techniques used to gain a foothold include targeted spear phishing and exploiting weaknesses on public-facing web servers. Foot holds gained through initial access may allow for continued access, like valid accounts and use of external remote services, or may be limited-use due to changing passwords. An extensive coverage on threat hunting has been done in a recent master's dissertation submitted to Rochester Institute of Technology [20] and simulation techniques have been discussed in [21]. For our exercise we used the technique of phishing and created a powershell script as shown in Fig. 8 and sent it to the user as an attachment. We created an encoded powershell launcher using Covenant and sent it via email. As per a survey 93% of attach organizations start with phishing and creating the initial access/foothold of

Fig. 8 Powershell script creation on C&C

the attacker. This launcher will use the powershell of the victim machine and hide itself in the background as soon as it runs on the machine.

4.4 Exploitation

Once the user clicks on the phishing mail, covenant gets a reverse shell back and can be seen in the Grunts. After the execution of the script we received a Grunt (remote session) of the victim machine as seen in Fig. 9 and now the victim machine is visible on our covenant tool to perform further actions on the machine that include privilege escalation, lateral movement and later AD access which is the end goal of this exercise. We interacted with the Grunt using interface and found its name using whoami.

4.5 Privilege Escalation

We tried to crack the username and password for the user using mimikatz [15] to dump user credentials from the memory as hashes and used offline methods to crack the NTLM hash returned for the user annuire as shown in Fig. 10.

Launchers
Grunts
Templates
Tasks
Taskings
Graph
Data
Users
Name
Hostname
User
Integrity
3d154b491b
WORKSTATION6
annuire
High
3d154b491b
(annuire) > whoami
WORKSTATION6\annuire

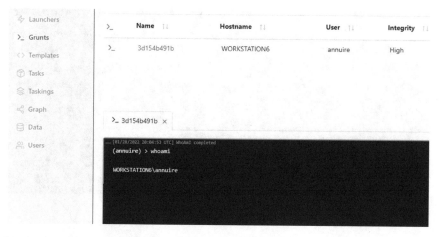

Fig. 9 Remote access of victim machine

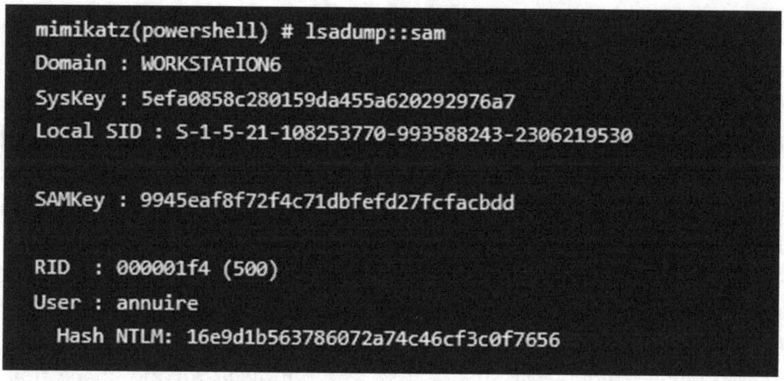

mimikatz(powershell) # lsadump::sam
Domain : WORKSTATION6
SysKey : 5efa0858c280159da455a620292976a7
Local SID : S-1-5-21-108253770-993588243-2306219530
SAMKey : 9945eaf8f72f4c71dbfefd27fcfacbdd
RID : 000001f4 (500)
User : annuire
Hash NTLM: 16e9d1b563786072a74c46cf3c0f7656

Fig. 10 Mimikatz hash dump for compromised user

After cracking the user credentials and domain information, we tested the system with Rubeus [16] for tracking kerberos tickets and eventually could manage complete control over the compromised network as a compromised administrator impersonated as seen in Fig. 11 using the c&c, towards attainment of our intended goal of the red teaming activity [17] during cyber-attacks automation and evaluation [18]. This may even be helpful for the intrusion detection process as discussed in the IBM research paper [19].

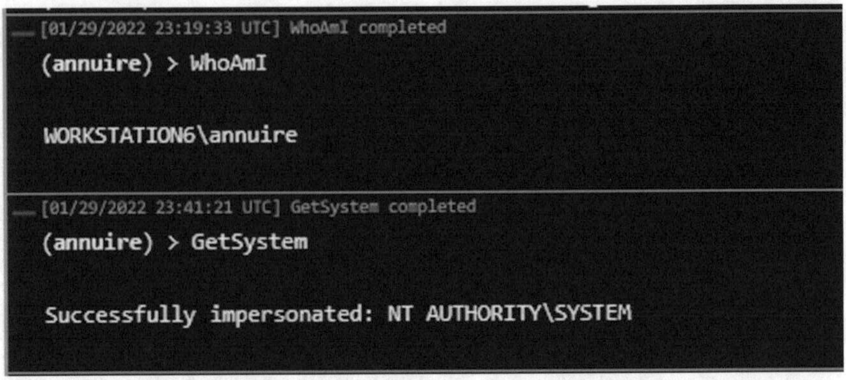

```
___ [01/29/2022 23:19:33 UTC] WhoAmI completed
    (annuire) > WhoAmI


    WORKSTATION6\annuire

___ [01/29/2022 23:41:21 UTC] GetSystem completed
    (annuire) > GetSystem


    Successfully impersonated: NT AUTHORITY\SYSTEM
```

Fig. 11 Successful impersonation of administrator account

5 Results

The visual representation of the red teaming activity in a simple network is not necessary, however in cases of complex networks, the process can be visualized with the help of a modified attack tree for red teams having paths labeled with differentiated information like weights or colors to indicate success or failure depending on the case. A modified red team attack tree with successful attack paths being represented in green and the unsuccessful attempts represented in red has been shown in Fig. 12.

The representation shown has been overly simplified with a small network assumption and low attack profiles with high success ratio, but in actual networks, the visualization can serve as a dynamic template for the organizational security policy makers while making strategic decisions on infrastructural investments and operations team management. The dual attack tree approach involving the blue team decision attack tree and the red team differentiated-labeling attack tree can provide both better comprehension of the crown jewel exposure, organizational security and better test coverage.

6 Conclusions and Future Directions

The attack tree approach to organizational network security is one of the most preferred choices of seasoned security managers. The dual applicability of the attack tree methodology in both blue teaming and red teaming activity holds immense potential for increasing the security test coverage and exposure quotient of the network crown jewels for an organization. The method of drawing an attack tree however is best suited to a manual rendering, drawing heavily from the domain expertise of

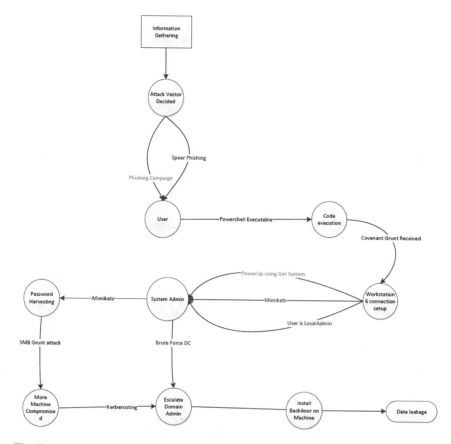

Fig. 12 The differentiated labeling red team attack tree

the security analysis team. Some solutions towards automated attack tree generation mechanisms can be considered for development by the open source security community.

References

1. Lallie HS, Debattista K, Bal J (2020) A review of attack graph and attack tree visual syntax in cyber security. Comput Sci Rev 35:100219. https://doi.org/10.1016/j.cosrev.2019.100219
2. "Computer Network Defense", Web.archive.org, 2011. [Online]. Available: https://web. archive.org/web/20160425120250/, https://www.sypriselectronics.com/information-security/ cyber-security-solutions/computer-network-defense/. Accessed 20 Mar 2022
3. Johnson R (2015) How your red team penetration testers can help improve your blue team. SC Magazine UK, 2015. [Online]. Available: https://web.archive.org/web/201605302 30034/, http://www.scmagazineuk.com/how-your-red-team-penetration-testers-can-help-imp rove-your-blue-team/article/431023/. Accessed 20 Mar 2022

4. Faircloth J (2017) Chapter 2—Reconnaissance. In: Faircloth J (ed) Penetration tester's open source toolkit (Fourth Edition), 4th edn. Syngress, Boston, pp 31–106
5. Schneier B (1999) Attack trees: modeling security threats. Dr. Dobbs J
6. Sheyner O (2004) Scenario graphs and attack graphs. Carnegie Mellon University, Ph.D.
7. Noel S, Jajodia S (2004) Managing attack graph complexity through visual hierarchical aggregation. In: Proceedings of the 2004 ACM workshop on Visualization and data mining for computer security. New York, USA, pp 109–118
8. Mauw S, Oostdijk M (2005) Foundations of attack trees. Inf Secur Cryptol ICISC 2006:186–198
9. Ray I, Poolsapassit N (2005) Using attack trees to identify malicious attacks from authorized insiders. In: Computer security—ESORICS 2005, pp 231–246
10. Steffan J, Schumacher M (2002) Collaborative attack modeling. In: Proceedings of the 2002 ACM symposium on applied computing, pp 253–259. https://doi.org/10.1145/508791.508843
11. Edge KS, Dalton II GC, Raines RA, Mills RF (2007) Using attack and protection trees to analyze threats and defenses to homeland security. MILCOM 2007, pp 1–7
12. Kordy B, Mauw S, Radomirovic S, Schweitzer P (2010) Foundations of attack–defense trees. In: LNCS, Springer, Heidelberg. available at http://satoss.uni.lu/members/barbara/papers/adt.pdf
13. Jhawar R, Kordy B, Mauw S, Radomirović S, Trujillo-Rasua R (2015) Attack trees with sequential conjunction. In: ICT systems security and privacy protection, pp 339–353
14. Entering a covenant: .NET command and control. Cobbr.io. [Online]. Available https://cobbr.io/Covenant.html. Accessed 9 Mar 2022
15. GitHub—gentilkiwi/mimikatz: A little tool to play with Windows security. GitHub. [Online]. Available https://github.com/gentilkiwi/mimikatz. Accessed 20 Mar 2022
16. GitHub—GhostPack/Rubeus: Trying to tame the three-headed dog. GitHub. [Online]. Available https://github.com/GhostPack/Rubeus. Accessed 20 Mar 2022
17. Blumbergs B (2019) Specialized cyber red team responsive computer network operations. Tallinn University of Technology, Ph.D.
18. Enoch SY, Huang Z, Moon CY, Lee D, Ahn MK, Kim DS (2020) HARMer: cyber-attacks automation and evaluation. IEEE Access 8:129397–129414. https://doi.org/10.1109/ACCESS.2020.3009748
19. Araujo F, Ayoade G, Al-Naami K, Gao Y, Hamlen K, Khan L (2021) Crook-sourced intrusion detection as a service. J Inf Secur Appl 61:102880. Available https://doi.org/10.1016/j.jisa.2021.102880
20. Bornholm B (2019) Network-based APT profiler. MSc, Rochester Institute of Technology
21. Moskal, Yang S, Kuhl ME (2017) Cyber threat assessment via attack scenario simulation using an integrated adversary and network modeling approach. J Defense Model Simul Appl Methodol Technol 15:13–29

A Propagation Model of Malicious Objects via Removable Devices and Sensitivity Analysis of the Parameters

Apeksha Prajapati

Abstract This paper presents an epidemic model that describes the spread of malicious objects in a network due to removable devices. Removable devices play a major role in spreading malicious objects in any network. An epidemic model is developed and analyzed both analytically and numerically. Equilibrium points, both endemic and malware free are obtained and also reproduction number is formulated. Global stability of endemic equilibrium is established analytically and also shown it graphically through simulation. Sensitivity of parameters are obtained and also supported by simulated results. The role of sensitivity analysis is to improve system reliability. To support all the qualitative results, simulation is carried out under certain set of the parameters and initial condition.

Keywords Epidemic model · Differential equation · Reproduction number · Stability · Simulation

1 Introduction

Even though the volume of malicious codes fell to a seven-year low in 2021, it started to rise in 2022. The conflict between Russia and Ukraine is one of the causes of the rise in malware attacks. *WannaCry* ransomware attack on May 2017 was global epidemic. About 200 thousand computers were infected in more than 150 countries [1]. Internet is now a huge depository of malware or malicious codes. Malware are basically software most of them are Trojan, Viruses, worm, spyware, ransomware etc. Together all are called malware or malicious codes. Each having special features that make them unique in the way they operate. In the past few years, malicious codes developed in the form of programs. These programs are simple code and they are able to destabilize the smooth function of networked computers.

A. Prajapati (✉)
Nirmala College, Ranchi, India
e-mail: prajapatiapeksha@gmail.com

© The Author(s), under exclusive license to Springer Nature Switzerland AG 2023
A. A. Abd El-Latif et al. (eds.), *Advances in Cybersecurity, Cybercrimes, and Smart Emerging Technologies*, Engineering Cyber-Physical Systems and Critical Infrastructures 4,
https://doi.org/10.1007/978-3-031-21101-0_6

Earlier, malicious codes caused not major damages to networked computer because their spread was very slow but now the time has changed. Over the years, the fast and advanced development of communication and technology, malicious codes have become a most important threat. Malicious codes spread over the network due to removable devices (pen drives, SD cards, External hard disk etc.) and internet. To have a better understanding of the dynamics of spread of malicious codes especially due to removable devices this paper explores the qualitative and quantitative analysis through epidemic model. The assumption while developing the model was interaction among the computer and removable devices. Removable devices play very significant role in spreading malicious codes into the network. Malicious codes also spread due to internet but our main focus for this work is the role of removable devices in spreading infection into the network.

Epidemic—models are the models to study disease prevalence. The epidemics model on malicious objects are inspired by the classical SIR model and infectious disease models [2, 3]. The compartmental model like SIS (Susceptible-Infectious-Susceptible), SEIRS (Susceptible-Exposed-Infectious-Recovered-Susceptible) have formulated to understand the spreading nature of malicious objects in mathematical way including their stability [4–6]. Various Models with Quarantine class, antidotal class, delays, bifurcation have been implemented in classical SIR model and analyzed for network robustness [7, 8]. The delay model has its own specialty which focuses on how model behaves in delays. Delay model are more realistic. The bifurcation analysis is implemented in SIR model to investigate network robustness [9–11].

Hernández et al., proposed an epidemic model SICR on WSN with carrier compartment. This model incorporated vaccination and reinfected processes. Some control measures are implemented depending on threshold value [12].

Zhu et al. [13], developed a model by taking into account the role of removable devices in computer network. The paper shows the interaction of two population class, computer network and removable devices. The stability analysis has also performed for the developed model [13].

This paper is an extension of Zhu et al., by dividing the computer network into a greater number of compartments to improve the understanding of infecting and spreading nature of malicious codes due to removable devices and it makes model more realistic. It is important to thoroughly study each and every parameter of the generated model in order to increase the safety and security of networked computers. To increase the system's dependability, the sensitivity of parameters was studied both analytically and visually for the developed model.

The following is a breakdown of the layout of this article. Section 2, describes all the state variables, parameters and the formulation of SEIR-$R_S R_I$ model. Equilibrium points, Reproduction number and stability of both malware free and endemic equilibrium has been investigated qualitatively in Sects. 3 and 4. Section 6 delves into the details of the simulation-based results in three categories: system behavior, stability, and influence of parameter. Finally, there is a concluding section.

2 Proposed SEIR-$R_S R_I$ Model

This section states the model assumption, variable, parameters and mathematical formulation. The total population of computer devices and removable devices are N and M respectively. S is the total number of Susceptible computers which are at risk of malware attack due to malicious objects. E is the total number of Exposed computers, which are infectious but not able to spread infection to others computers or to the removable devices. I is the total number of infected computers, which are infectious and able to spread infection to others. R is the total number of recovered computers. The total number of susceptible-removable devices are R_S. These are the removable devices that are not infected yet could catch an infection if connected to an infected computers. R_I represents the total number of infected-removable devices. These are the infected devices that have the potential to transfer malicious objects to computers in a network.

There are various parameters involved in the model which connects all the state variables. b_1 and b_2 denote the birth rate of networked computers and removable devices respectively. γ_1 and γ_2 are recovery rate of networked computers and removable devices respectively. β and α are Infectivity contact rate of networked computers and removable devices respectively. μ_1 and μ_2 represent the death rate of networked computers and removable devices respectively. ρ_1 is the Infection rate in exposed class of networked computers. The assumption of the model are as follows.

The total population of the computer network and removable devices are constant.

Initially, all removable devices are vulnerable. When removable devices and infected computers come into contact, the removable devices become infectious. The force of infection for infectious and susceptible computer is given by βSI. If an infected removeable device is linked to a computer with antivirus software after recovery, it may become vulnerable. The recovered force of infected device is $\gamma_2 R R_I$.

Antivirus software may be installed on the computers to safeguard them. Figure 1 shows the flow diagram of the model which leads the system of differential equations.

The differential equations for the model can be written as

$$\left. \begin{array}{l} \frac{dS}{dt} = b_1 - \beta SI - \alpha R_I S - \mu_1 S \\ \frac{dE}{dt} = \beta SI + \alpha S R_I - (\mu_1 + \rho_1) E \\ \frac{dI}{dt} = \rho_1 E - \mu_1 I - \gamma_1 I \\ \frac{dR}{dt} = \gamma_1 I - \mu_1 R \\ \frac{dR_S}{dt} = b_2 - \alpha R_S I + \gamma_2 R R_I - \mu_2 R_S \\ \frac{dR_I}{dt} = \alpha R_S I - \mu_2 R_I - \gamma_2 R R_I \end{array} \right\} \tag{1}$$

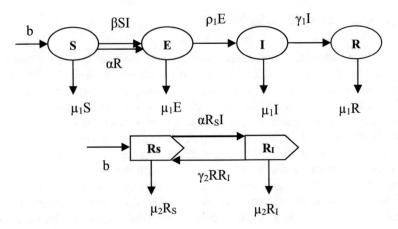

Fig. 1 Flow diagram of the model

3 Equilibrium Point and Reproduction Number

The equilibrium points lead to study the behavior of the dynamical system and it also shows the long-term behavior of different states and based on that the security measure can be achieved.

For equilibrium points,

$$\frac{dS}{dt} = 0, \frac{dE}{dt} = 0, \frac{dI}{dt} = 0, \frac{dR}{dt} = 0, \frac{dR_S}{dt} = 0, \frac{dR_I}{dt} = 0$$

The system of Eq. (1) has two types of equilibrium point, malware-free (E^{MF}) equilibrium points and endemic equilibrium point (E^{EE}).

$$E^{EE} = \left(1 - (1 + a_1 + a_2)I^*, a_2I^*, a_1I^*, 1 - \frac{a_3I^* - \beta}{\alpha(1 - I^*(1 + a_1 + a_2))}, \frac{a_3I^* - \beta}{\alpha(1 - I^*(1 + a_1 + a_2))} \right)$$

where, $a_1 = \frac{\gamma_1}{\mu_1}, a_2 = \frac{\gamma_1 + \mu_1}{\rho_1}, a_3 = \beta(a_2 + \alpha + 1) + a_2(\mu_1 + \rho_1)$

I^* can be found from the equation $m_1 I^2 + m_2 I - m_3 = 0$

Solving the above equation gives two values of I^*

i.e., $I^* = \frac{-m_2 \pm \sqrt{m_2^2 + 4m_1 m_2}}{2m_1}$, ignoring the negative value of I^*

So,

$$I^* = \frac{-m_2 + \sqrt{m_2^2 + 4m_1 m_2}}{2m_1}, \text{ provided } m_2^2 + 4m_1 m_2 > m_2^2$$

$$m_1 = \gamma_2 a_1 a_3 + a_3 \alpha + \alpha^2(1 + a_1 + a_2), m_2 = \mu_2 a_3 - \gamma_2 a_1 \beta - \alpha \beta - \alpha^2, m_3 = \beta \mu_2$$

Malware-free equilibrium point is given as,

$$E^{MF} = \left(\frac{b_1}{\mu_1 + \beta}, 0, 0, 0, \frac{b_2}{\mu_2}, 0 \right)$$

The reproduction number is defined as number of secondary infections produces when an infection is introduced in a fully susceptible population. The reproduction number is calculated from next generation matrix. For set of Eq. (1)

$$V = \begin{pmatrix} \rho_1 + \mu_1 & 0 \\ -\rho_1 & \gamma_1 + \mu_1 \end{pmatrix}, F = \begin{pmatrix} 0 & \alpha R_I + \beta S_0 \\ 0 & 0 \end{pmatrix}$$

$$V^{-1} = \begin{pmatrix} \frac{1}{(\rho_1 + \mu_1)} & 0 \\ \frac{\rho_1}{(\rho_1 + \mu_1)(\gamma_1 + \mu_1)} & \frac{1}{(\gamma_1 + \mu_1)} \end{pmatrix}$$

where F and V represent the rate of infection and the rate of infection by compartment to compartment.

$$FV^{-1} = \begin{pmatrix} \frac{\rho_1(\alpha R_I + \beta S_0)}{(\rho_1 + \mu_1)(\gamma_1 + \mu_1)} & \frac{(\alpha R_I + \beta S_0)}{(\gamma_1 + \mu_1)} \\ 0 & 0 \end{pmatrix}$$

Eigen values are $\frac{\rho_1(\alpha R_I + \beta S_0)}{(\rho_1 + \mu_1)(\gamma_1 + \mu_1)}$ and 0

The reproduction number is given as the spectral radius of FV^{-1} [14].

So, for networked computers, reproduction number is $R_{01} = \frac{\rho_1(\alpha R_I + \beta S_0)}{(\rho_1 + \mu_1)(\gamma_1 + \mu_1)}$.

In the absence of infected removable devices, the reproduction number is $\frac{\rho_1 \beta S_0}{(\rho_1 + \mu_1)(\gamma_1 + \mu_1)}$.

For removable devices, from second equation of (1), we have,

$$\alpha R_S I - \mu_2 R_I - \gamma_2 R R_I < 0$$

In case of no-infection in Computer networks we have $I = 0$.

i.e., $\mu_2 R_I - \gamma_2 R R_I < 0$

Hence for removable devices' reproduction number is $R_{02} = \frac{\gamma_2 R}{\mu_2}$

Reproduction numbers is given as,

$$R_0^2 = R_{01} \times R_{02} = \frac{\gamma_2 R}{\mu_2} \times \frac{\rho_1(\alpha R_I + \beta S_0)}{(\rho_1 + \mu_1)(\gamma_1 + \mu_1)}$$

4 Stability Analysis: Local and Global

This section presents the stability of malware free and endemic equilibrium point.

Theorem 1 The malware-free equilibrium is locally stable if $R_0 < 1$.

Proof The above theorem can be proved in two parts. First for only connected computers and the second part is for overall network including computers and portable or removable devices.

The Jacobian of the computer network is given as,

$$J^{AF} = \begin{bmatrix} -\mu_1 & 0 & -\beta S_0 & 0 \\ 0 & -(\mu_1 + \rho_1) & \beta S_0 & 0 \\ 0 & \rho_1 & -(\mu_1 + \gamma_1) & 0 \\ 0 & 0 & \gamma_1 & -\mu_1 \end{bmatrix}$$

For eigen values, $\left| J^{AF} - \lambda I \right| = 0$

i.e., $$\begin{bmatrix} -(\mu_1 + \lambda) & 0 & -\beta S_0 & 0 \\ 0 & -(\mu_1 + \rho_1 + \lambda) & \beta S_0 & 0 \\ 0 & \rho_1 & -(\mu_1 + \gamma_1 + \lambda) & 0 \\ 0 & 0 & \gamma_1 & -(\mu_1 + \lambda) \end{bmatrix} = 0$$

The characteristic equation of the above matrix is given as

$$(\mu_1 + \lambda)(\mu_1 + \lambda)(\lambda^2 + \lambda(\mu_1 + \rho_1 + \mu_1 + \gamma_1) + ((\mu_1 + \rho_1)(\mu_1 + \gamma_1) - \beta S_0 \rho_1) = 0 \tag{2}$$

The above equation has the roots $-\mu_1, -\mu_1$ and roots of the quadratic equation

$$(\lambda^2 + \lambda(\mu_1 + \rho_1 + \mu_1 + \gamma_1) + ((\mu_1 + \rho_1)(\mu_1 + \gamma_1) - \beta S_0) = 0$$

All the roots of the above equation have negative roots if $\frac{\rho_1(\beta S_0)}{(\rho_1 + \mu_1)(\gamma_1 + \mu_1)} < 1$, i.e., $R_{01} < 1$.

All the roots of the Eq. (2) have negative real part; hence the malware free state in computer network is locally asymptotically stable.

Infection would not spread among removable devices if $\mu_2 R_I - \gamma_2 R R_I < 0$, i.e., $R_{02} < 1$.

The infection will not spread in a network consisting both computers and removable devices when, $R_{01} < 1$ and $R_{02} < 1$ i.e., $R_0 < 1$.

Theorem 2 If $R_0 > 1$, then the unique endemic equilibrium is globally stable.

Proof The global stability of the endemic equilibrium will be studied by geometric method in this section [15, 16]. The global

stability of endemic equilibrium point is shown in the domain $\Delta =$
$\{(S, E, I, R, R_S, R_I) : S \geq 0, E \geq 0, I \geq 0, R \geq 0, R_S \geq 0, R \geq 0,$
$S + I + E + R \leq 1; R + R_I \leq 1\}$

Consider the autonomous dynamical system:
$\dot{x} = f(x)$, where f: $G \rightarrow R_n$, $G \subset R_n$ is an open set and simply connected and f
$\in C^1(G)$.

Let x* be a unique endemic equilibrium point. There exists a compact absorbing
subset K of G then x* is globally stable if it satisfies the additional Bendixson criteria
given as

$$\bar{q} = limsup_{t \to \infty} \, sup_{x_0 \in K} \frac{1}{t} \int_0^t \mu(G(x(s, x_0))) \, ds < 0, \text{ where } D = A_f A^{-1} + A \frac{\partial f^{[2]}}{\partial x} A^{-1}$$

The matrix A_f is obtained by replacing each entry Aij of A by its derivative in
the direction of f, $A_{ij}f$, and μ is the Lozinski̇ı measure of Q with respect to a vector
norm $|\cdot|$ in R_n.

The Jacobian of order 4 is given as,

$$J = \begin{bmatrix} -(\beta I + \alpha R_I + \mu_1 + \rho_1) & \beta(1 - E - 2I - R) - \alpha R_I & -(\beta I + \alpha R_I) & -\alpha(1 - E - I - R) \\ \rho_1 & -(\gamma_1 + \mu_1) & 0 & 0 \\ 0 & \gamma_1 & -\mu_1 & 0 \\ 0 & \alpha(1 - R_I) & -\gamma_2 R_I & -(\alpha I + \gamma_2 R + \mu_2) \end{bmatrix}$$

Further, $J^{[2]} = \frac{\partial f^{[2]}}{\partial x}$ is the second compound Jacobian matrix is given as,

$$J^{[2]} = \begin{bmatrix} -a_{11} & 0 & 0 & \alpha R_I + \beta I & \alpha(1 - E - I - R) & 0 \\ \gamma_1 & -a_{22} & 0 & \beta(1 - E - 2I - R) & 0 & \alpha(1 - E - I - R) \\ \alpha(1 - R_I) & \gamma_2 R_I & -a_{33} & 0 & \beta(1 - E - 2I - R) & -(\alpha R_I + \beta I) \\ 0 & -\gamma_1 - \mu_1 & 0 & -a_{44} & 0 & 0 \\ 0 & 0 & -\gamma_1 - \mu_1 & \alpha(1 - R_I) & -a_{55} & 0 \\ 0 & 0 & 0 & -\alpha(1 - R_I) & \gamma_1 & -a_{66} \end{bmatrix}$$

where,

$$a_{11} = (\beta I + \alpha R_I + 2\mu_1 + \rho_1 + \gamma_1), a_{22} = (\beta I + \alpha R_I + 2\mu_1 + \rho_1)$$

$$a_{33} = (\beta I + \alpha R_I + 2\mu_1 + \mu_2 + \rho_1 + \gamma_1 + \gamma_2 R + \alpha I)$$

$$a_{44} = (2\mu_1 + \gamma_1)$$

$$a_{55} = (\mu_1 + \mu_2 + \gamma_1 + \gamma_2 R + \alpha I), a_{66} = (\mu_1 + \mu_2 + \gamma_2 R + \alpha I)$$

For the matrix D in the Bendixion criteria, characterize a diagonal matrix A as

$$A = diag\left(1, \frac{R_I}{I}, \frac{R_I}{I}, \frac{R_I}{I}, \frac{R_I}{I}, \frac{R_I}{I}\right)$$

and $A_f A^{-1} = diag\left(0, \frac{R_I}{I}\left(\frac{I}{R_I}\right)_f, \frac{R_I}{I}\left(\frac{I}{R_I}\right)_f, \frac{R_I}{I}\left(\frac{I}{R_I}\right)_f, \frac{R_I}{I}\left(\frac{I}{R_I}\right)_f, \frac{R_I}{I}\left(\frac{I}{R_I}\right)_f\right)$

So D is given as

$$D = D_f D^{-1} + D J^{[2]} D^{-1}$$

where,

$$D = \begin{bmatrix} -a_{11} & 0 & 0 & \alpha R_I + \beta I & \alpha(1-E-I-R) & 0 \\ \gamma_1 & \left(\frac{R_I}{I}\right)_f \frac{I}{R} - a_{22} & 0 & \beta(1-E-2I-R) & 0 & \alpha(1-E-I-R) \\ \alpha(1-R_I) & \gamma_2 R_I & \left(\frac{R_I}{I}\right)_f \frac{I}{R} - a_{33} & 0 & \beta(1-E-2I-R) & -(\alpha R_I + \beta I) \\ 0 & -\gamma_1 - \mu_1 & 0 & \left(\frac{R_I}{I}\right)_f \frac{I}{R} - a_{44} & 0 & 0 \\ 0 & 0 & -\gamma_1 - \mu_1 & \alpha(1-R_I) & \left(\frac{R_I}{I}\right)_f \frac{I}{R} - a_{55} & 0 \\ 0 & 0 & 0 & -\alpha(1-R_I) & \gamma_1 & \left(\frac{R_I}{I}\right)_f \frac{I}{R} - a_{66} \end{bmatrix}$$

Which can be expressed in the form of a block matrix as

$$D \quad = \quad \begin{bmatrix} D_{11} & D_{12} \\ D_{21} & D_{22} \end{bmatrix}, \quad \text{Where,} \quad D_{11} \quad = \quad [a_{11}]; D_{12} \quad =$$

$$\left[0 \ 0 \ \frac{R_I}{I}(\alpha R_I + \beta I) \ \alpha(1-E-I-R) \ 0\right];$$

$$D_{21} = \begin{bmatrix} \frac{R_I}{I} \\ \alpha(1-R_I) \\ 0 \\ 0 \\ 0 \end{bmatrix}$$

$$D_{22} = \begin{bmatrix} \left(\frac{R_I}{I}\right)_f \frac{I}{R} - a_{22} & 0 & \beta(1-E-2I-R) & 0 & \alpha(1-E-I-R) \\ \gamma_2 R_I & \left(\frac{R_I}{I}\right)_f \frac{I}{R} - a_{33} & 0 & \beta(1-E-2I-R) & -(\alpha R_I + \beta I) \\ -\gamma_1 - \mu_1 & 0 & \left(\frac{R_I}{I}\right)_f \frac{I}{R} - a_{44} & 0 & 0 \\ 0 & -\gamma_1 - \mu_1 & \alpha(1-R_I) & \left(\frac{R_I}{I}\right)_f \frac{I}{R} - a_{55} & 0 \\ 0 & 0 & -\alpha(1-R_I) & \gamma_1 & \left(\frac{R_I}{I}\right)_f \frac{I}{R} - a_{66} \end{bmatrix}$$

To calculate Lozinskii measure of matrix D, first step is to calculate sup {g1, g2} where g1 and g2 are defined as,

$$g_1 = \mu(D_{11}) + |D_{12}| = -(\beta I + \alpha R_I + 2\mu_1 + \rho_1 + \gamma_1) + \frac{R_I}{I}(\alpha R_I + \beta I) + \alpha(1-E-I-R)$$

$$g_2 = \mu_0(D_{22}) + |D_{21}| = \left(\frac{R_I}{I}\right)_f \frac{I}{R_I} - 2\mu_1$$

where $|D_{12}|$, $|D_{21}|$ are matrix norms w.r.t. the l1 vector norm, and μ denotes the Lozinskii Measure w.r.to the l1 [15–17]. To obtain $\mu(D_{22})$, add the absolute value of the off-diagonal elements to the diagonal one in each column of D_{22} and then take the maximum of two sums.

Now $\frac{R_I}{I}\left(\frac{I}{R_I}\right)_f = \frac{R_I}{I}\left(\frac{IR_I{}'-R_II'}{R_I{}^2}\right) = \frac{R_I{}'}{R_I} - \frac{I'}{I}$ and so.

$\frac{R_I{}'}{R_I} = \alpha I \frac{R_S}{R_I} - \gamma_2 R - \mu_2$ and $\frac{I'}{I} = \rho_1 \frac{E}{I} - \mu_1 - \gamma_1$

Hence g_1 and g_2 reduce as,

$$g_1 = (\beta I + \alpha R_I + 2\mu_1 + \rho_1) + \frac{R_I}{I}(\alpha R_I + \beta I) + \alpha(1 - E - I - R) - \rho_1 \frac{E}{I} + \frac{I'}{I}$$

$$g_2 = \frac{I'}{I} - \frac{R_I{}'}{R_I} - 2\mu_1$$

$$= \frac{I'}{I} - \alpha I - \mu_2 - \gamma_2 R - 2\mu_1$$

Now, $\mu(D) \leq \sup\{g1, g2\}$

$$\leq \frac{I'}{I} + sup\{-(\mu_1 + \mu_2 + \rho_1 + \gamma_1), \beta I, \alpha R_I, \gamma_2 R, \alpha(1 - E - I - R)\}$$

which gives $\mu(D) \leq \frac{I'}{I} - (\mu_1 + \mu_2 + \rho_1 + \gamma_1)$ and so

$$\int_0^t \mu(D)dt < log\,I(t) - (\mu_1 + \mu_2 + \rho_1 + \gamma_1)t$$

Hence, $q = \frac{1}{t}\int_0^t \mu(D)dt < \frac{1}{t}\log I(t) - (\mu_1 + \mu_2 + \rho_1 + \gamma_1) < 0$ for all initial values of S, E, I, R, R_S, R_I which belongs to Δ. The condition $q < 0$ is fulfilled and so it proves the global stability of endemic equilibrium. This condition also proves the local stability of the endemic equilibrium.

5 Sensitivity Analysis

The sensitivity of the parameters is investigated in this part in order to determine how the model reacts when the parameter values are changed. The sensitivity analysis is carried out by computing the sensitivity indices of the basic reproduction number R_0.

The normalized forward sensitivity index of a variable with respect to a parameter is used to conduct the sensitivity analysis.

The reproduction number is

$$R_0{}^2 = \frac{\gamma_2 R}{\mu_2} \times \frac{\rho_1(\alpha R_I + \beta S_0)}{(\rho_1 + \mu_1)(\gamma_1 + \mu_1)}$$

Now the sensitivity index for different parameters are as follows

$$\frac{\partial R_0/R_0}{\partial \beta/\beta} = \frac{\beta}{R_0}\frac{\partial R_0}{\partial \beta} = \frac{\beta}{R_0}\left(\left(\frac{\gamma_2 R}{\mu_2}\right)^{1/2} \times \frac{1}{2}\left(\frac{\rho_1 S_0}{(\alpha R_I + \beta S_0)(\rho_1 + \mu_1)(\gamma_1 + \mu_1)}\right)^{1/2}\right) > 0$$

$$\frac{\partial R_0/R_0}{\partial \alpha/\alpha} = \frac{\alpha}{R_0}\frac{\partial R_0}{\partial \alpha} = \frac{\alpha}{R_0}\left(\left(\frac{\gamma_2 R}{\mu_2}\right)^{1/2} \times \frac{1}{2}\left(\frac{\rho_1 R_I}{(\alpha R_I + \beta S_0)(\rho_1 + \mu_1)(\gamma_1 + \mu_1)}\right)^{1/2}\right) > 0$$

$$\frac{\partial R_0/R_0}{\partial \gamma_1/\gamma_1} = \frac{\gamma_1}{R_0}\frac{\partial R_0}{\partial \gamma_1} = -\frac{\gamma_1}{R_0}\left(\left(\frac{\gamma_2 R}{\mu_2}\right)^{1/2} \times \frac{1}{2}\frac{(\rho_1(\alpha R_I + \beta S_0))^{1/2}}{(\mu_1 + \rho_1)(\mu_1 + \gamma_1)^{3/2}}\right) < 0$$

$$\frac{\frac{\partial R_0}{R_0}}{\frac{\partial \gamma_2}{\gamma_2}} = \frac{\gamma_2}{R_0}\frac{\partial R_0}{\partial \gamma_2} = \frac{\gamma_2}{R_0}\left(\frac{1}{2}\left(\frac{R}{\gamma_2\mu_2}\right)^{\frac{1}{2}} \times \left(\frac{\rho_1(\alpha R_I + \beta S_0)}{(\rho_1 + \mu_1)(\gamma_1 + \mu_1)}\right)^{\frac{1}{2}}\right) > 0$$

$$\frac{\partial R_0/R_0}{\partial \rho_1/\rho_1} = \frac{\rho_1}{R_0}\frac{\partial R_0}{\partial \rho_1} = \frac{\rho_1}{R_0}\times\left(\frac{\gamma_2 R}{\mu_2}\right)^{1/2} \times \frac{\mu_1(\alpha R_I + \beta S_0)^{1/2}}{2(\rho_1)^{1/2}(\rho_1 + \mu_1)^{3/2}(\gamma_1 + \mu_1)^{1/2}} > 0$$

From the above calculations one can see that, the positive sensitive parameters of the model are α, β, ρ_1, and γ_2 whereas γ_1 is the negative sensitive parameters. This can also be seen in the next section from Figs. 7, 8, 9 and 10 which shows how parameters influence the number of infections in a network.

6   Simulations Based Analysis

This section verifies the qualitative analysis discussed in Sects. 4 and 5. The simulation is carried out in MATLAB with ode 45 suite. The initial values of the variables are S = 0 .93, E = 0, I = 0.07, R = 0, R_S = 0.90, R_I = 0.1.

The values of the parameters for the Figs. 2, 3, 4, 7, 8, 9 and 10 are b_1 = 0.001, b_2 = 0.01, β= 0.001, α=0.1 γ_1 = 0.003, γ_2 = 0.02, ρ_1 = 0.03, μ_1 =0.02, μ_2 =0.001.

For Figs. 5 and 6 values of the parameters are b_1 = 0.1, b_2 = 0.2, β = 0.071, α=0.04 γ_1 = 0.1, γ_2 = 0.005, ρ_1 = 0.1, μ_1 =0.003, μ_2 = 0.21

6.1   System Behavior with Respect to Time

Figure 2 shows the behavior of different population class with time. The susceptible removable devices (Rs) become infected in a short period of time and then become susceptible again. This shows the genuine scenario where removable devices are first susceptible then it will become infected in the network and due to recovery tools

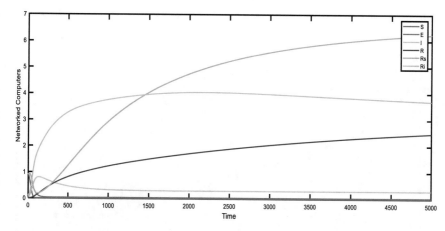

Fig. 2 Behavior of different class with respect to time

(antivirus or formatting the device) they become susceptible again. The population of the infectious computers (I) within the network begins to decline after 100 unit of time but the infection persists in the network. This shows the actual scenario because presence of removable devices makes networked computer at risk.

6.2 Stability Analysis

Figures 3 and 4 show the global stability of endemic equilibrium point for one set of parameters whereas Figs. 5 and 6 show the stability and asymptotic behavior of the system for another set of parameters.

6.3 Impact of Parameters

The simulations in this section depicts the extent to which malicious objects can propagate. It also shows how can the vast majority of computers be infected in a short period of time. Comparative simulations for infectious computers vs time have been run with various parameter settings of α, ρ_1, γ_1 and β.

Impact of α—Figure 7 shows 80% computers got infected in 125 unit of time when α is 0.1 whereas the infection reduced into 65% at 425 unit of time for $\alpha = 0.01$. The spread of infection is directly proportional to α. Higher value of α means, high infection of malicious codes in the network.

Impact of ρ_1—Figure 8 shows 90% computers got infected in 64 unit of time when $\rho_1 = 0.3$ whereas the infection reduced to 65% at 428 unit of time when $\rho_1 = 0.03$.

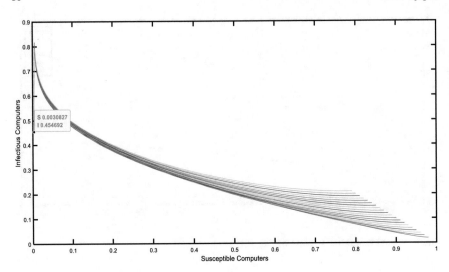

Fig. 3 Phase plane between susceptible and infectious class

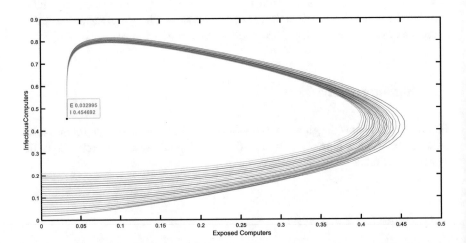

Fig. 4 Phase plane between exposed and infectious class

This figure also explains that the network broke down due to spread of malicious code in very short period of time when $\rho_1 = 0.3$.

Impact of γ_1—Figure 9 shows 80% computers got infected in 128 unit of time when $\gamma_1 = 0.003$ whereas the infection reduced to 30% at 81 unit of time when $\gamma_1 = 0.03$. It can be seen from the below simulated result that the higher value of γ_1 is responsible for rapid spread of malicious code in the network.

Impact of β—Figure 10 shows higher the value of β, higher will be the infection among computers in the network.

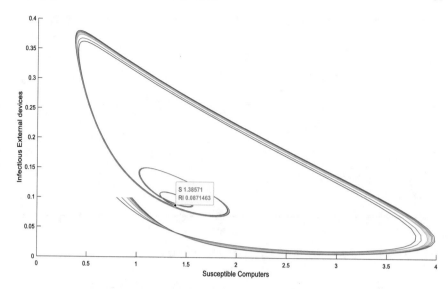

Fig. 5 Phase plane between S and R_I

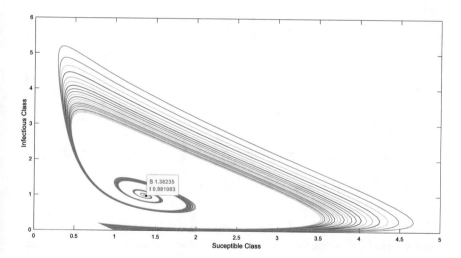

Fig. 6 Phase plane between S and I

7 Conclusion

This paper discussed the qualitative and quantitative aspects on the role of removable devices in a computer network and explained how the malicious objects spread due to removable devices. It also described how certain parameters play an important role to restrict the spread of malicious codes in the network. The influence of various parameters on the spread of malicious codes is depicted in various figures of Sect. 6.

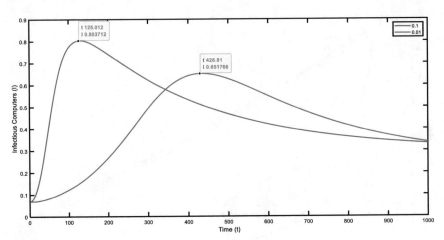

Fig. 7 Flow of Infectious computers with time for two different values of α

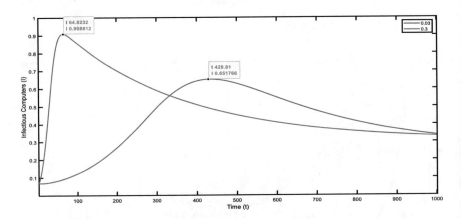

Fig. 8 Infectious class with respect to time for different values of ρ1

These illustrate how a high value for one parameter causes high spread of malicious codes, while a lower value for another parameter may control the spread of malicious objects. Parameters like infectivity contact rate and infectious rate in exposed class can reduce the spread of malicious objects and also it can control the spreading time in the network. Thus, by adjusting the parameters and employing various recovery tools such as antivirus and firewalls, the infection may be delimited for the network's smooth operation.

Implementing control strategies in exposed class can definitely reduce the spread of malicious codes and by maintaining some set of parameters it can also control the infection period of the network. The future work may include the analysis and implementation of different control strategies at exposed class.

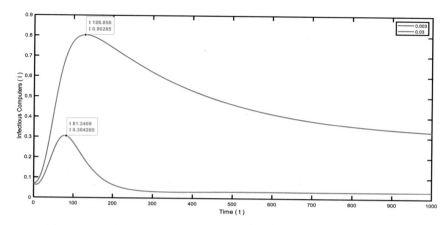

Fig. 9 Infectious class with respect to time for different values of γ_1

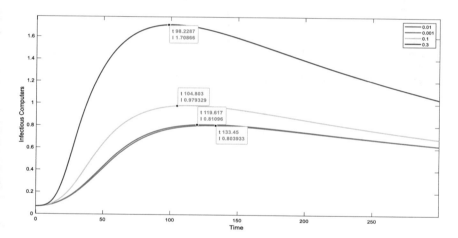

Fig. 10 Infectious computer versus time for different values of β

References

1. Maxat, Vassilios V (2019) Wannacry ransomware: analysis of infection, persistence, recovery prevention and propagation mechanisms. J Telecommun Inf Technol 1:113–124
2. Kermack W, McKendrick A (1927) Contributions to the mathematical theory of epidemics. Proceedings of the royal society of London. Series A 115:700–721
3. Hethcote HW (2000) The mathematics of infectious diseases. SIAM Rev 42(4):599–653
4. Kim J, Radhakrishnan S, Jang J (2006) Cost optimization in SIS model of worm infection. ETRI J 28(5):692–695
5. Chen TM, Jamil N (2006) Effectiveness of quarantine in worm epidemics. Cost optimization in sis model of worm infection in communications, ICC'06. IEEE, 52142–2147
6. Han X, Tan Q (2010) Dynamical behavior of computer virus on internet. Appl Math Comput 217(6):2520–2526

7. Dong T, Liao X, Li H (2012) Stability and hopf bifurcation in a computer virus model with multistate antivirus. Abstr Appl Anal. https://doi.org/10.1155/2012/841987
8. Liu J, Bianca C, Guerrini L (2016) Dynamical analysis of a computer virus model with delays. Discret Dyn Nat Soc. https://doi.org/10.1155/2016/5649584
9. Zhang ZH, Zhao T (2019) Bifurcation analysis of an e-SEIARS model with multiple delays for point-to-group worm propagation. Adv Diff Eq 228 (2019). https://doi.org/10.1186/s13662-019-2164-7
10. Batistela CM, Piqueira JRC (2019) SIRA computer viruses propagation model: mortality and robustness. Int J Appl Comput Math 4(5):1–9. https://doi.org/10.1155/2019/6467104
11. Piqueira JRC, Vasconcelos AAD, Gabriel CECJ, Araujo VO (2008) Dynamic models for computer viruses. Comput Secur 27(7):355–359. https://doi.org/10.1016/j.cose.2008.07.006
12. Hernández GJD, Martín AR (2019) A mathematical model for malware spread on WSNs with population dynamics. Physica A. https://doi.org/10.1016/j.physa.2019.123609
13. Zhu Q, Yang X, Ren J (2012) Modeling and analysis of the spread of computer virus. Commun Nonlinear Sci Numer Simul 17(12):5117–5124. https://doi.org/10.1016/j.cnsns.2012.05.030
14. Hurford A, Cownden D, Day T (2009) Next-generation tools for evolutionary invasion analyses. J R Soc Interface 7:561–571. https://doi.org/10.1098/rsif
15. Li M, Muldowney JS (1994) On Bendixson's criterion. J Differ Equ 106:27–39
16. Li M, Muldowney JS (1996) A geometric approach to global-stability problems. SIAM J Math Anal, 1070–1083. https://doi.org/10.1137/S0036141094266449
17. Prajapati A (2020) Mathematical analysis of effectiveness of security patches in securing wireless sensor network. In: Sahana S, Bhattacharjee V (eds) Advances in computational intelligence. Advances in Intelligent Systems and Computing, vol 988. Springer, Singapore. https://doi.org/10.1007/978-981-13-8222-2_12

Improving Deep Learning Model Robustness Against Adversarial Attack by Increasing the Network Capacity

Marco Marchetti and Edmond S. L. Ho

Abstract Nowadays, we are more and more reliant on Deep Learning (DL) models and thus it is essential to safeguard the security of these systems. This paper explores the security issues in Deep Learning and analyses, through the use of experiments, the way forward to build more resilient models. Experiments are conducted to identify the strengths and weaknesses of a new approach to improve the robustness of DL models against adversarial attacks. The results show improvements and new ideas that can be used as recommendations for researchers and practitioners to create increasingly better DL algorithms.

Keywords Machine learning · Deep learning · Security · Measurement · Perturbation methods · Robustness

1 Introduction

Security in Deep Learning (DL) is becoming increasingly important due to the extensive presence of its application in our daily life [1]. DL is also bringing benefits in a lot of aspects of the computing field, such as solving complex problems efficiently, great understanding of unstructured data, reduction of costs and high-quality results, etc. However, DL still faces great disadvantages, such as the need for a huge amount of data and the computational intensity [2] constitute important cons for the use of Deep Learning.

This implies DL applications are having an increasing demand for resources, therefore outsourcing techniques are required, such as MLaaS (Machine Learning as a Service). Here, data is sent to the cloud and the DL algorithm is there executed,

M. Marchetti · E. S. L. Ho (✉)
Department of Computer and Information Sciences, Northumbria University, 2 Ellison Place, Newcastle upon Tyne NE1 8ST, United Kingdom
e-mail: e.ho@northumbria.ac.uk

M. Marchetti
e-mail: m.marchetti@northumbria.ac.uk

© The Author(s), under exclusive license to Springer Nature Switzerland AG 2023
A. A. Abd El-Latif et al. (eds.), *Advances in Cybersecurity, Cybercrimes, and Smart Emerging Technologies*, Engineering Cyber-Physical Systems and Critical Infrastructures 4, https://doi.org/10.1007/978-3-031-21101-0_7

freeing the users from running heavy programs on their machines. However, this outsourcing of data creates data privacy and security concerns because there is a DL algorithm exposure to adversarial attacks [3].

To this regard, Xue et al. [4] clearly sum up the reasons why DL can be attacked. This includes: the outsourcing of the training phase, the integration into the network of third parties pre-trained models and the lack of data validations for data coming from untrusted users or third parties. Despite the fact that these working paradigms bring undoubtedly advantages, it might also rise security issues.

Essentially, adversarial attacks mainly target MLaaS or DML (Distributed Machine Learning) systems [5], for their nature of relying on the Internet. In particular, in a distributed situation the thing that, by expanding, generates more probable threats is the attack surface, i.e. the sum of the different points where an attacker can try to enter data or extract data from a software environment [6]. Additionally, it needs to be pointed out that even the most famous Deep Learning frameworks, such as TensorFlow, Caffe and Torch, present vulnerabilities. Xiao et al. [7] have studied the risks caused by those and their impact on common DL applications, particularly voice recognition and image classification. To tackle the problem of guaranteeing security while outsourcing potentially sensitive data, Ghodsi and Garg have built SafetyNet [8] which implements a framework to enable an untrusted server (the cloud in our case) to provide a short mathematical proof of the correctness of the tasks being performed for the client (MLaaS). By this, the DNN prevents any poisoning attacks by detecting them.

Adversarial attacks can be of two kinds [9]: a poisoning attack or an evasion attack. The first type of attack affects the training phase of the learning process by manipulating the training data [10] with the use of the so-called poisoning examples. Whereas, an adversarial attack on the interface phase (or predictions, see Appendix A) of the learning process is called evasion attack, where an attacker manipulates the test data or the real-time inputs to the model in order to produce false results [11]. Usually, the examples used to fool a DL model at interface time are called adversarial examples, noticed for the first time by Szegedy et al. [12], are defined as follows:

$$x^* = x + arg \min\{\|\delta\| : f(x + \delta) = t\} \tag{1}$$

In other words, an adversarial example (or adversarial sample) x^* is created by adding an imperceptible perturbation δ to the sample x correctly classified. δ is computed by approximating the optimisation problem (defined in the Eq. 1) in an iterative way until the crafted adversarial sample is classified by the DL classifier $f(.)$ in targeted class t.

While attacks attempt to force the target DL models to misclassify using adversarial examples, defense techniques tend to strengthen the resilience of DL models against adversarial examples, while preserving the performance of DL models on legitimate instances (i.e. non adversarial samples). In fact, based on what kind of attack they contrast, Tariq et al. [2] classify the defense techniques as against evasion attacks, which can use detection of adversarial examples, adversarial training or defensive distillation. Adversarial training was introduced for the first time by

Papernot et al. [13]. Defensive distillation refers to a DL that produces a set of confidence levels for each training example. Whereas, defenses against poisoning attacks usually eliminate the extreme values that fall outside the relevant group as described in the framework proposed by Steinhardt et al. [14].

To evaluate the performance of defence methods, Qayyum et al. [9] list some principles such as defending against the adversary, testing the worst-case robustness and measuring the performance towards human-level abilities. In addition, they propose some common evaluation recommendations as well, including using both targeted and untargeted attacks, performing ablation (i.e. removing some defence components and testing again), diversifying test settings, evaluating defence on broader domains and analysing the transferability attacks.

In this paper, we evaluate the methods for improving the robustness of DL models against adversarial attacks. In particular, DeepSec [15] provides us with a unique platform to analyse the security of a DL model. Despite the fact that this platform already includes a series of attacks and defense techniques that are subsequently evaluated, we are interested in improving the robustness of DL models based on the recommendations in [16]. Extensive experimental results are presented to demonstrate the effectiveness of increasing the capacity of DL models to improve the robustness against adversarial attacks.

2 Methodology

The attacks and defenses in DeepSec [15] are essentially PGD (Projected Gradient Descent) [16] and PAT (PGD Adversarial Training) [16]. However, the DeepSec implementation of PGD and PAT are different from [16] and, therefore, it loses some key aspects of the project.

In particular, [16] employ a ResNet model with wider layers in order to study the impact of the network capacity and they set $\epsilon = 8$. Those are essential parts of the new approach they propose that are not considered in DeepSec [15].

Based on this observation, we propose to modify DeepSec accordingly to evaluate whether the optimisation problem and the relative solution proposed by [16] can improve the robustness of deep learning models against adversarial attacks. In addition, we focus on the CIFAR-10 [17] dataset because existing methods have already achieved a state-of-the-art accuracy near 100% on the MNIST [18] dataset.

At this point, we could have focused on developing a particular defence mechanism for a specific type of attack. However, we can take a different approach as in [16], and create a unique mechanism to focus on both attacks and defense for any adversarial example. In the following sections, we will first review the basics of creating robust DL models.

2.1 Creating Robust Deep Learning Models

Madry et al. proposed a new approach in [16], i.e. the robustness to adversarial attacks is seen under the perspective of a saddle point optimisation problem (or min-max). Min-max optimisation is a decision rule commonly used in AI for minimizing the possible loss for a worst-case (max loss) scenario. Not only does this technique allows for DL models robust to any class of attacks, but it provides us with a unique theoretical framework that includes both defenses and attacks.

In particular, Madry et al. [16] define the problem as follows:

$$\min_{\theta} \rho(\theta), \; where \; \rho(\theta) = \mathbb{E}_{(x,y)-\mathcal{D}}[\max_{\delta \in s} L(\theta, x + \delta, y)] \tag{2}$$

In order to obtain this, the authors start by considering a standard classification task that employs a data distribution $\mathcal{D}$ over pairs of samples $x \in R^d$ and the corresponding labels $y \in [k]$. In addition, they assume that they are given a loss function, defined as $L(\theta, x, y)$, where $\theta \in R^p$ is the set of parameters for the model. The goal is to find the parameters θ that minimise the risk $\mathbb{E}_{(x,y)-\mathcal{D}}[L(\theta, x, y)]$. Empiric risk minimisation (ERM) is a successful method to find the classifier with a small population risk.

Madry et al. [16] further define the attack model as: for each input x, a set of perturbations $S \subseteq R^d$ that formalise the manipulation power of the adversarial attack will be introduced. Recall the definition of adversarial examples from Eq. 1 and we incorporate that in the population risk $\mathbb{E}_{\mathcal{D}}[L]$, we will obtain the saddle point problem described in Eq. 2.

This robust optimisation problem can be viewed as an inner maximisation problem and an outer minimisation problem. The first one aims to create an adversarial version of a given data point x that achieves a high loss. Whereas, the second one's goal is to find model parameters so that the adversarial loss given by the adversarial is minimised. As a result, when the parameters θ yield a (nearly) vanishing risk, the corresponding model is perfectly robust to attacks specified by our attack model. In other words, the unified view on attacks and defenses give an answer to both the question of how to produce effective adversarial examples and how to train a model free from adversarial samples.

On the attack side, Madry et al. [16] focus on PGD (Projected Gradient Descent), which they claim to be a 'universal' adversary. While on the defence side, they use a training dataset augmented with adversarial samples created by PGD. In addition, they argue that if a network is trained to be robust against PGD adversaries, then it becomes robust against a wide range of other attacks as well. The formula can be seen below:

$$x^{t+1} = \prod_{x+s}(x^t + \alpha \; sgn(\nabla_x L(\theta, x, y))) \tag{3}$$

From here, we can see that PGD is a multi-step variant of FGSM (Fast Gradient Sign Method):

$$x + \epsilon \; sgn(\nabla_x L(\theta, x, y)) \tag{4}$$

Therefore, applying PGD solves the inner maximisation problem. As for the outer minimisation, i.e. finding model parameters that minimise the 'adversarial loss', this is done by applying SGD (Stochastic Gradient Descent) using the gradient of the loss at adversarial examples during training. These techniques employed in [16] result in successful optimisation of the saddle point problem and ultimately in the training of a robust DL model.

In other words, as we saw, the attack side of the problem is represented by the PGD, whereas on the defence side the authors employ the adversarial training technique called PAT (PGD Adversarial Training), which retrains the model on a dataset that includes the adversarial samples created by the PGD attack. Adversarial training is generally thought to be one of the most effective defence techniques, further contributing to the importance of the methods presented in [16].

3 Experimental Results

In the experiments, we evaluate the effectiveness of the DeepSec platform [15], different adversarial attacks (CW2 and PGD) and defenses (NAT and PAT) on benchmark datasets MNIST [18] and CIFAR10 [17].

3.1 PGD Attack and PAT Defense

For PGD and PAT, the experiments were run on CIFAR-10 [17]. We started with the unmodified model, i.e. a ResNet, which reports an accuracy of 90.01%. As a second step, we randomly select the clean candidate examples that will be used in the PGD attack, and 9001 samples were selected. Next, the parameters ($\epsilon = 0.1$ and $\epsilon_{iter} = 0.01$) suggested by the authors were used for creating the adversarial examples. As suggested by the authors. This returned a misclassification ratio (MR) of 100%.

Using these adversarial examples generated, the PGD Adversarial Training (PAT) is executed with $\epsilon = 0.3137, number\ of\ steps = 7$ and $step\ size = 0.007843$. This results in an accuracy of the retrained (i.e. trained on the dataset augmented with adversarial examples) model of 81.11%.

Finally, the evaluations of the attack and the defence technique are reported in the PGD ($\epsilon = 0.3137$) column in Table 1 and the PAT ($\epsilon = 0.1$) column in Table 2, respectively.

Table 1 Experimental results of the utility metrics of attacks

Metric	CW2	CW2-DeepSec	PGD ($\epsilon = 0.3137$)	PGD-DeepSec	PGD ($\epsilon = 8$)
MR	99.8% (ConvNet on MINST)	99.7% (ConvNet on MINST)	100% (ResNet on CIFAR10)	100% (ResNet on CIFAR10)	99.6% (Wide ResNet on CIFAR10)
ACAC	0.320	0.326	1.000	1.000	0.956
ACTA	0.312	0.318	0.000	0.000	0.002
ALDp					
L0	0.428	0.342	0.978	0.979	0.995
L1	1.492	1.746	3.380	3.682	29.228
L2	0.447	0.528	0.100	0.100	0.975
ASS	0.731	0.925	0.782	0.827	0.000
PSD	12.616	14.086	149.487	165.721	1331.845
NTE	0.001	0.003	1.000	1.000	0.978
RGB	0.003	0.004	1.000	1.000	0.978

Table 2 Experimental results of the utility metrics of defences

Metric	NAT	NAT-DeepSec	PAT ($\epsilon = 0.1$)	PAT-DeepSec	PAT ($\epsilon = 8$)
ACC (Raw model)	99.36% (ConvNet on MINST)	99.27% (ConvNet on MINST)	90.01% (ResNet on CIFAR10)	85.95% (ResNet on CIFAR10)	94.16% (Wide ResNet on CIFAR10)
ACC (defence-enhanced model)	99.27%	99.51%	81.11%	80.23%	40.25%
CAV	−0.09%	0.24%	−8.90%	−5.72%	−49.76%
CRR	0.25%	0.44%	3.92%	6.60%	2.61%
CSR	0.34%	0.20%	12.82%	12.32%	52.37%
CCV	0.26%	0.17%	16.58%	13.87%	31.58%
COS	0.0009	0.0006	0.0683	0.0572	0.1345

3.2 The Wide ResNet Model

As mentioned in Sect. 2, the DeepSec implementation of PGD and PAT lack some key aspect highlighted by Madry et al. [16]. In particular, Madry et al. [16] demonstrated the ResNet [19] model can be successfully trained against strong adversaries by 1) increasing the network capacity and 2) constructing adversarial examples with $\epsilon = 8$.

Inspired by the recommendations, we included a new model, called *Wide ResNet*, by modifying the original ResNet as follows:

- Changing the layer size: the values of layer 1, 2 and 3 and the Fully Connected (FC) are multiplied to a wide factor (10 in our experiments).

- Changing the ϵ from 0.1 (default in DeepSec) to 8 and the ϵ_{iter} to 0.875. The latter is due to the fact that Madry et al. [16] run the algorithm for 7 iterations, therefore ϵ is divided by the number of iterations: $8/7 = 0.875$.
- For the PAT, the new parameters are: $\epsilon = 8$, *number of steps* $= 10$ and the *step size* $= 2$ as suggested in [16].

We follow the tests explained in Sect. 3.1 and the results are reported in the PGD ($\epsilon = 8$) column in Table 1 and PAT ($\epsilon = 8$) column in Table 2. The accuracy resulting obtained by the raw *Wide ResNet* model is 94.16%, which is comparable to the 95.2% obtained in [16].

Then, after the candidate selection, we ran the PGD attack with the modified parameters, i.e. $\epsilon = 8$ and $\epsilon_{iter} = 0.875$. For the PAT test, the modified parameters are:$\epsilon = 8$, *number of steps* $= 10$ and *step size* $= 2$, as in [16]. The final evaluation accuracy was 40.06% as presented in the PAT ($\epsilon = 8$) column in Table 2.

3.3 Experimental Setup

For running the experiments we used a PC equipped with an NVIDIA GeForce 1080 GPU and Ubuntu as the operating system.

3.4 Evaluation Metrics

We follow the literature to evaluate the performance of the adversarial attacks and defenses. Specifically, 9 metrics (Table 3) were used for evaluating the adversarial attacks as in [15] and 5 metrics (Table 4) were used for evaluating the defence performance as in [16].

4 Results and Discussions

The results of the experiments give us the evaluation utility metrics that are shown in Tables 1 and 2. While DeepSec [15] focuses on comparing the metrics of different attacks and defences to determine which one is the best, we try to analyse the impact of the modification we brought, namely the wider ResNet and the higher value of ϵ.

4.1 Attack Utility Metrics

In the first two columns in Table 1, we reported the results obtained from the corresponding values presented in [15]. This is done with the aim of determining to what extent the random component present in every DL algorithm will create a difference in results.

Despite not being exactly the same, the various utility metrics seem to be in a very similar range, therefore, not compromising their meaning. For instance, we can notice how the Carlini-Wagner (CW2) algorithm presents a very strong attacking ability, confirmed by the close to 100% MR. In addition, the relatively low ACTC confirm a higher resilience of the CW2 attack to other models, such as a defence-enhanced one. Therefore, confirming its good attack ability.

CW2 also presents a good imperceptibility, as proven by the low value of ALD and ASS. However, it needs to be noticed the relatively high PSD, which is one of the best metrics to evaluate imperceptibility [21]. This means that the attack is quite perceptible by humans.

Another disadvantage of CW2 seems to be its robustness, indeed it presents very low values in its robustness metrics (NTE, RGB and RIC). Finally, the computational cost of CW2 is quite high. Despite not being the worst, CW2 is not the best for the amount of resources it needs. This similarity of results can also be noticed when comparing our second experiment on a PGD attack and the values reported in [15]. In fact, both the experiments conducted find that PGD is one of the best attack methods available, with a state-of-the-art misclassification rate. In addition, PGD generated adversarial examples are way more robust to pre-processing than CW2. However,

Table 3 Metrics for evaluating adversarial attacks

Misclassification Ratio (MR)	The percentage of adversarial samples that are successfully misclassified
Average Confidence of Adversarial Class (ACAC)	The average prediction confidence towards the incorrect class
Average Confidence of True Class (ACTC)	Evaluate to what extent the attacks escape from the ground truth, by averaging the prediction confidence of true classes for adversarial examples
Average Lp Distortion (ALDp)	The average normalized Lp (with $p = 0, 1, \infty$) norm distortion for all successful adversarial examples. The smaller ALDp is, the more imperceptible the adversarial examples are
Average Structural Similarity (ASS)	i.e. the average SSIM similarity [20] (alternative to $Lp\ norm$) between all successful adversarial examples and their original samples. Therefore, the greater ASS is, the more imperceptible the adversarial samples are
Perturbation Sensitivity Distance (PSD) [21]	Evaluates the human perception of perturbations. The smaller it is, the more imperceptible the adversarial examples are
Noise Tolerance Estimation (NTE)	Calculates the gap between the probability of misclassified class and the maximum probability of all other classes. The higher NTE is, the more robust adversarial samples are
Robustness to Gaussian Blur (RGB)	Measures how much a robust adversarial example maintain its misclassification effect after RGB is applied, which is used in pre-processing. The higher RGB is, the more robust adversarial examples are
Computation Cost (CC)	The average runtime employed by the attacker to generate an adversarial example, i.e. the attack cost

Table 4 Metrics for evaluating defenses

Classification Accuracy Variance (CAV)	Since the most important metric for evaluating a DL model is its accuracy, and a defence technique should maintain this as much as possible on normal samples
Classification Rectify/Sacrifice Ratio (CRR/CSR)	CRR is defined as the percentage of testing samples that are misclassified by the model previously but correctly classified by the defence-enhanced model. Inversely, CSR is the percentage of testing samples that are correctly classified by the original model but misclassified by the defence-enhanced one. Therefore, CAV is equal to the difference between CRR and CSR
Classification Confidence Variance (CCV)	Even though the accuracy might remain the same, sometimes the confidence of correctly classified samples can significantly decrease
Classification Output Stability (COS)	Measure the classification output stability between the original model and the defence-enhanced model, we use JS divergence [22] to calculate the similarity of their output probability. We average the JS divergence between the output of original and defence-enhanced model on all correctly classified testing examples

despite presenting a great imperceptibility (due to low ALD and ASS values), the PSD metric is extremely high, especially if compared to other attacks. This could result in an easy to detect adversarial examples for humans.

Finally, the Computational Cost is very low, especially when compared with other attacks. If we analyse the impact of increasing the dimension of the layers (*Wide ResNet*) and changing the value of ϵ from 0.3 to 8, we can notice a series of differences. In general, the values seem to be just slightly worse, with an exception on the Average Structural Similarity (ASS) and on PSD. The first one seems to be more of a computational mistake rather than a plausible value, indeed such a low value is not reported by any type of attack. Whereas, the latter looks to be a worrying value which could make adversarial examples really easy to detect by a human operator.

4.2 Defence Utility Metrics

As for the defence utility metrics, we tend to obtain values that present a more noticeable difference from those reported in [15]. In fact, while generally obtaining better accuracy on the raw models as reported in Table 2, we notice a drop in accuracy on the defence-enhanced model that is more noticeable than the one obtained in [15]. However, this is in line with the results obtained in [16], i.e. a 47.1% accuracy on the standard model and 50% on the wide one. It, therefore, DeepSec [15] has achieved better results.

Starting from the Naive Adversarial Training (NAT) defence technique, we notice how this algorithm increases the accuracy of the defence-enhanced model. In fact, according to the CRR value, 44% of the samples misclassified by the raw model are classified correctly by the new model. While 20% (CSR) of the samples are correctly classified by the raw model, but misclassified by the defence-enhanced one. Therefore, creating an increase in accuracy, as proven by the positive value of CAV (CAV=CRR-CSR).

However, this benefit brought by the NAT technique does not seem to be produced by the PGD Adversarial Training. In fact, in all the results relative to PAT we can notice a negative CAV, which represents a decrease in the accuracy of the new model in comparison to the original one.

It needs to be pointed out that the NAT technique was used on the MNIST dataset, which is simpler than the CIFAR10 one. As a result, most of the values that look less performing in Table 2 are to be ascribed to this higher complexity that generates a higher instability in the prediction of testing samples. In fact, other defence methods on MNIST have similar results to the NAT technique and PAT is among the best performing defence techniques on CIFAR10.

As for the change we brought to the model, we can notice a remarkable increase in the accuracy of the raw model (up to 94.16%), due to a wider neural network. This gain in network capacity is essential to have more room to protect us against misclassification. However, the variation of epsilon seems to indicate a significant loss in the accuracy of the model to which the defence technique was applied.

Lastly, despite not being included in the defence utility metrics, the Computational Cost of PAT was a metric worth mentioning. In fact, the retraining on the adversarial augmented dataset required our machine for approximately 5 days or around 7000 minutes. This highlights a considerable weakness of this algorithm, i.e. its high demand for resources in order to obtain a robust model.

5   Conclusions

In this research, the results indicate that increasing ϵ to 8, as used by [16], has generated worse metrics in the experiment condition set by DeepSec [15]. However, operating on a wider ResNet has allowed us to gain more accuracy to be used to protect against misclassification caused by adversarial attacks and, therefore, create a more robust model.

Finally, as for evaluation, Carlini and Wagner [23] propose an alternative method. In particular, they identify weaknesses in one of the most effective defence methods, i.e. defensive distillation. They develop three new attacks algorithms in order to be used as a benchmark for the effectiveness of a robust model and this is an interesting future direction to further evaluate the performance of adversarial attack and defense.

References

1. Ho ESL (2022) In: Abd El-Latif AA, Abd-El-Atty B, Venegas-Andraca SE, Mazurczyk W, Gupta BB (eds) Data security challenges in deep neural network for healthcare IoT systems. Springer, Cham, pp 19–37. https://doi.org/10.1007/978-3-030-85428-7_2
2. Tariq MI, Memon NA, Ahmed S, Tayyaba S, Mushtaq MT, Mian NA, Imran M, Ashraf MW (2020) A review of deep learning security and privacy defensive techniques. Mobile Inf Syst 2020:6535834. https://doi.org/10.1155/2020/6535834
3. Xu G, Li H, Ren H, Yang K, Deng RH (2019) Data security issues in deep learning: attacks, countermeasures, and opportunities. IEEE Commun Mag 57(11):116–122. https://doi.org/10.1109/MCOM.001.1900091
4. Xue M, Yuan C, Wu H, Zhang Y, Liu W (2020) Machine learning security: threats, countermeasures, and evaluations. IEEE Access 8:74720–74742. https://doi.org/10.1109/ACCESS.2020.2987435
5. Chen Y, Mao Y, Liang H, Yu S, Wei Y, Leng S (2020) Data poison detection schemes for distributed machine learning. IEEE Access 8:7442–7454. https://doi.org/10.1109/ACCESS.2019.2962525
6. Manadhata PK, Wing JM (2011) An attack surface metric. IEEE Trans Software Eng 37(3):371–386. https://doi.org/10.1109/TSE.2010.60
7. Xiao Q, Li K, Zhang D, Xu W (2018) Security risks in deep learning implementations. In: 2018 IEEE security and privacy workshops (SPW), pp 123–128. https://doi.org/10.1109/SPW.2018.00027
8. Ghodsi Z, Gu T, Garg S (2017) Safetynets: verifiable execution of deep neural networks on an untrusted cloud. In: Proceedings of the 31st international conference on neural information processing systems. NIPS'17, pp 4675–4684. Curran Associates Inc., Red Hook, NY, USA
9. Qayyum A, Usama M, Qadir J, Al-Fuqaha A (2020) Securing connected and autonomous vehicles: challenges posed by adversarial machine learning and the way forward. IEEE Commun Surv Tutorials 22(2):998–1026. https://doi.org/10.1109/COMST.2020.2975048
10. Biggio B, Nelson B, Laskov P (2012) Poisoning attacks against support vector machines. In: Proceedings of the 29th international coference on international conference on machine learning. ICML'12, pp 1467–1474. Omnipress, Madison, WI, USA
11. Biggio B, Corona I, Maiorca D, Nelson B, Šrndić N, Laskov P, Giacinto G, Roli F (2013) Evasion attacks against machine learning at test time. In: Blockeel H, Kersting K, Nijssen S, Železný F (eds) Machine learning and knowledge discovery in databases. Springer, Berlin, Heidelberg, pp 387–402
12. Szegedy C, Zaremba W, Sutskever I, Bruna J, Erhan D, Goodfellow IJ, Fergus R (2014) Intriguing properties of neural networks. In: Bengio Y, LeCun Y (eds) 2nd international conference on learning representations, ICLR 2014, Banff, AB, Canada, April 14-16, 2014, Conference Track Proceedings. arXiv:1312.6199
13. Papernot N, McDaniel PD, Jha S, Fredrikson M, Celik ZB, Swami A (2015) The limitations of deep learning in adversarial settings. arXiv:1511.07528
14. Steinhardt J, Koh PW, Liang P (2017) Certified defenses for data poisoning attacks. In: Proceedings of the 31st international conference on neural information processing systems. NIPS'17. Curran Associates Inc., Red Hook, NY, USA, pp 3520–3532
15. Ling X, Ji S, Zou J, Wang J, Wu C, Li B, Wang T (2019) Deepsec: a uniform platform for security analysis of deep learning model. In: 2019 IEEE symposium on security and privacy (SP), pp 673–690. https://doi.org/10.1109/SP.2019.00023
16. Madry A, Makelov A, Schmidt L, Tsipras D, Vladu A (2018) Towards deep learning models resistant to adversarial attacks. In: International conference on learning representations
17. Krizhevsky A, Nair V, Hinton G (2009) Cifar-10 (canadian institute for advanced research)
18. LeCun Y, Cortes C (2010) MNIST handwritten digit database
19. He K, Zhang X, Ren S, Sun J (2016) Deep residual learning for image recognition. In: 2016 IEEE conference on computer vision and pattern recognition (CVPR), pp 770–778. https://doi.org/10.1109/CVPR.2016.90

20. Wang Z, Bovik AC, Sheikh HR, Simoncelli EP (2004) Image quality assessment: from error visibility to structural similarity. IEEE Trans Image Process 13(4):600–612. https://doi.org/10.1109/TIP.2003.819861
21. Liu A, Lin W, Paul M, Deng C, Zhang F (2010) Just noticeable difference for images with decomposition model for separating edge and textured regions. IEEE Trans Circuits Syst Video Technol 20(11):1648–1652. https://doi.org/10.1109/TCSVT.2010.2087432
22. Dagan I, Lee L, Pereira F (1997) Similarity-based methods for word sense disambiguation. In: 35th annual meeting of the association for computational linguistics and 8th conference of the European chapter of the association for computational linguistics. Association for Computational Linguistics, Madrid, Spain, pp 56–63. https://doi.org/10.3115/976909.979625
23. Carlini N, Wagner D (2017) Towards evaluating the robustness of neural networks. In: 2017 IEEE symposium on security and privacy (SP). IEEE Computer Society, Los Alamitos, CA, USA, pp 39–57. https://doi.org/10.1109/SP.2017.49

Intelligent Detection System for Spoofing and Jamming Attacks in UAVs

Khadeeja Sabah Jasim, Khattab M. Ali Alheeti, and Abdul Kareem A. Najem Alaloosy

Abstract Unmanned aerial vehicles (UAVs) have recently gained popularity due to their extensive uses in parcel delivery, wildlife conservation, agriculture, and the military. However, security worries with UAVs are developing, as UAV nodes are becoming appealing targets for assaults due to rapidly growing volumes and inadequate inbuilt security. This paper proposes an intelligent security system for UAVs that harness machine learning to detect cybersecurity attacks. It determines whether the signals coming to the UAV are benign, or offensive using a UAV attacks dataset containing two types of attacks: GPS spoofing and Jamming. In order to improve security in UAV networks, this research shows how machine learning methods may be utilized to categorize benign and malicious signals. Finally, the accuracy rate, recall, F1-score, precision, and confusion matrix of the tested ML algorithms are compared for efficacy. Compared to all other ML classifiers, the decision tree model performed well, with a maximum accuracy rate of 99.86% in detecting various attacks.

Keywords Decision tree · Intelligent Detection System (IDS) · Machine learning · Gradient Boosting · K-nearest neighbour · Cybersecurity attacks · UAV

1 Introduction

Unmanned aerial vehicles (UAVs) are lightweight robots [1] that may fly autonomously or with pilot assistance from a ground station or via remote control [2], as shown in Fig. 1. Drones or UAVs are often employed for secure communication, packet delivery, and aerial surveillance. Communication signals exchanged between

K. S. Jasim
Computer Sciences Department, College of Computer and Information Technology, University of Anbar, Anbar, Iraq
e-mail: kha20c1005@uoanbar.edu.iq

K. M. Ali Alheeti (✉) · A. K. A. Najem Alaloosy
Computer Networking Systems Department, College of Computer and Information Technology, University of Anbar, Anbar, Iraq
e-mail: co.khattab.alheeti@uoanbar.edu.iq

© The Author(s), under exclusive license to Springer Nature Switzerland AG 2023
A. A. Abd El-Latif et al. (eds.), *Advances in Cybersecurity, Cybercrimes, and Smart Emerging Technologies*, Engineering Cyber-Physical Systems and Critical Infrastructures 4,
https://doi.org/10.1007/978-3-031-21101-0_8

Fig. 1 A simplified form of
drone [5]

UAVs and ground stations in a risk zone may be lost or damaged due to cyber-attacks such as spoofing and Jamming [3, 4].

UAVs systems are commonly used GPS to land on the intended location. Even though the GPS can operate the UAVs, it may be exposed to radiofrequency interference [6]. Jamming [7] and GPS spoofing [8] are two principal vulnerabilities that pose a serious threat to military and civilian GPS users. In GPS spoofing, the spy transmits bogus signals and changes the predefined position or creates a fake position to prevent the UAV from completing its objective. As a result, spoofing a civilian or military UAV is possible by changing the predefined trajectory without the user's knowledge [9]. While Jamming is defined as broadcasting radio signals at a higher frequency than the actual signals while hiding the original signals with noisy signals to disrupt them [10]. However, Suppose the drones receive erroneous information. In that case, this leads to many problems, such as the plane's loss, crashing, or carrying out its mission incorrectly due to the wrong information. So Securing communications in UAV networks is essential [11]. Several IDS have been developed to provide this goal and detect network or system threats by monitoring signals [12], analyzing for inconsistent accidents that violate system security rules, and identifying them as unauthorized access from allowed items or malicious [13].

Many IDS have been developed in the past two decades; most of them are in machine learning, such as Naïve Bayes [14], Random Forests [15], support vector machine [16], Logistic regression [17], Decision tree [18] and deep learning [19].

This paper proposes an intelligent security system for UAVs that harness machine learning to detect cybersecurity attacks. It determines whether the signals coming to the UAV are benign, or offensive using a UAV attacks dataset containing two types of attacks: GPS spoofing and Jamming.

The main contribution of this paper is represented in the following:

- Design and build an Intelligent Detection System to detect attacks on UAV signals.
- Machine learning was used to classify the signals as either authentic or deceptive.
- Several ML models are used to select the appropriate models for the proposed work.
- The dataset was divided into different sets for testing and training to choose the best splitting for the proposed system in execution time and accuracy.
- Design security system lightweight to maintain privacy data.
- Several performance measures analysis to evaluate the proposed system: the confusion matrix, accuracy, execution time, precision, recall, and F_score error rate.

This paper is structured as follows: Sect. 2 will present related works, and Sect. 3 will present an overview of machine learning. The Research methodology will be described in Sect. 4. The results and discussion will be presented in Sect. 5, and finally, the conclusion and future works will be presented in Sect. 6.

2 Related Works

UAVs have been widely used in various fields in recent years, the most important of which are military, remote monitoring, rescue, and others [20]. The importance of the areas in which UAVs are used attracted the attention of researchers to present several studies to improve safety in the UAV network and discover the different types of attacks that target the drone and cause damage to it. Below, some previous studies on detecting jamming and spoofing attacks on drones are summarized.

Liu et al. [21], Suggested a GPS-Probe spoofing detection algorithm by taking advantage of Air Traffic Control (ATC) messages transmitted by drones by analyzing the timestamps and strength of messages received for ATC messages. GPS-Probe creates a machine-learning framework for antenna location assessment and GPS spoofing detection. The proposed system achieved precision and accuracy of 85.3% and 81.7%, respectively, and 91.5% and 89.7% at best.

Greco et al. [22], Propose a system for detecting Jamming using machine learning with two techniques: decision tree (DT) and multi-layer perceptions (MLP). The two techniques were trained for the part of the data set designated for training. Researchers noticed that the decision tree outperformed by a large percentage, where its accuracy rate reached 93%, while MLP achieved 86.7%. In this work, the researchers used one machine learning algorithm, and it was possible to use another algorithm to achieve higher efficiency.

Majidi et al. [23], The authors suggested a prediction discrepancy based on an innovative particle filter(PDIPF) technique to deal with GPS spoofing attempts that resulted in unknown rapid changes in system state variables. PDIPF technique was used to adjust the compensation of the GPS spoofing effect. In the context of GPS spoofing assaults, the proposed algorithm reduces the effects of GPSspoofing mistakes and calculates the actual position of the UAV. One of the limitations of this work is that it was evaluated by redundancy and accuracy only.

Shafique et al. [24], Proposed a new intrusion detection system to detect GPS spoofing attacks on drones using several machine learning algorithms. K_flod was used before implementing the machine learning algorithms to increase the system's accuracy. After implementation, a Sport vector machine (with polynomial) achieved the highest ratio among all machine learning algorithms.

Sedjelmaci et al. [25], Suggest an intrusion detection system using a support vector machine(SVR) to distinguish malicious behaviour from ordinary. The proposed system detects cybersecurity attacks on drones and focuses on spoofing, Jamming, black holes, and gray holes. It achieved a high rate for detecting attacks, with an accuracy rate of 93%. One of the essential issues with rule-based detection systems is that many rules may be required for each variation of an attack, which is a limiting limitation, especially in the rapidly changing network attack types.

Mitchell et al. [26] proposed an adaptive (IDS)system by incorporating a conduct specification for identifying harmful UAVs into the UAV system. The system determines whether the UAV operates under the influence of various attacks or is operating normally. Despite this, the system failed in the face of cyber-attacks on UAVs.

Panice et al. [27], The authors suggested a method using a Support Vector Machine for detecting GPS spoofing attacks on UAV systems based on state estimate analysis. This approach uses the error dispersion between the inertial navigation system and the GPS of UASs. The authors discovered that this distribution shifts in the presence of a GPS spoofing attack. However, In a long-duration attack, this approach suffers from severe performance loss.

Our work aims to increase the accuracy in detecting attacks, higher than all previous studies, while at the same time reducing the execution time and the rate of false alarms. In addition, the proposed work is simple and lightweight.

3 An Overview of Machine Learning

ML is a sort of artificial intelligence that learns new things through algorithms, and it focuses on teaching algorithms how to complete specific activities on their own. Machine learning employs both existing data knowledge and experience in the intrusion detection domain to teach a system to cluster or classify future unknown data [28]. The proposed IDS use three ML models: a decision tree, Gradient Boosting, and K-nearest neighbour, to choose the best model for implementing the proposed system.

- Gradient Boosting (GB): is a machine learning technique utilized in various applications, including regression and classification. Creates a powerful specular reflection-boosting model by combining weak prediction models, often decision trees. The boosting method produces basic models sequentially. The accuracy of the predictions is increased by creating several models in order and focusing on training examples that are difficult to estimate. The hard-to-estimate examples appear more frequently in the training data during the boosting process than correctly estimated models [29].
- K-nearest neighbour (KNN): is a more straightforward machine learning method that classifies objects based on their neighbours' superiority polling [30] and assigns them to the nearest
  Neighbour class. KNN can be used with statistical techniques as an intrusion detection system classifier, and it can be achieved easily by utilizing local data to generate a highly adaptive attitude. This approach, however, necessitates a high computation cost and a considerable distance calculation amount [31].
- A Decision Tree (DT): is a type of supervised machine learning (ML) approach used for categorization by a collection of if–then rules to improve readability.

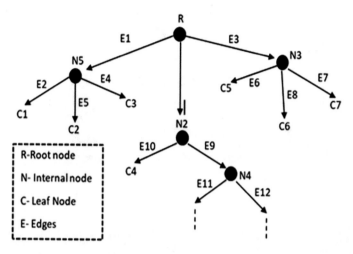

Fig. 2 Representation graphical of DT classifier [33]

DT consists of two types of nodes known as decision nodes and leaf nodes, as shown in Fig. 2, and DT can forecast a target or class by creating a training model based on inferred decision rules from training data. The DT's significant benefit is decreased decision-making ambiguity, transparency, and comprehensive analysis [32].

4 Research Methodology

This paper suggested IDS to detect spoofing and jamming attacks on the UAV network. The proposed system uses ML algorithms to classify a dataset into three classes (Jamming, benign, spoofing). Classification using three ML algorithms (DT, GB and KNN) to classify the data set. Finally, the performance of the proposed system is evaluated and its efficiency checked by calculating a set of necessary performance measures.

The general structure of the proposed system is divided into four phases as follows:

1. Dataset Source phase.
2. Dataset Preprocessing phase.
3. Splitting and Classifying the Dataset Phase.
4. Performance Metric Phase.

The details of these phases are illustrated in Fig. 3.

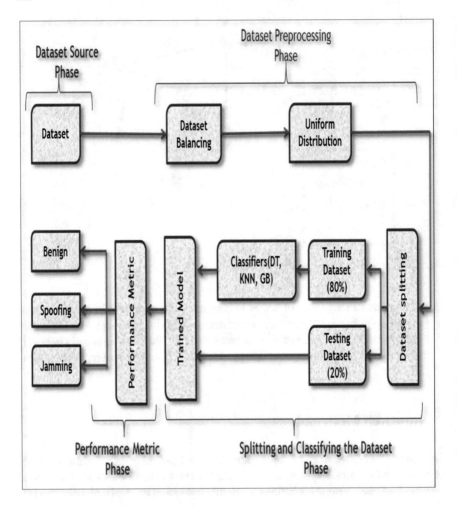

Fig. 3 The architecture of the proposed system

4.1 Dataset Source Phase

We used the UAV attack dataset to implement the proposed system [34]. The data is collected through three flights: a secure flight, a flight under a jamming attack, and a flight under a GPS spoofing attack. Data is extracted from the drone after each flight. This dataset contains three types of classes: benign, spoofing, jamming, and 9 features are timestamp, valid, timestamp_sample, hover_thrust, hover_thrust_var, accel_innov, accel_innov_var, accel_innov_test_ratio, accel_noise_var, and label. The dataset contains 7516 records. A sample of the dataset is shown in Fig. 4.

timestamp	timestamp_sample	hover_thrust	hover_thrust_var	accel_innov	accel_innov_var	accel_innov_test_ratio	accel_noise_var	valid	label
617413972	617412837	0.2095902	9.58E-05	0.31506634	5.963904	0.0018494	5.5784016	0	Benign
617434488	617432834	0.20897473	9.06E-05	0.653568	5.900586	0.008043479	5.531206	0	Benign
617513933	617512823	0.208286	7.29E-05	-0.94025874	5.702376	0.017226474	5.3675117	0	Benign
617614157	617612802	0.21362934	5.91E-05	-0.9329307	5.463999	0.017698877	5.227068	0	Benign
617713857	617712785	0.21369384	5.07E-05	-1.8655807	5.267211	0.07341839	5.0975423	0	Benign
617814061	617812767	0.21687685	4.47E-05	-2.5681896	5.1188354	0.1431662	4.993591	0	Benign
617913957	617912750	0.2206844	3.99E-05	-0.38517416	5.097782	0.003233631	4.9405475	0	Benign
618014043	618012750	0.2218348	3.60E-05	-0.13365293	4.905634	0.000404594	4.7608695	0	Benign
618113789	618112714	0.2274239	3.29E-05	-0.20589328	5.2694936	0.000893867	5.1377316	0	Benign
618213938	618212697	0.22858205	3.05E-05	-0.72188425	5.192738	0.011150547	5.068033	0	Benign
618313781	618312680	0.22941162	2.85E-05	-0.67718047	5.064627	0.010060484	4.948289	0	Benign
618413823	618412662	0.22922115	2.67E-05	-3.1224446	4.9567604	0.21854912	4.904588	0	Benign
618514271	618512645	0.23265384	2.52E-05	-3.414241	4.9971733	0.25919193	4.9572616	0	Benign
618613863	618612627	0.23286858	2.38E-05	-1.6704195	4.9467006	0.0626748	4.8527822	0	Benign
618713839	618712610	0.2331284	2.26E-05	-1.2863055	4.825719	0.038096383	4.7294164	0	Benign
618813737	618812595	0.23494358	2.14E-05	-6.2996583	4.670584	0.9441049	4.868124	0	Benign
618913812	618912575	0.23674074	2.04E-05	0.37284613	4.9504666	0.003120115	4.87113	0	Benign
619013627	619012557	0.23690142	1.95E-05	1.71316	5.0538573	0.06452535	5.0225663	0	Benign
619114577	619112570	0.23621617	1.87E-05	2.2113557	4.9370513	0.11005431	4.927731	0	Benign
619213693	619212525	0.236805	1.80E-05	-0.5911134	4.8156066	0.008062098	4.7334886	0	Benign
619313940	619312505	0.23603314	1.76E-05	2.3523812	4.709668	0.13055173	4.728821	0	Benign

Fig. 4 Sample of the UAV attacks dataset

4.2 Dataset Preprocessing Phase

Preprocessing the data set is an important step to solve the problems that this dataset suffers from because machine learning algorithms require a pure and balanced dataset to get the highest accuracy and the least execution time.

After obtaining the dataset, we noticed that it suffers from unbalance of the classes of the dataset, so the technique of balancing the dataset is used. In addition, it suffers from bias for a particular class at the expense of the rest of the classes, which requires a uniform distribution of records for the dataset to give equal training opportunities for all classes in training.

The preprocessing steps for the dataset are as follows:

- Uniform distribution step: Ranking the classes randomly to ensure that all classes are included in training or testing is the process of randomly arranging the records of a dataset to provide equal opportunities for all dataset classes in training or testing without bias toward a particular class.
- Dataset balancing step: The classes of the dataset were balanced using ADASYN. The ADASYN equalizes the classes in the dataset to make the classes equal/close in the number of records by over/under samples. As a result, ADASYN technology improves data distribution in two ways: Reducing the data set's imbalance by decreasing the high classes or increasing the few categories. In the proposed system, we used the second way.

4.3 Splitting and Classifying the Dataset Phase

After completing the data set preprocessing, the dataset is separated into two parts: the training data set and the test data set. As explained below:

- Training Dataset: This dataset is used to train the machine learning algorithms used in the proposed system.
- Test Dataset: This dataset is used to evaluate the trained model.

The proposed system will separate the dataset into several splits and choose the best split for the dataset.

After applying the data preprocessing and dataset separation process, The dataset is ready to enter the classification phase, and the three ML algorithms (DT, GB, KNN) are applied to the proposed system.

In the classification phase, 8 of the data set's full features are considered input to the ML algorithms, and the Label feature of the classes is considered the output of the classification phase.

The classification phase is divided into two phases: training and testing. The training phase creates a high-accuracy trained model that shows how the features are related to the label, so this phase is considered essential in building the proposed system to improve performance and reduce false alarms. The testing phase begins after completing the training phase. In this phase, the trained model is tested with an unknown dataset of the trained model (test dataset). The trained model must predict the precise class of each test sample.

4.4 Performance Metric Phase

Many performance metrics are used to evaluate the proposed IDS. We will evaluate the ML model using the Confusion Matrix. The confusion matrix determines the number of incorrect or correct predictions, and it consists of four types of error alarms [35]. Below, the four types are explained:

- True Positive (TP): Positive classes in a dataset classify as the positive class and are defined as:

$$\text{TP Rate} = \frac{TP}{(FN + TP)} * 100\% \tag{1}$$

- False Positive (FP): Positive classes in a dataset classify as the negative class and are defined as:

$$\text{FP Rate} = \frac{FP}{(FP + TN)} * 100\% \tag{2}$$

- True negative (TN): negative classes in a dataset classify as the negative class and are defined as:

$$\text{TN Rate} = \frac{TN}{(FP + TN)} * 100\% \tag{3}$$

- False Negative (FN): negative classes in a dataset classify as the Positive class and are defined as:

$$\text{FN Rate} = \frac{FN}{(FN + TP)} * 100\% \tag{4}$$

We also evaluated the proposed IDS based on several other performance evaluation metrics: recall, error rate, precision, accuracy, and F-measure. Below is an explanation of each type with its equations [36].

- Accuracy: Percentage of correct predictions (positive and mmm negative) during model testing and defined as:

$$\text{Accuracy Rate} = \frac{(TN + TP)}{(FN + TP + TN + FP)} * 100\% \tag{5}$$

- Precision: Represents the number of correctly classified cases. Formally defined as representing the number of correctly positive classified cases divided by positive predictions. The worst precision is when the resultant value is 0.0, while the best rate equals 0.1. Formally defined as:

$$\text{Precision Rate} = \frac{TP}{(FP + TP)} * 100\% \tag{6}$$

- Recall: Represent correct positive predictions is divided by the sum number of positives. The worst recall is when the resultant value is zero, while the best rate equals 1. The recall is also called True positive rate or Sensitivity. Formally defined as:

$$\text{Recall Rate} = \frac{TP}{(FN + TP)} * 100\% \tag{7}$$

- F-measure (F1): Represent Precision and Recall are in a delicate balance. Formally defined as:

$$\text{F1 Rate} = \frac{(2 * \text{Recall} * \text{Precision})}{(\text{recall} + \text{Precision})} * 100\% \tag{8}$$

- Error Rate: The number of false predictions (FP + FN) is divided by the sum of the dataset. The worst error rate is when the resulting value is 1.0, whereas the best is when the value is 0.0. Formally defined as:

$$\text{Error Rate} = \frac{(FN + FP)}{(FN + TP + TN + FP)} * 100\% \tag{9}$$

5   Results and Discussion

This section will explain the results obtained and evaluate the efficiency of the proposed system. Accuracy, precision, recall, and F1 are used to evaluate the performance of the proposed system.

Table 1 shows machine learning algorithms' accuracy and execution time based on Several splits of the data set for training and testing.

Based on the results shown in Table 1, the best data splitting in terms of time and accuracy is 80% for training and 20% for testing. The higher the test percentage, the higher the execution time with slight accuracy changes. The proposed system is based on the best split, and all the results mentioned below depend on this percentage.

Table 1 Comparing the efficiency of machine learning algorithms based on the dataset split

Model	Training (%)	Testing (%)	Accuracy (%)	Execution time (s)
KNN	80	20	97.16	0.72
	70	30	97.10	0.87
	60	40	96.71	0.99
GB	80	20	98.87	4.47
	70	30	98.12	5.26
	60	40	98.49	5.70
DT	80	20	99.86	0.66
	70	30	99.86	0.86
	60	40	99.74	0.87

Table 2 Evaluate the performance of machine learning algorithms

Model	Accuracy (%)	Precision (%)	Recall (%)	F1 (%)
KNN	97.16	97	97	97.33
GB	98.87	99	99	99.66
DT	99.86	99.93	99.93	99.92

Table 3 Error rate of machine learning algorithms

Model	KNN (%)	GB(%)	DT(%)
Error rate	2.84	1.13	0.14

Table 2 shows proposed model results (precision, recall, and F1) when implementing machine learning algorithms (K-nearest neighbour, Decision tree, and Gradient Boosting).

In addition, the error rate is extracted for all the ML algorithms used in the proposed system. Table 3 shows the error rate of machine learning algorithms.

The results obtained show that all algorithms achieved a high percentage with varying efficiency. KNN achieved a good percentage in classifying normal and abnormal behaviours, as the accuracy rate reached 97.16%, while the accuracy and recall were 97%, with an increase in the F1 percentage by 97.33%. GB was higher than KNN, achieved 98.87% for accuracy, 99% for precision and Recall, and for F1, it was 97.33%. As for the decision, the tree was fared excellently in the speed of execution and classification of behaviours, with accuracy close to 100%, precision, recall, and f1 being 100%, As shown in Fig. 5.

As for the error rate, the decision tree also achieved significant superiority. Its percentage approached zero, which is the best value for the error rate, where it was equal to 0.14, while in KNN and GB, it was 2.84 and 1.13, respectively.

Confusion matrix results were also extracted to determine each algorithm's false and positive predictions rate in the proposed system.

Fig. 5 The performance evaluation metric

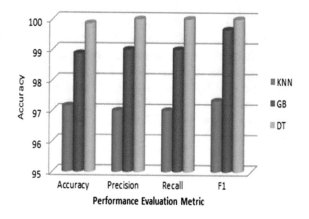

Table 4 Confusion matrix of machine learning algorithms

Model	TP (%)	TN (%)	FP (%)	FN (%)
KNN	97.162	98.560	1.439	2.837
GB	98.878	99.436	0.563	1.121
DT	99.861	99.930	0.0692	0.138

Table 5 Comparison of proposed IDS with the existing IDS

Author's	Model	Main Target	Accuracy (%)
This paper	Decision Tree	Detection Spoofing and Jamming	99.86
[37]	STL and SVM	Detection Spoofing and Jamming	92

Table 4 shows (FP, FN, TP, TN) all machine learning algorithms used in this system.

In machine learning, the confusion matrix indicates that the model is perfect when the value of TN and TP is enormous while the value of FP and FN is close to or equal to zero, and this is what the decision tree achieved in the proposed IDS, where the number of FP and FN predictions was equal to 10, which is a small percentage compared to the number of enormous test data.

A decision tree is proposed as the best algorithm to implement the proposed IDS based on the results.

After the results are obtained, the proposed IDS is compared with [37], which has this paper's same goal in Table 5.

A comparison of the results in Table 5 shows the superiority of the proposed IDS over the IDS in the paper [37]

6 Conclusion and Future Works

This paper proposes IDS using machine learning in UAV networks to classify ordinary and malicious signals received by UAVs. Three machine learning algorithms are used to determine the most efficient algorithm. The proposed system achieved high efficiency in identifying normal behaviour, Jamming, and spoofing, as the decision tree achieved a high percentage in all performance evaluation tools and the confusion matrix. In addition to that, it achieved the lowest error rate. Moreover, when comparing the proposed IDS with previous work, the proposed IDS were better than the previous work.

In future works, the proposed system can use other machine learning algorithms such as a random forest or sport vector machine or use a second dataset.

References

1. Jasim KS, Alheeti KMA, Alaloosy AKN (2021) A review paper on secure communications in FANET. In: 2021 International conference of modern trends in information and communication technology industry (MTICTI). IEEE, pp 3–9
2. Kwon W, Park JH, Lee M, Her J, Kim SH, Seo JW (2020) Robust autonomous navigation of unmanned aerial vehicles (UAVs) for Warehouses' inventory application. IEEE Robot Autom Lett 5(1):243–249. https://doi.org/10.1109/LRA.2019.2955003
3. Attacks SJ (2019) SS symmetry a Dyna-Q-based solution for UAV networks against smart jamming attacks. Symmetry (Basel)
4. Davidson D, Jellinek R, Ristenpart T, Controlling UAVs with sensor input spoofing attacks
5. Shahmoradi J, Talebi E, Roghanchi P, Hassanalian M (2020) A comprehensive review of applications of drone technology in the mining industry. Drones 4(3):1–25. https://doi.org/10.3390/drones4030034
6. Bada M, Eddine D, Lagraa N, Abdelaziz C, Imran M, Shoaib M (2021) A policy-based solution for the detection of colluding GPS-Spoofing attacks in FANETs. Transp Res Part A 149(May):300–318. https://doi.org/10.1016/j.tra.2021.04.022
7. Yahuza M et al (2021) Internet of drones security and privacy issues: taxonomy and open challenges. IEEE Access 9:57243–57270. https://doi.org/10.1109/ACCESS.2021.3072030
8. Arteaga SP, Hernandez LAM, Perez GS, Orozco ALS, Villalba LJG (2019) Analysis of the GPS spoofing vulnerability in the drone 3DR solo. IEEE Access 7:51782–51789. https://doi.org/10.1109/ACCESS.2019.2911526
9. Karpe RV (2020) Software defined radio based global positioning system jamming and spoofing for vulnerability analysis. In: International conference on electronics and sustainable communication systems, no. Icesc, pp 881–888
10. Manesh MR, Kenney J, Hu WC, Devabhaktuni VK, Kaabouch N (2019) Detection of GPS spoofing attacks on unmanned aerial systems. In: IEEE annual consumer communications & networking conference (CCNC). https://doi.org/10.1109/CCNC.2019.8651804
11. Arnosti SZ, Pires RM, Branco KRLJC (2017) Evaluation of cryptography applied to broadcast storm mitigation algorithms in FANETs. 2017 International conference unmanned aircraft system ICUAS 2017, pp 1368–1377.https://doi.org/10.1109/ICUAS.2017.7991377
12. Sultana N, Chilamkurti N, Peng W, Alhadad R (2019) Survey on SDN based network intrusion detection system using machine learning approaches. Peer-to-Peer Netw Appl 12(2):493–501. https://doi.org/10.1007/s12083-017-0630-0
13. Ali Alheeti KM, Al-Jobouri L, McDonald-Maier K (2013) Increasing the rate of intrusion detection based on a hybrid technique. 2013 5th computer science electronic engineering conference CEEC 2013—conference proceedings, pp 179–182. https://doi.org/10.1109/CEEC.2013.6659468
14. AHA, Sundarakantham K (2019) Machine learning based intrusion. 2019 3rd international conference trends electronic informatics, Icoei, pp 916–920
15. Farnaaz N, Jabbar MA (2016) Random forest modeling for network intrusion detection system. Procedia Comput Sci 89:213–217. https://doi.org/10.1016/j.procs.2016.06.047
16. Whelan J, Sangarapillai T, Minawi O, Almehmadi A, El-Khatib K (2020) Novelty-based intrusion detection of sensor attacks on unmanned aerial vehicles. Q2SWinet 2020—proceedings 16th ACM sympium QoS security wireless mobblie networks, pp 23–28. https://doi.org/10.1145/3416013.3426446
17. Upadhyay K (2021) Network intrusion detection system based on machine learning 25(4):12445–12451
18. Shrestha R, Omidkar A, Roudi SA, Abbas R, Kim S (2021) Machine-learning-enabled intrusion detection system for cellular connected uav networks. Electron 10(13):1–28. https://doi.org/10.3390/electronics10131549
19. Chou D, Jiang M (2022) A survey on data-driven network intrusion detection. ACM Comput Surv 54(9):1–36. https://doi.org/10.1145/3472753

20. Sedjelmaci H, Senouci SM, Messous MA (2016) How to detect cyber-attacks in unmanned aerial vehicles network? 2016 IEEE global communication conference GLOBECOM 2016—proceedings. https://doi.org/10.1109/GLOCOM.2016.7841878

21. Liu G, Zhang R, Wang C, Liu L (2019) Synchronization-free GPS spoofing detection with crowdsourced air traffic control data. Proceedings—IEEE international conference moblie data manag. Mdm, pp 260–268. https://doi.org/10.1109/MDM.2019.00-49

22. Greco C, Pace P, Basagni S, Fortino G (2021) Jamming detection at the edge of drone networks using multi-layer perceptrons and decision trees. Appl Soft Comput 111:107806. https://doi.org/10.1016/j.asoc.2021.107806

23. Majidi M, Erfanian A, Khaloozadeh H (2020) Prediction-discrepancy based on innovative particle filter for estimating UAV true position in the presence of the GPS spoofing attacks. https://doi.org/10.1049/iet-rsn.2019.0520

24. Shafique A, Mehmood A, Elhadef M (2021) Detecting signal spoofing attack in UAVs using machine learning models. IEEE Access 9:93803–93815. https://doi.org/10.1109/ACCESS.2021.3089847

25. Sedjelmaci H, Senouci SM, Ansari N (2018) A hierarchical detection and response system to enhance security against lethal cyber-attacks in UAV networks. IEEE Trans Syst Man Cybern Syst 48(9):1594–1606. https://doi.org/10.1109/TSMC.2017.2681698

26. Mitchell R, Chen I (2014) Adaptive intrusion detection of malicious unmanned air vehicles using behavior rule specifications. IEEE Trans Syst 44(5):593–604

27. Panice G, Luongo S, Gigante G, Pascarella D, Di Benedetto C, Vozella A (2017) A SVM-based detection approach for GPS spoofing attacks to UAV, pp 7–8

28. Hamid Y, Sugumaran M, Balasaraswathi V (2016) IDS using machine learning—current state of art and future directions. Br J Appl Sci Technol 15(3):1–22. https://doi.org/10.9734/bjast/2016/23668

29. Trivedi NK, Simaiya S, Lilhore UK, Sharma SK (2020) An efficient credit card fraud detection model based on machine learning methods. Int J Adv Sci Technol 29(5):3414–3424

30. Asaju LB, Shola PB, Franklin N, Abiola HM (2017) Intrusion detection system on a computer network using an ensemble of randomizable filtered classifier, K-nearest …. Ftst Journal.Com 2(1):550–553. http://ftstjournal.com/uploads/docs/21BArticle39.pdf

31. Al-Abrez SM, Alheeti KMA, Alaloosy AKAN (2020) A hybrid security system for unmanned aerial vehicles. J Southwest Jiaotong Univ 55(2):1–9. https://doi.org/10.35741/issn.0258-2724.55.2.1

32. Praveen Kumar D, Amgoth T, Annavarapu CSR (2019) Machine learning algorithms for wireless sensor networks: a survey. Inf Fusion 49:1–25. https://doi.org/10.1016/j.inffus.2018.09.013

33. Ab V (2018) Fault diagnosis of wind turbine structures using decision tree learning algorithms with big data. Saf Reliab Soc a Chang World

34. Whelan J, Sangarapillai T, Minawi O, Almehmadi A, El-Khatib K (2020) UAV attack dataset_IEEE DataPort

35. Alheeti KMA, Gruebler A, McDonald-Maier KD (2015) An intrusion detection system against black hole attacks on the communication network of self-driving cars. Proceedings—2015 6th international conference emerging security technology EST 2015, pp 86–91. https://doi.org/10.1109/EST.2015.10

36. Hossin M, Sulaiman MN (2015) A review on evaluation metrics for data classification evaluations. Int J Data Min Knowl Manag Process 5(2):1–11. https://doi.org/10.5121/ijdkp.2015.5201

37. Arthur MP (2019) Detecting signal spoofing and jamming attacks in UAV networks using a lightweight IDS. CITS 2019—proceeding 2019 international conference computer information telecommunity system, pp 1–5. https://doi.org/10.1109/CITS.2019.8862148

AST-Based LSTM Neural Network for Predicting Input Validation Vulnerabilities

Abdalla Wasef Marashdih, Zarul Fitri Zaaba, and Khaled Suwais

Abstract Due to the increased popularity of Web-based applications, input validation problems are becoming more common. Two input validation issues to be aware
of are SQL injection (SQLi) and Cross-Site Scripting (XSS). Vulnerability prediction
methods based on machine learning have lately increased in favor in the field of Web
security. Due to the simplicity and efficiency of such procedures, they are becoming
more popular. They usually make use of sophisticated graphs drawn from source code
or highly proficient regex patterns to accomplish their goals. Essentially, tokenization is a technique of breaking down source code into a set of tokens to determine
the vulnerability of a program's structure and flow using neural network methods.
This paper proposed a model for predicting input validation vulnerabilities based on
Abstract Syntax Tree (AST) and Long Short-Term Memory (LSTM) algorithm. The
programs are translated into an AST structure, which is followed by a reduction in
the number of nodes that are not linked to the program's vulnerabilities. The token
sequence generated from AST is used as input for the LSTM model, which is used
to understand the flow of vulnerabilities in the source code. The proposed model's
accuracy was 96.44%, which was higher than the related studies.

Keywords Input validation vulnerability · XSS · SQLi · Minimal SSA · Deep
learning · Tokenization

A. W. Marashdih · Z. F. Zaaba (✉)
School of Computer Sciences, Universiti Sains Malaysia, 11800 Pulau Pinang, Malaysia
e-mail: zarulfitri@usm.my

K. Suwais
Faculty of Computer Studies, Arab Open University, Riyadh, Saudi Arabia

1 Introduction

Web apps are currently among the most extensively used platforms for displaying data and launching digital products. Nevertheless, due to intrinsic security concerns, programming vulnerabilities may be widely exploited. The Open Web Application Security Project (OWASP) [1] lists SQLi and XSS as major web application vulnerabilities. SQLi stands for introducing SQL instructions for the website inputs to modify SQL query execution. The XSS vulnerability works by attaching a malicious code to web URLs to fool people towards visiting dangerous links. Assuring that internet apps are safe is a constant endeavor by security firms, programmers, and white-hat hackers [2].

Machine Learning (ML) has recently been hailed as the most effective technology for finding security vulnerabilities. For application code processing, and even the output if it has any defects, further approaches include Natural Language Processing (NLP) [3] and neural networks [4, 5]. Some of the concerns, such as dependency between code parts over time [6, 7], might have a negative impact on the efficiency of these methods [6, 7]. The identification of vulnerabilities is highly dependent on the interactions that arise between various factors. It is also possible that a considerable gap exists between parts that are conceptually connected. The identification and use of a variable name in a program, for example, may have a number of unnecessary codes that are interspersed throughout the code. Another major concern is the 'out of vocabulary (OoV)' such as variable identifiers and methods, which cause a shift in lexical structure.

In this study, we proposed a method for tokenizing the program codes into a series of tokens using the AST. Then, in predicting the vulnerabilities, we used the LSTM classifier to understand the tokens stream. The proposed model was evaluated and shown to be successful on Hypertext Preprocessor (PHP) web technology [8]. The remaining sections of the paper are organized as follows. Deep learning, tokenization, and input validation issues (including SQLi and XSS) are discussed in detail in Sect. 2. The related works are discussed in Sect. 2.5. Section 3 examines the model's structure and implementation. Section 4 contains a summary of the experimental findings and an evaluation of the findings. Section 6 concludes with a discussion of possible future directions.

2 Background and Related Works

2.1 Input Validation Vulnerabilities

The reliability of a single threat model should not be relied upon for all input sources [9]. Unverified data must be validated before being submitted into the website. The most popular attack types are XSS and SQLi. The malicious code execution sites differentiate these attacks.

XSS [10] involves the injection of a malicious JavaScript code, enabling the attacker to run scripts on the target browser. These scripts may cause damage to web applications when they are included in them. The injection of the script is carried out in such a way that it seems harmless to an unwary person. Last but not least, a trusted domain runs this script that has been hacked by the unauthorized code [11].

Database languages such as SQL are the most frequently used in the world, and they are used to connect with database systems [12]. Nowadays, dynamic query creation generally involves mixing text to form SQL instructions. While the inquiry is being conducted, such instructions are constructed based on input data collected from external sources. Due to the fact that queries may be adjusted to take into account certain limitations supplied by users, this design enables attackers to interfere. To boost security, developers choose either parametrized query generation methods or the use of stored procedures. If the correct steps aren't taken, the code can still be used and hacked.

2.2 Long Short Term Memory (LSTM)

The model of a neural network has neurons and their connections [13]. Feature definition is not necessary for DNN models since learning is possible. DNN models include several layers, where the data is passed progressively between them, transforming it. These layers help the model understand important and complex needs by using simple and vague requirements [14].

An LSTM model is presented to overcome the RNN's failure to process data streams, which is caused by a lack of long-term memory. LSTM is well-known for being a very successful model for NLP applications and memory capacity; as a result, it is a widely used model.

There are gates in the LSTM that allow for the addition and deletion of features. The input gate, the output gate, and the forget gate are the three gates that make up this circuit. When information is no longer needed, the forget gate determines what has to be forgotten. A decision is made by the input gate as to which of the new pieces of information should be saved. LSTM output activation is calculated by the output layer, which determines which information is required for calculation.

2.3 Abstract Syntax Tree

The Abstract Syntax Tree (AST) depicts the program's overall organization. Parentheses, brackets, and the semicolon are among the delimiters and punctuation marks that are removed by this program [15]. This implies that, for example, an if-statement consists just of the text of the statement and not of the parenthesis or braces around the statement. The fact that an AST is made up of of necessary nodes means that the tree itself is minimal, making it simpler to work with and understand. ASTs are often used by compilers to improve the efficiency with which they work.

2.4 Tokenization

Tokenization is the process of turning source code into a sequence of tokens that are then used by neural network algorithms to find vulnerable code scripts. Code tokenization is currently considered an innovative auditing technique. The tokenizer parses the source code using NLP. Without formatting or reliance on variable names, the pattern retains essential significance. Meanwhile, the generated tokens from the program code may use the learning algorithm to incorporate vector space. Tokenization also avoids the requirement for complex structures.

2.5 Related Works

Manual web application code audits resulted in inefficiency and a lack of accuracy in the results. Massive deployments need a significant amount of work [16]. Predefined prediction instructions are used by the bulk of current PHP code vulnerability auditing solutions. They use ordinary code node matching against a large rule set that may be greatly enhanced [17]. Machine learning for code verification and optimization may be integrated with code vectorization [18, 19].

The Abstract Syntax Tree (AST) and Control Flow Graph (CFG) are two of the most commonly used technologies and traditional analysis approaches today.They contribute to the enhancement of abstraction program logic and also parallelization results, which become considered adequate for use as input for machine learning models and are given to the models. A code vectorization techniquethat relies on AST and CFG was developed by Michael et al. [20], and it has the advantage of helping to minimize noise caused by the diverse program styles used by different developers. However, there is a drawback with this approach in that they all transform the original structure of the source code into a new structure, which requires more ways to parse the new data structure and makes the researchers' jobs more difficult and complicated.

Text-mining techniques are utilized to identify program files that have been recognized as a collection of terms. Walden et al. [21] used a text mining approach to extract a set of tokens that may be used to predict XSS vulnerabilities. The researchers' experimental results showed that their strategy was better than software metrics features by a wide margin.

A TAP tokenizer was presented by Fang et al. [22], and it is recognized as a suitable solution for tokenizing PHP programs. They used the PHP built-in function *function_get_all()* to get the program tokens. Through the data flow analysis and the rules from the authors, auditing of different vulnerabilities was done on the Software Assurance Reference Dataset (SARD) [23].

3 Methodology

This section describes our proposed methodology for anticipating input validation vulnerabilities, which involves processing the program under test and converting it to AST. To begin the tokenization process, the AST is traversed using forward slicing to produce a collection of tokens. During the embedding phase, the generated token sequences are utilized as input for the neural network model (LSTM). The LSTM layers' outputs are fed into a dense layer, which predicts the vulnerabilities.

3.1 Abstract Syntax Tree

Abstract syntax trees (ASTs) are used to visualize how programs work. These trees demonstrate how code may be conceptually split into its grammatical components. The tree is generic in that it doesn't include all of the details of how to create a particular application. They merely demonstrate how program components are stacked in order to produce the final program.

A node in an abstract syntax tree may be divided into two categories. Nodes representing operators like as statements and method calls are represented by inner nodes, while leaf nodes are represented by operands such as constants and unique identifiers. Figure 1 displays the abstract syntax tree for the previous example code in Listing 1, which serves as an illustration.

```php
<?php
$x = 10;
$y = 0;
if ($x == 5) { $y = 3; }
else { $y = x; }
print $y;
```

Listing 1 PHP code example

3.2 Tokenization

The AST node sequence should be fed into the neural network by turning the node's content into a token sequence. Each node's type is used to build a structure stream that is generated by slicing [24]. Forward slice has been used since it always starts from the first node. It will initially record each node's type in a token array. After

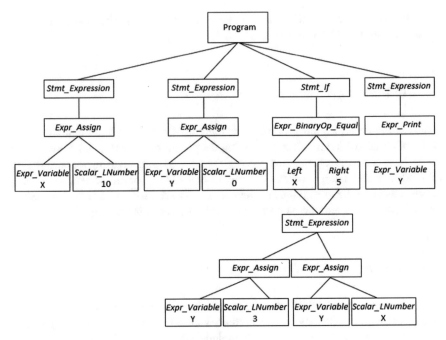

Fig. 1 Abstract Syntax Tree for code in Listing 1

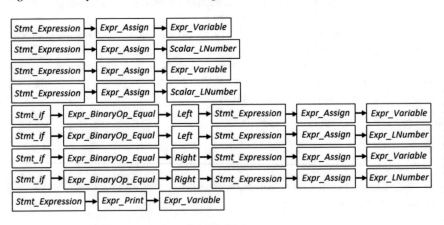

Fig. 2 The generated data stream for code in Listing 1

slicing, all the tokens were concatenated to create token streams for the program. Figure 2 illustrates the slicing streams generated from Listing 1 code.

To eliminate the interference of unnecessary information from the generated data streams. We removed the assignment of constant variables (like *LNumber*) and operators (like *Left* and *Right*) that didn't make sense for the vulnerabilities flow in the programs.

3.3   Neural Network Model

This research uses LSTMs as a neural network model for predicting the vulnerabilities. The LSTM model has five layers:

1. Input Layer: After tokenization, data content is represented by vectors in this layer.
2. LSTM Layer: This layer may send each input to the LSTM unit to get the hidden layer's output. It solves the long-term dependence issue and learns the vulnerabilities' abstract aspects.
3. Dense Layer: Adding this layer of noise to the model will help to make it less likely that the LSTM layer will store patterns that aren't important.
4. Dropout Layer: This layer avoids overfitting from occurring.
5. Output Layer: The *Softmax* classifier outputs the classification result y.

4   Evaluation

4.1   Dataset

Numerous research [6, 7, 22] have used the Software Assurance Reference Dataset (SARD) [23] to evaluate algorithms for web application vulnerability prediction. The specifications of the SARD datasets that were utilized to evaluate the proposed model are listed in Table 1.

4.2   Experimental Environment

As for the LSTM neural network, many tests were done to compare the parameters and choose the suitable values that would make the graph more understandable. We came up with these values: The vector dimension is 256, the LSTM units are 128 and the epochs are 30.

Table 1 SARD dataset for XSS and SQLi

Vulnerability	# of Safe	# of Unsafe	Total
XSS	5728	4352	10080
SQLi	8232	912	9144

Using the publicly accessible Google Collaboratory,[1] we were able to train our model. Keras, a deep learning package, has also been used in conjunction with Tensorflow in the backend of the system.

4.3 Evaluation Metrics

We utilized a confusion matrix to track both wrong and right predictions. True positives (TNs) may help decide whether the approach correctly detects the program to be vulnerable; true negatives (TPs) can assist in evaluating if the method correctly identifies the program to be invulnerable. False positives (FPs) may help identify whether the technique wrongly deems the program to be vulnerable, whereas false negatives (FNs) can assist in identifying whether the method wrongly identifies the program to be invulnerable. The optimum tool has the greatest TN and TP values while having low FP and FN rates. Five metrics are used to evaluate the effectiveness of the proposed model: accuracy, recall, precision, F1 score, and confusion matrix. The following are their calculations:

$$Precision = \frac{TP}{TP + FP} \tag{1}$$

$$Recall = \frac{TP}{TP + FN} \tag{2}$$

$$F1 - score = \frac{2 * (Precision * Recall)}{Precision + Recall} \tag{3}$$

$$Accuracy = \frac{TP + TN}{TP + TN + FP + FN} \tag{4}$$

Confusion matrix: The table is employed for defining classification performance.

4.4 Results

This section discusses the proposed method outcomes and compares them to those of three comparable approaches. To assess each method's performance, the structure of the proposed LSTM algorithm were maintained and we applied the produced tokens from each method to the same algorithm (LSTM).

token_get_all(), a PHP built-in function, is used to parse the program code into tokens [25]. TAP [7] retrieves tokens from PHP code using the PHP function (*token_get_all()*), and then does data flow analysis to find critical code lines with

---

[1] https://colab.research.google.com/notebooks/welcome.ipynb.

Table 2 Comparison of different tokenization methods

Method	Precision (%)	Recall (%)	F1 (%)	Accuracy (%)
Walden et al. [21]	81.36	66.62	75.12	85.12
Fang et al. [22]	85.59	90.27	88.96	92.89
token_get_all() [25]	69.4	70.65	72.80	82.14
Our method	**95.72**	**94.42**	**94.53**	**96.44**

function calls. Walden et al. [21] presented text mining metrics, which begin by tokenizing the program code using the *token_get_all()* function. Additionally, their tokenizer analyzes the token set to remove spaces and remarks. The results of each method are shown in Table 2. The performance indicators show the method's ability to predict the vulnerabilities.

The Walden et al. [21] method produced a low vulnerability recall of 66.62%. On the other hand, the accuracy of predicting vulnerabilities is higher, at 81.36%. The generated tokens via the token get all() function had a lower precision of 69.4% and a 70.65% ability to predict vulnerability, which is slightly better than the Walden et al. [21] method. For both vulnerable and safe classes, the Fang et al. [22] method produced better results than the previous methods, with a recall value of 90.27% and a precision value of 85.59%. With a 94.42% recall and a 95.72% precision, the proposed model improved the ability to predict vulnerabilities.It also has a better accuracy rate (96.44%) than the previous models for predicting XSS and SQLi vulnerabilities.

The results of Walden et al. [21] method achieved low recall estimation of the vulnerability with 66.62%. However, the precision of the predicting the vulnerabilities is higher with 81.36. The generated tokens from *token_get_all()* function achieved the lower precision results with 69.4%, and the ability to predict the vulnerability using this function is 70.65%, which is slitly better than Walden et al. [21] method. On the other hand, Fang et al. [22] tokenizer achieved a good results in regards of the recall value with 90.27, and a precision value 85.59 for both vulnerable and safe classes. The outcomes of the proposed method have an improved in the ability to predict the vulnerabilities with a 94.42% recall and 95.72% precision. Additionally, the proposed method's accuracy (96.44%) is greater than that of previous methods for predicting XSS and SQLi vulnerabilities.

Figure 3 shows the confusion matrices that each method produced as a result of the studies. It should be highlighted that the majority of the approaches were successful in predicting XSS vulnerabilities. Walden et al. [21] method predicted 95% and 86% of true labels for XSS and SQLi, respectively. The *token_get_all()* function, on the other hand, predicted 87% XSS and 70% SQLi. Fang et al [22] improved the XSS vulnerability prediction results by 95% and the SQLi vulnerability prediction results by 86%. Our AST tokenization method has a 99% chance of correctly predicting XSS vulnerabilities and a 93% chance of correctly predicting SQLi vulnerabilities. Overall, the proposed method outperformed other methods and made it easier to find XSS and SQLi in PHP source code.

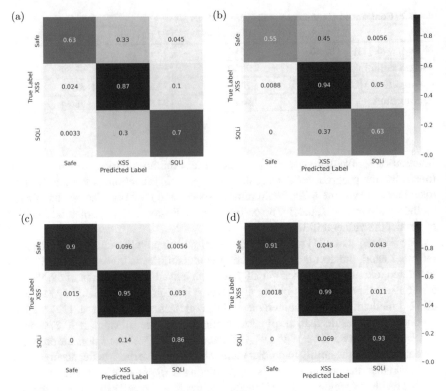

Fig. 3 Confusion matrix of different tokinzation method

5 Discussion

The proposed AST-tokenization method starts by converting the source code into an AST structure, followed by storing the set of tokens in the AST-pattern stream and removing the tokens that are not related to the vulnerability in the stream. An LSTM neural network is performed as a classifier to address the benign and vulnerable samples. Compared to other token methods, the proposed AST-tokenization method reduces the vocabulary size in the token streams and keeps the tokens that are related to the vulnerability pattern in the programs. The accuracy of the proposed method is 4% higher than the data flow tokens method by Fang et al. [5], and up to 12% higher than the PHP built in function (*token_get_all()*) [25] and data mining method [21].

6 Conclusion

A deep neural network model was suggested in this paper for identifying input validation vulnerabilities in PHP source code. Converting the programs into AST and slicing the tree allows tracking vulnerabilities across programs and removing nodes that are unrelated to the vulnerabilities. The proposed method's accuracy rate was 96.44% in the test dataset. This highlights the need to identify the vulnerability pattern in the programs. In future work, investigating the context of vulnerabilities in source code and assigning each vulnerability a unique token will provide additional information for the model to understand the vulnerability flow in programs. Furthermore, the proposed model now focuses on two categories of vulnerabilities (XSS and SQLi). It might, however, be expanded to include more vulnerabilities.

Acknowledgment The authors would like to thank Universiti Sains Malaysia and Arab Open University, Saudi Arabia for supporting this study.

References

1. OWASP: Owasp top 10-2021. https://owasp.org/Top10/. Accessed 10 Mar 2022
2. CVE: Cve details. https://www.cvedetails.com/browse-by-date.php. Acessed 12 January 2022
3. Medeiros I, Neves N, Correia M (2016) Dekant: a static analysis tool that learns to detect web application vulnerabilities. In: Proceedings of the 25th international symposium on software testing and analysis, pp 1–11
4. Li Z, Zou D, Xu S, Jin H, Zhu Y, Chen Z (2022) SySeVR: a framework for using deep learning to detect software vulnerabilities. In: IEEE Trans Dependable Secure Comput 19(4):2244–2258. https://doi.org/10.1109/TDSC.2021.3051525
5. Li Z, Zou D, Xu S, Chen Z, Zhu Y, Jin H (2022) VulDeeLocator: a deep learning-based fine-grained vulnerability detector. In: IEEE Trans Dependable Secure Comput 19(4):2821–2837. https://doi.org/10.1109/TDSC.2021.3076142
6. Li X, Wang L, Xin Y, Yang Y, Tang Q, Chen Y (2021) Automated software vulnerability detection based on hybrid neural network. Appl Sci 11(7):3201
7. Li C, Wang Y, Miao C, Huang C (2020) Cross-site scripting guardian: a static XSS detector based on data stream input-output association mining. Appl Sci 10(14):4740
8. Odeh AH et al (2019) Analytical and comparison study of main web programming languages-asp and PHP. TEM J 8(4):1517–1522
9. Li X, Xue Y (2011) A survey on web application security. Nashville, TN USA 25(5):1–14
10. Martin MC, Lam MS (2008) Automatic generation of XSS and SQL injection attacks with goal-directed model checking. In: USENIX security symposium, pp 31–44
11. Gupta S, Gupta BB (2017) Cross-site scripting (XSS) attacks and defense mechanisms: classification and state-of-the-art. Int J Syst Assur Eng Manag 8(1):512–530
12. Shar LK, Tan HBK (2012) Defeating SQL injection. Computer 46(3):69–77
13. Alom MZ, Taha TM, Yakopcic C, Westberg S, Sidike P, Nasrin MS, Hasan M, Van Essen BC, Awwal AA, Asari VK (2019) A state-of-the-art survey on deep learning theory and architectures. Electronics 8(3):292
14. Goodfellow I, Bengio Y, Courville A (2016) Deep learning. MIT Press
15. Klingström S, Olsson P (2020) Type inference in PHP using deep learning. LU-CS-EX
16. Yamaguchi F, Lindner F, Rieck K (2011) Vulnerability extrapolation: assisted discovery of vulnerabilities using machine learning. In: Proceedings of the 5th USENIX conference on Offensive technologies, p 13

17. Lingzi X, Zhi L (2015) An overview of source code audit. In: 2015 international conference on industrial informatics-computing technology, intelligent technology, industrial information integration. IEEE, pp 26–29
18. Choi YH, Liu P, Shang Z, Wang H, Wang Z, Zhang L, Zhou J, Zou Q (2020) Using deep learning to solve computer security challenges: a survey. Cybersecurity 3(1):1–32
19. Liu S, Lin G, Han QL, Wen S, Zhang J, Xiang Y (2019) Deepbalance: deep-learning and fuzzy oversampling for vulnerability detection. IEEE Trans Fuzzy Syst 28(7):1329–1343
20. Alon U, Zilberstein M, Levy O, Yahav E (2019) code2vec: learning distributed representations of code. In: Proceedings of the ACM on programming languages 3(POPL), pp 1–29
21. Walden J, Stuckman J, Scandariato R (2014) Predicting vulnerable components: software metrics vs text mining. In: 2014 IEEE 25th international symposium on software reliability engineering. IEEE, pp 23–33
22. Fang Y, Han S, Huang C, Wu R (2019) Tap: a static analysis model for PHP vulnerabilities based on token and deep learning technology. PloS One 14(11):e0225196
23. Database NV. Nvd—statistics search. https://web.nvd.nist.gov/view/vuln/statistics. Accessed 30 Apr 2021
24. Weiser M (1984) Program slicing. IEEE Trans Softw Eng 4:352–357
25. PHP: token_get_all. https://www.php.net/manual/en/function.token-get-all.php. Accessed 15 Jan 2022

IoT Trust Management as an SIoT Enabler Overcoming Security Issues

Assiya Akli⊕ and Khalid Chougdali

Abstract Nowadays, security in the IoT world is a very popular area because it affects everyone's privacy and safety. The risk is increasing with Social IoT, involving intelligent devices interacting with each other dynamically and autonomously. However, while trying to limit interactions only to objects known as secure, we take the risk of service disruption. This means that traditional techniques are in general not applicable or inefficient in this context. In this paper, we present trust management as a promising paradigm that can deal with security concerns in IoT. We draw up various techniques and parameters implemented in this context and we describe the different challenges of managing SIoT trust.

Keywords IoT · SIoT · Security · Trust management · Trust characteristics · Trust metrics

1 Introduction

In the IoT world, everyday things interact with each other via internet, and most of these communications occur without human intervention.

These interactions are the basis of SIoT (Social Internet of Things), which include also the relationship between the users/owners of the object, and their relationship with objects.

In such open world, devices behavior can be dangerous if they are infected, or are subject to the owner's mis-behaving. Thus, to ensure the devices interaction security, a minimum level of trust is required.

A. Akli (✉) · K. Chougdali
Engineering Sciences Laboratory, National School of Applied Sciences, Ibn Tofail University, Kenitra, Morocco
e-mail: assiya.akli@uit.ac.ma

K. Chougdali
e-mail: khalid.chougdali@uit.ac.ma

© The Author(s), under exclusive license to Springer Nature Switzerland AG 2023
A. A. Abd El-Latif et al. (eds.), *Advances in Cybersecurity, Cybercrimes, and Smart Emerging Technologies*, Engineering Cyber-Physical Systems and Critical Infrastructures 4,
https://doi.org/10.1007/978-3-031-21101-0_10

In human every day's life, people need to communicate and interact with each other. There are people they know and trust/untrust based on their own experience or on what they have heard about other person's experiences. But often they may have seen some persons for the first time and then need to take a decision. Humans have learned to make such decisions based on other parameters such as appearance, speaking manner or meeting point.

This concept of trust is more and more explored in the SIoT field, as it helps dealing with many similar situations, taking in many cases the subjective aspect into account. And research works are increasingly focusing on this concept.

In this paper, we overview the trust management, and its different properties in Sect. 2. We present the importance of this field through the study of previous scientific research in Sect. 3. We look over the most used calculation models and methods for trust-based security in the fourth section. In the fifth section, we expand the constraints researchers are facing in dealing with such systems. We end up by a conclusion of this work.

2 Trust Definition and Characteristics

Trust, as a concept, is very complicated and hardly measurable. It concerns many parameters such as the reliability, the availability, the ability of the device, its strength, and goodness [13]. Making the decision to trust an interactor or not, is the result of the trust management processes. Trust management is identified by the best ways to establish, ensure, and maintain trust [13] As an alliance joining a trustee to a trustor for mutual benefit, with mitigation of the possible risk in this relationship, trust has many characteristics(grouped as DATCAS), as we present on Fig. 1.

In fact, trust is:

- Dynamic (flexible): a device A can trust another device B at certain time or circumstances, and untrust it on others. Policies can be used for this. (Trust is updated periodically or every time conditions change)
- Asymmetric: a device A may trust a device B, while the device B don't trust A (as if A is a vehicle behind another vehicle B, A can trust B about any coming problems in the road, but B can't trust A)
- Transitive: if a device A trust a device B and the device B trust another device C, it can influence the trust of A to C (This characteristic is important especially when the device A has no previous experience with C. It can ask its trustee B about his recommendation)
- Context dependent: a device A can trust a device B relatively in a certain context and untrust it on others (ex: it can trust temperature sensor about a heat measure, but not about a network information)
- Approximative: trust is never 100% and includes risk mitigation (By definition trust is a degree of belief, it can be expressed by a percentage or a number in a continuous or discontinuous range)

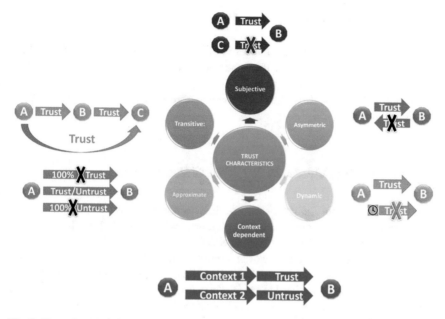

Fig. 1 Trust characteristics

– Subjective: a device A can trust a device B, while another device C does not trust it. It depends on the point of view of each A and C's standpoint about the device B. (as in the case where A and B has the same owner)

The trust sources are based on self-observation's history and experience (Direct), or transitives (Indirect) represented by reputation, recommendation, rating, or feedback. We mostly consider indirect sources when the nodes have no previous experience.

Joining direct and indirect sources (Hybrid) is commonly used to get more relevant trust evaluation.

3 IoT Trust Management in Literature

3.1 Trust Importance in IoT Security

In the literature relative to IoT security, researches name trust as an essential aspect even though it's a relatively new paradigm. We cite here some examples.

Riahi Sfar et al. [16] use cognitive and systemic approach to present an overview of IoT security issues, they classify researches into 4 areas: access control, identification, privacy, and trust. They suggest, namely, conceiving and implementing trust methods which deal with the heterogeneous and changing IoT network environments, as a good direction for researches.

Binti et al. analyse research activities in IoT security from 2016 to 2018 [3]. They observe that publications on IoT trust management are increasing, and affirm that it is a promising technique to enhance IoT security.

We can conclude that trust is an aspect which deserves to be inspected in any SIoT security research.

3.2 Increasing Interest in IoT Trust Management

To get an idea about the interest of the scientific community in this aspect we conduct a statistical study. We query the most important and widely used databases in this field, namely: IEEE, Scopus and Web of Science.

In the process of choosing keywords, we notice that we must combine the expression "internet of things" with the abbreviation "IoT" to have comprehensive results. However, adding the keywords (M2M OR "machine to machine") has no remarkable added value. In the same way, we exclude other keywords as "physical internet".

On another hand, the removal of the keywords relative to security leads to other interpretations of the word "trust", leading to inaccurate results. Substituting the keyword "trust" by "trust management" or "trustworthiness" helps to put right the intended meaning of "trust".

The keywords expression (IOT OR "internet of things") AND ("trust management" OR trustworthiness), gives us relevant results which we sort by year to visualize the evolution.

As shown in Fig. 2, there is a growing number of research in IoT trust management in the last years. The results in terms of enhancing IoT security are promising. Therfore, in SIoT context trust is a paradigm which worth to be explored.

3.3 Trust Models Classification

Najib et al. classify, in their survey, trust models according to five parameters [13]: metrics, sources, algorithms, architecture, and propagation.

In a service-oriented IoT systems context, Guo et al. survey trust computation models [6]. They compare researches according to trust calculation methods in each dimension in a classification tree. They use five dimensions: composition, propagation, aggregation, formation, and update

Abdelghani et al. [1] classify trust management methods in SIoT considering their composition, propagation, aggregation, and update algorithms.

Pourghebleh et al. categorized trust management methods based on four main classes [15]: recommendation, prediction, policy, and reputation.

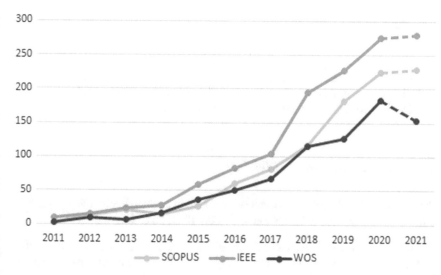

Fig. 2 Research interests in IoT trust management

Table 1 Comparing current survey to cited ones

	SIoT concern	Trust metrics details	Aggregation methods expantion
Najib et al. [13]		✓	~
Guo et al. [6]		✓	✓
Abdelghani et al. [1]	✓	~	
Pourghebleh et al. [15]		✓	✓
Current study	✓	✓	✓

In our survey we aim to study the interest of the trust management in dealing with SIoT security issues. Therefore, we focus on the details of the composition and the aggregation aspects in order to facilitate the choices which fit with SIoT particularities.

As we can see in Table 1, Abdelghani et al. [1] are concerned in SIoT and cite trust composition and aggregation methods without details. While the other surveys don't involve Social Internet of things.

In order to highlight suitable trust parameters for SIoT context, we expose, in the following part of this work, the most important metrics and calculation methods.

4 Trust Management

The trust management consists of evaluating the trust, propagating the evaluation and continuously updating this one.

Building trust management evaluation is based on adjusting multiple parameters (metrics) and combining them via an aggregation method.

4.1 Trust Composition (Metrics)

The definition of relevant parameters is a key step in trust evaluation. We can choose into many metrics. Figure 3 illustrate their classification as follows:

1. **QoS Based Trust**: it evaluates the capability to provide the service and qualifies the performance via several metrics as: packet delivery ratio, success rate, energy consumption and task completion capability.
2. **Social Trust**: it calculates properties that are based on devices social relationship:

   - Honesty: during direct contacts the node use anomaly detection rules to discover suspicious or dishonest experiences, and evaluate the honesty of every device to recognize malicious nodes
   - Cooperativeness: refers to readiness to help different devices, uncooperative node is not necessarily malicious, it acts just for its own advantage
   - Community interest: is mostly seen as related to friendship, expressing the level of common interest
   - Recommendation: represent proposals of the near devices
   - Reputation: is a global rating of the device evaluating the opinion based on previous experiences
   - Centrality: the multitude of relationships and transactions, it shows device's group efficiency
   - Knowledge: is the data given by trustee to assess its reliability
   - Owners relationship: objects owned by the same person influence each one another's reputation

4.2 Trust Aggregation

Depending on the context, a single trust metric can be important for trust evaluation. Whereas, in general, multiple attributes are often required to define more satisfying trust assessment.

To conclude an aggregated trust value, we need to combine many trust metrics. Many techniques are used in the literature:

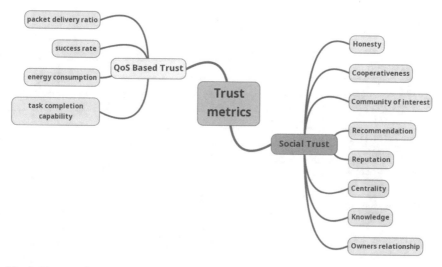

Fig. 3 Trust metrics

- **Weighted sum (static or dynamic) or weighted mean**
  is usually used to combine different properties, adjustment of the weights acts out
  the relevance of each property. Penalty coefficient weights can also be applied to
  express untrustworthiness [14].
- **Regression based**
  As used in [10, 18], the regression is a statistical method of modeling. It helps to
  estimate the correlation between different parameters.
- **Bayesian inference**
  Simple model to continuously update the trust estimation. It is based on the statistical method which uses Bayes' theorem to recalculate probability, each time
  where new informations are available. It has been used for example in Gao and
  Liu [5] to develop a dynamic method for trust management in medical wireless
  networks.
- **Fuzzy logic**
  In fuzzy logic, we use different logic values between 0 and 1 where 0 is corresponding to "completely false" and 1 means "completely true". This logic is used to
  generate trust measures between "untrusted" and "completely trustworthy". This
  method has been widely investigated as in Mitchell et al. [12], Wu and Zhou [19].
- **Particle swarm**
  An intelligent optimization method inspired by colony organization where every
  organism performs a small task to achieve a complex one. Chakraborty proposes
  this method in Chakraborty and Datta [4], as it is convenient for decentralized and
  self-organized systems, it is used in Sun et al. [17] as well.
- **Machine learning based**
  As the AI branch which provides the ability to learn automatically from data via
  performant algorithms, machine learning is nowadays widely used to make effi-

Table 2 Trust aggregation methods

	Weighted average	Regression based	Bayesian inference	Fuzzy logic	Particle swarm	ML based
Nitti et al. [14]	x					
Wang et al. [18]		x				
Li et al. [10]		x				
Gao et al. [5]			x			
Mitchell et al. [12]				x		
Wu et al. [19]				x		
Chakraborty et al. [4]					x	
Sun et al. [17]					x	
Jayasinghe et al. [8]						x
Jinarajadasa et al. [9]						x
Jayasinghe et al. [7]						x
Abdelghani et al. [2]						x
Masmoudi et al. [11]						x

cient decisions and accurate predictions, based on data and without human intervention or explicit instructions. Many research works implement ML algorithms to evaluate devices trustworthiness.

Table 2 summarizes the implementation of the cited aggregation methods in literature.

5 Implementing Trust System Challenges

Every system that aims to guarantee an appropriate level of trust and security in SIoT communications must respect some constraints related to IoT characteristics:

- It has to be a lightweight solution with minimal resource consumption
- Take in count different hardware specifications and standards
- Deal with mobility and multiple interactions and protocols.

5.1 Metrics and Method Selection

The metrics used for the trust composition need to be generic, easy and legitimate for collection. These characteristics are more available in QoS properties, which explains why this type is the most used in literature. However, for an efficient trust assessment, social properties are essential parameters to consider. In addition, the diversity in the IoT world means that many metrics are exclusive to certain IoT contexts.

To choose a management method, we need a trade-off between performance which can be measured via numerous parameters, and constraints related to IoT character-istics.

5.2   Trust Category Issues

We often tend to decide on trustworthiness in a deterministic way to avoid unsat-isfactory experiences. However, in the IoT context which changes frequently and unpredictably, we can hardly implement a totally deterministic solutions.

1. Deterministic

   - Policy-based: A set of policies conclude the trust evaluation based on predefined parameters and scenarios
   - Certificate system: Trust is relative to authentication with digital signature and verified by a public key

2. Non deterministic

   - Recommendation-based: Helps to decide about the trustworthiness of a node, even if there was no interaction with it in the past. It requires defining the reliable recommenders and the method of trust calculation. It may be an explicit recommendation or a transitive one.
   - Reputation based: It utilizes the collected rating and feedback
   - Prediction-based: When there is no prior experiences or relative information, prediction methods are useful to decide about the trustworthiness of new devices
   - Social network based: In SIoT, objects are in communities where they can share experiences, ratings and opinions about one another

   Thus, without non-deterministic trust management methods, devices will not be able to provide a service or complete a task: "nothing ventured, nothing gained".

5.3   Trust-Related Attacks

Trust management system must be aware of any attacks that can corrupt the results. As it is an interactive procedure in IoT context, trust management is vulnerable to many types of these attack:

- **Self-promoting attack**: a device provide good recommendation to promote itself as trustworthy node
- **Bad-Mouthing Attack**: giving bad recommendation to harm other devices by soiling their reputation

- **Whitewashing attack**: a disappearance for enough time allows a malicious object to wash away its bad reputation
- **Opportunistic Service attack**: a hostile device provides services to acquire the trust
- **Collision attack**: a false recommendation is made by a group of devices to boost or ruin entities reputation
- **Sybil attack**: a node creates multiple fake identities, it uses them to give multiple recommendations and control devices reliability
- **On off attack**: the device behave well during a period to become reliable, then it uses the trust it had to proceed an attack

An efficient solution should be insusceptible to these trust-related attacks.

6 Conclusion

In this paper we highlight the trust management importance for IoT security and the raising focus of scientific research on this field. We deduce that trust management is a promising way to trade off between security challenges and the open ambitious prospects in the SIoT world. In order to help choosing the most important parameters for any future system design, we survey trust specifications and we describe the most popular trust calculation methods and metrics. And finally we point out issues and challenges which should be considered for the development of an effective trust management system in SIoT context.

References

1. Abdelghani W, Zayani CA, Amous I, Sèdes F (2016) Trust management in social internet of things: a survey. In: Dwivedi YK, Mäntymäki M, Ravishankar M, Janssen M, Clement M, Slade EL, Rana NP, Al-Sharhan S, Simintiras AC (eds) Social media: the good, the bad, and the ugly. Springer International Publishing, Cham, pp 430–441
2. Abdelghani W, Zayani CA, Amous I, Sèdes F (2018) Trust evaluation model for attack detection in social Internet of Things. In: CRISIS 2018—13th international conference on risks and security of Internet and systems, Arcachon, France, pp 48–64, October 2018. https://hal.archives-ouvertes.fr/hal-02296115
3. Binti M, Noor M, Hassan WH (2019) Current research on Internet of Things (IoT) security: a survey. Comput Netw 148:283–294. https://doi.org/10.1016/j.comnet.2018.11.025
4. Chakraborty T, Datta SK (2017) Application of swarm intelligence in Internet of Things. In: 2017 IEEE international symposium on consumer electronics (ISCE), pp 67–68. https://doi.org/10.1109/ISCE.2017.8355550
5. Gao Y, Liu W (2014) Betrust: a dynamic trust model based on Bayesian inference and Tsallis entropy for medical sensor networks. J Sens 2014:649392
6. Guo J, Chen IR, Tsai JJ (2017) A survey of trust computation models for service management in Internet of Things systems. Comput Commun 97:1–14

7. Jayasinghe U, Lee GM, Um TW, Shi Q (2019) Machine learning based trust computational model for IoT services. IEEE Trans Sustain Comput 4(1):39–52. https://doi.org/10.1109/TSUSC.2018.2839623
8. Jayasinghe U, Lee HW, Lee GM (2017) A computational model to evaluate honesty in social Internet of Things. Proc ACM Symp Appl Comput Part F1280:1830–1835. https://doi.org/10.1145/3019612.3019840
9. Jinarajadasa G, Rupasinghe L, Murray I (2018) A reinforcement learning approach to enhance the trust level of MANETs. In: 2018 national information technology conference, NITC 2018, pp 1–7. https://doi.org/10.1109/NITC.2018.8550072
10. Li Z, Li X, Narasimhan V, Nayak A, Stojmenovic I (2011) Autoregression models for trust management in wireless ad hoc networks. In: 2011 IEEE global telecommunications conference—GLOBECOM 2011, pp 1–5. https://doi.org/10.1109/GLOCOM.2011.6133993
11. Masmoudi M, Abdelghani W, Amous I, Sèdes F (2020) Deep learning for trust-related attacks detection in social Internet of Things. In: Chao KM, Jiang L, Hussain OK, Ma SP, Fei X (eds) Advances in e-business engineering for ubiquitous computing. Springer International Publishing, Cham, pp 389–404
12. Mitchell J, Rizvi S, Ryoo J (2015) A fuzzy-logic approach for evaluating a cloud service provider. In: 2015 1st international conference on software security and assurance (ICSSA), pp 19–24. https://doi.org/10.1109/ICSSA.2015.014
13. Najib W, Sulistyo S, Widyawan (2019) Survey on trust calculation methods in Internet of Things. Procedia Comput Sci 161:1300–1307. https://doi.org/10.1016/j.procs.2019.11.245
14. Nitti M, Girau R, Atzori L (2014) Trustworthiness management in the social Internet of Things. IEEE Trans Knowl Data Eng 26(5):1253–1266. https://doi.org/10.1109/TKDE.2013.105
15. Pourghebleh B, Wakil K, Navimipour NJ (2019) A comprehensive study on the trust management techniques in the Internet of Things. IEEE Internet of Things J, 1 (2019). https://doi.org/10.1109/jiot.2019.2933518
16. Riahi Sfar A, Natalizio E, Challal Y, Chtourou Z (2018) A roadmap for security challenges in the Internet of Things. Digital Commun Netw 4(2). https://doi.org/10.1016/j.dcan.2017.04.003
17. Sun Z, Zhang Z, Xiao C, Qu G (2018) D-S evidence theory based trust ant colony routing in WSN. China Commun 15(3):27–41. https://doi.org/10.1109/CC.2018.8331989
18. Wang Y, Lu YC, Chen IR, Cho JH, Swami A, Lu CT (2014) Logittrust : a logit regression-based trust model for mobile ad hoc networks
19. Wu Z, Zhou Y (2016) Customized cloud service trustworthiness evaluation and comparison using fuzzy neural networks. In: 2016 IEEE 40th annual computer software and applications conference (COMPSAC), vol 1, pp 433–442. https://doi.org/10.1109/COMPSAC.2016.86

An Overview of the Security Improvements of Artificial Intelligence in MANET

Hafida Khalfaoui⬤, Abderrazak Farchane, and Said Safi⬤

Abstract Mobile Ad hoc Networks (hereinafter, MANETs) are formed by nodes that communicate with wireless mediums without resorting to pre-existing network infrastructure. In MANETs, terminals can act as both end systems and intermediate nodes as routers. MANETs present a good solution for lowering communication costs but are more susceptible to security attacks because of node mobility, wireless transmission links and dynamic topology. Artificial Intelligence (AI) techniques are recently bestowed to tackle these security issues. In this paper, we present a detailed view of issues in MANET security. Then, we define some security solutions presented in the literature and their shortcomings, explaining how AI techniques can be exploited to improve the security of routing protocols and connectivity between dynamic mobile nodes in MANETs.

Keywords MANET · Security · Artificial intelligence

1 Introduction

Since the 1950s artificial intelligence (AI) research has been carried out in mathematics and computer science domains. Initially, AI research tried to address relatively simple issues that are difficult for humans and consume their time. Over time, research has shifted towards developing algorithms for solving pattern recognition problems (such as face and speech recognition) that are relatively easy for humans to solve but difficult to articulate clearly. After years of neglect, many fields of AI research have

H. Khalfaoui (✉) · A. Farchane · S. Safi
Laboratory of Innovation in Mathematics, Polydisciplinary Faculty, Applications and Information Technologies, Sultan Moulay Slimane University, Beni Mellal, Morocco
e-mail: hafida.khalfaoui@usms.ac.ma

© The Author(s), under exclusive license to Springer Nature Switzerland AG 2023
A. A. Abd El-Latif et al. (eds.), *Advances in Cybersecurity, Cybercrimes, and Smart Emerging Technologies*, Engineering Cyber-Physical Systems and Critical Infrastructures 4,
https://doi.org/10.1007/978-3-031-21101-0_11

seen significant recovery thanks to technological advances in computing power and increased availability of "big data" [1].

Recently, advancements in telecommunications, intelligent devices and computational systems have opened up new opportunities to solve networking problems. AI is widely used in this field to streamline traditional approaches. It can improve the security and efficiency by receiving information from a variety of sources, improving user awareness, and predicting to prevent potential attacks [2]. In this work, we select MANET as a type for wireless networks, explaining how AI technology was used to improve it.

The concept of MANET is not novel, dating back to the DARPA Packet Radio Network project in 1972 [3]. After that, Burchfield et al. explained the functions and structure of a Packet radios station in [4]. Actually, MANET is defined as a collection of mobile devices like smartphones, laptops and PDAs that communicate via a wireless medium without fixed infrastructure or centralized administration [5, 6]. This technology aims to reduce costs and allow network usage in harsh conditions such as disaster management, military, vehicle computing, etc.

MANET's network devices can use different technologies to communicate (e.g., Bluetooth, Zigbee, Ultrawideband, WiFi and 802.11 variants...) [7]. Each technology has specific characteristics like theoretical bit rate, frequency, range and power consumption, as summarized in Table 3 of [3]. The limitation of these characteristics and the specifics of MANET, like the lack of centralized management and dynamic topology, make these networks more vulnerable to attacks than wired networks. Mobile nodes are low-resource devices that can't secure or defend themselves, leaving them open to hacking and compromise [8].

Traditional approaches, such as access control mechanisms and encryption, have significant drawbacks in effectively safeguarding networks and systems from more sophisticated attacks like denial of service. Furthermore, most systems based on such methodologies cannot adapt to changing malicious behaviors. This paper's main contribution is to outline the issues with current MANET security solutions and encourage the use of AI to provide more robust security.

The outline of this paper is shown as follows: Section 2 explains security attacks in MANET. Section 3 discusses some security methods of MANET that exist in the literature, followed by a discussion of problems of existing security solutions in Sect. 4. Section 5 spots the light on the evolution of AI towards MANET security, showing their advantages in this domain in Sect. 6. In the last section, we conclude and give some perspectives.

2   Security Attacks in MANET

All nodes in the MANET are free to join and quit the network, which is known as an open network boundary. Because the communication medium is wireless, each node will receive packets within its wireless range and transfer them to its neighbors until they arrive at the destination. Due to these features, each node can readily

Table 1 The classification of various type of attacks [12]

Attack type	Source	Behavior
Eavesdropping, selfish misbehavior, monitoring	Interne	Passive
Traffic analysis, selfishness	Externe	Passive
Blackhole, interceptions, sybil, sinkhole, session hijacking, resoure exhaustion, link spoofing, grayhole, wormhole, information disclosure	Externe	Active
Jellyfish, SYN flooding, replication, packet, man-in-the middle, byzantine flooding, DoS, link withholding, replay attack, resource consumption, impersonation, traffic jamming	Interne	Active

access packets of other nodes or insert fault packets into the network. As a result, protecting MANET from malicious nodes and behaviors has become one of the most critical challenges in MANET [9]. Additionally, privacy, availability, integrity, and non-repudiation are all aspects of security that must be considered. Security also refers to detecting potential attacks and a system's vulnerability to illegal access that can cause altering, releasing or denying of data. Attacks can be generally defined into two categories:

- **Passive attacks** refer to attacks that get access to resource networks but do not alter their content. The attacker monitors certain connections to gather traffic information without injecting bogus data that could endanger the system's resources or normal network operations. This type of assault is not severe, but it compromises confidentiality, aids the attacker in obtaining information and creates a bridge to other dangerous attacks [10].
- **Active Attacks** mean manipulation or alteration of data transferred in the network. It can limit network operations, change, inject packets or take advantage of many network features.

Attacks are also classified into external and internal. For the first type, attacks are utilized by illegitimate persons who want to enter the network. If they gain access, they will abuse it by sending faked packets, causing the entire network to go down. For the second type, attacks usually occur within the network by selfish nodes, and it is difficult to predict them [11]. The Table 1 shows the classification of various attacks according to the source and behavior of the attacker.

3   Existing MANET Security Algorithms

MANET has many applications, but they all have their own set of issues. The main source of these problems is its decentralized architecture and mobile topology, which causes unpredictable changes in routing information. According to research, MANET has two sorts of security measures. The first type concerns protecting routing schemes, while the second concerns data security schemes [13].

3.1   Protecting Routing Schemes

Routing is one of the most difficult tasks in MANET because of the challenges cited above. It is desirable to create a routing protocol that is both efficient and adaptive to changing network conditions. A routing protocol must also support various services for multiple applications in MANET environment, which is dynamic, flexible to changes in network architecture, and vulnerable to security attacks [14].

Many surveys [15, 16] have been conducted to evaluate the routing process and strategies used by MANETs. The majority of authors in the literature, like Abolhasan et al. [17], divided MANET routing protocols into three categories: reactive routing protocols used for threat identification and speedier response times, proactive routing schemes focusing on preventing adversaries from invoking intruders, and hybrid routing protocols that take the best of both reactive and proactive methods and combines them to produce excellent results. There are other essential routing classifications in MANETs, such as Hierarchical, Geographical, Power-Aware and Multicast Routing. In Syed and Shahzad [14], for example, multiple existent MANETs routing protocols were discussed, with one example for each category of a routing technique being examined.

3.2   Data Security Schemes

Security in MANET will not be complete without these fundamental concepts: authentication, confidentiality, integrity, non-repudiation and availability. Protecting network resources and data information should regard authentication as the first line of defense against attacks. Malicious attackers might simply enter the network without authentication and use their available resources, acquire access to critical information and disrupt the operation of other nodes. Therefore, it is necessary to have a mechanism for preventing an outsider from performing an attack in the network. In our previous work [18], we presented a survey on various distinct MANET authentication mechanisms. When looking at the existing algorithms in the literature, it's evident that every author attempts to turn the authentication process into a decentralized system. Blockchain technology is a distributed database in which each node in the network participates in the authentication method. This technology ensures the node's privacy and the integrity of the data transferred by signing all transactions.

Some authors in [19] combine authentication with confidentiality mechanism. They are interested in encrypting data with a symmetric key after the node has been successfully authenticated because symmetric keys are simple, utilize less memory, consume less power, and occupy less memory. Other authors [20] developed a method for changing DSR routing protocol and gaining data integrity by protecting the discovery phase of the routing protocol from a Man-In-The-Middle attack.In this latter, The attacker captures all packets and then removes or modifies them. But according to the integrity service, only authorized nodes can originate, edit or delete packets.

Secure communication is an arduous task in MANET due to its specific characteristics. In Mandal et al. [21], authors spotlight different types of Intrusion Detection Systems (IDS) such as Host IDS (HIDS), Application-based IDS and Network IDS. They are mechanisms that use cryptography and access control technologies for detecting and preventing security attacks.

Security of MANET, in general, must consider the two previous aspects: protection of data and routing protocols. Moreover, numerous methods have been developed to combine security data with routing protocols, making it difficult to determine the most optimal and efficient solution that can be applied in different situations [22].

4 Problems in Existing MANET Security Solutions

We present below a general idea about some challenges of existing MANET security approaches in the literature [9, 23], which are as follows:

- **Processing Time**: This constraint relates to the time required for each technique to discover dangerous nodes and secure the network against all malicious nodes. Because of the dynamic topology of the network, routes are continuously changing, which necessitates rerouting and increases the time taken on the security approach. As a result, MANET flexibility is affected by long processing times.
- **Cooperative nodes**: In some contexts, the defeating approach can not discover cooperative malicious nodes, but it locates them one by one in different executions, which takes a long time.
- **Key distribution**: In MANET, key distribution is complex due to the lack of centralized control unit. Malicious nodes can obtain keys by intercepting packets or via a Man-In-The-Middle attack. The widest solution for this problem is using clusters to exchange keys in the network.
- **Packet overhead**: Means the additional packets generated by the source node to detect and delete unauthorized nodes. However, the high packet overhead caused by some defeating techniques, increases congestion, collision and packet loss due

to the wireless medium. It also lengthens the processing time and consumes more energy in the nodes.

- **Frequency knowledge**: In light of this constraint, each node must know the transmission frequency. Because all nodes in MANET are free to join or leave the network, keeping track of all frequencies is difficult.
- **Energy consumption**: In MANET, nodes have limited battery power. Consequently, the security approaches must save as low as possible energy consumption to keep the optimal network's lifetime.

5 Artificial Intelligence's Approaches in MANET Security

Artificial intelligence (AI) has grown in popularity thanks to its recent advancements. As cited in Haenlein and Kaplan [24], AI is defined as "a system's ability to interpret external data correctly, to learn from such data and to use those learnings to achieve specific goals and tasks through flexible adaptation." Additionally, Advances in hardware and algorithms have substantially expedited the broad application of AI in all simulation operations. Forecasting activities such as weather and natural environment forecasts, design and management work, and database construction and processing are representative examples of AI [25]. AI is built on mathematical, biological, computational, linguistic and engineering concepts. Figure 1 depicts some AI branches. Concerning the security aspect, AI arrives with new techniques that aim to procreate an intelligent network, defend the nodes and resolve the challenges and threats stalking MANETs and their protocols. This technology has been explored to transform network nodes into smart nodes that can make brilliant decisions on their own, much like humans [26].

According to different studies, e.g. Shivashankar and Shivakumar [13], involving the security approaches of AI in the MANET, the populated branches that have been reported to be utilized for securing network system are briefly discussed as follow:

- **Machine learning**:
  In [27], Luong et al. use machine learning to investigate a new method for detecting flooding attacks. This latter is a type of DoS attack in which malicious nodes broadcast fake packets in the network to absorb resources and disrupt network operations. The main idea of this approach is based on the route discovery history information of each node represented by a route discovery frequency vector (RDFV) to distinguish normal behavior from the abnormal one. The route discovery histories demonstrate that nodes belonging to the same class have similar characteristics and behaviors. RDFV is used with the Flooding Attack Detection Algorithm (FADA) and the AODV protocol for creating a novel prevention routing mechanism of flooding attacks in a MANET environment. Simulation results reveal that FADA achieves a higher misbehavior detection ratio (over 99 %). Furthermore, this approach is efficient compared to the standard AODV protocol.

Fig. 1 Some branches of AI

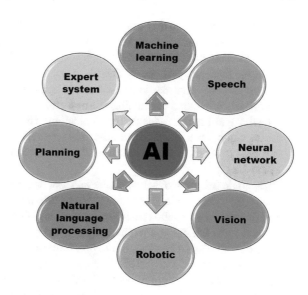

Also, it enhances network performance in terms of higher packet delivery ratio, lower end-to-end delay and reduced routing overhead.

Deep learning is a subclass of machine learning in which algorithms are built and function similarly to machine learning algorithms. Still, there are multiple layers in deep learning, each of which provides a distinct interpretation of the data it feeds on Laqtib et al. [28]. Sbai and El Boukhari [29] have exploited the deep learning model to propose an intrusion detection system (IDS) data flooding or UDP flooding attack in MANETs using the dataset CICDDoS2019: it is purely dedicated to DDoS attacks, with a large number of transaction network records. According to the environment of MANETs, the obtained results can achieve exciting performance (Accuracy, Precision, Recall and F1-score(F1-Score is higher, the machine learning model is better)).

– **Neural netwok**:

Clustering becomes especially important in highly dense networks and for MANET security due to its dynamic nature. Chatterjee and Saha [30] established an improved protocol for their previous work. They benefit from Artificial Neural Networks (ANNs) to develop a clustering algorithm using weight-based parameters to select cluster heads. ANNs are computing systems that are inspired by the biological neural networks that constitute human brains [31]. ANN is used to select cluster heads and train a model to predict the scale of the weights necessary for various network topologies utilizing four inputs: mobility, packet drop, energy and the number of neighbor nodes.

Kaushik [32] has also utilized ANNs to identify and detect the black hole node inside the network. It's an active attack that sends incorrect routing information to the victim node tables, resulting in fraudulent route entries. This attack can consume data packets intended to be transmitted to the destination. As a result, data packets are dropped in the network, lowering its performance [33]. Kaushik selects the best path with an efficient Ad-Hoc on Demand Distance Vector (AODV) routing protocol. This protocol can identify the routes, but it does not have information about the legitimacy of nodes. The fitness function of the Firefly Algorithm (FA) is used to determine the node features in order to solve this problem. FA is a metaheuristic method for finding the best path that is inspired by firefly flashing behavior and bioluminescent communication. The experiment has analyzed that the throughput has increased compared to previous works, and reduction has been observed in terms of energy consumption, packet loss rate and delay.

– **Fuzzy logic**:
Fuzzy logic (FL) is a robust mathematical tool that proved its applicability in IDSs. A multistemmed logic allows for defining intermediate values between conventional assessments such as true/false, yes/no, high/low, etc. These values like fairly tall and swift are defined mathematically and prepared by PCs, aiming to use a more human way of thinking in PC programming [34].
Ghasemnezhad and Ghaffari [35] created a novel protocol called Fuzzy Logic-Based Reliable Routing Protocol (FRRP). It regards network nodes as intelligent agents in charge of data routing and transfer. This approach was created to consider the following parameters while choosing an acceptable route and the next hop: available bandwidth, battery energy, hop count and the mobility's degree of wireless nodes for minimizing cost and network resource consumption. Thanks to FL, the solidity and stability of the FRRP had higher efficiency than the AODV routing protocol.
Routing protocols are required in the MANET environment to deal with attacks such as a black hole. Hiremath et al. [33] have developed a novel method for detecting and preventing the cooperative black hole attack on MANETs. This method uses an adaptive fuzzy inference system (FIS) for MANET using fuzzy rule bases. A node's fuzzy inference can represent and analyze imprecise or incomplete information about its neighbors, allowing it to choose which nodes to send data packets. Simulation experiments show that the adaptive FIS effectively and competently detects and prevents various attacks in MANETs.

The limitations of the approaches discussed above are the high time cost and memory resource consumption to perform the predictions. Moreover, the normal behavior of nodes is modeled based on the audit data collected throughout a normal operation. When the learning phase ends, the algorithm detects only known types of intrusion learned in this phase. Therefore, if the traffic changes, the detection becomes less reliable and the algorithm has to be reset in order to adapt to the new traffic.

Table 2 shows the summary of AI-based security methods in MANET discussed above.

Table 2 Some existing AI-based security methods in MANET

References	Problem	Solution	Advantages	Limitations
Luong et al. [27]	IDS of Flooding attacks	Machine learning (RDFV with DATA and AODV)	Higher misbehavior detection ratio. Higher packet delivery ratio Reduced routing overhead Lower end-to-end delay	Not resistant to dynamic attack
Sbai and El Boukhari [29]	IDS of data flooding or UDP flooding attack	Deep learning	High performances (Accuracy, Precision, Recall and F1-score)	Only resistant to flooding attacks
Chatterjee and Saha [30]	Clustering	Artificial Neural Networks (ANNs)	Creating model to predict the scale of the weights necessary to select cluster heads for various networks	Large requirement of storage and time processing
Kaushik [32]	IDS of Black hole attacks	Artificial Neural Network, Firefly Algorithm with AODV	Reduction of packet loss rate, delay and energy consumption	Only resistant to Black hole attack
Ghasemnezh- ad and Ghaffari [35]	Routing	Fuzzy Logic-Based Reliable Routing Protocol (FRRP)	Minimizing cost and network resources consumption	Done only for a single unicast routing protocol
Hiremath et al. [33]	IDS of Black hole attacks	Fuzzy inference system (FIS)	Detection and prevention of the cooperative black hole attack	Detect only Black hole attack

6 Advantages of AI in Security of MANET

In the field of MANET security, AI offers dependable solutions. A self-manageable network administration is the key to AI-based networking, which is the future where very little human intervention is required. As a result of the two key approaches, the error rate will decrease considerably. The first is using machines to manage networks in place of humans. Second, when a system learns from experience, AI will make the system more resilient day by day [36]. Because of the following reasons, AI approaches have several advantages in MANETs security:

1. **Improvement of IDSs**: AI contributes to the advancement of many detection systems that are more accurate and able to detect adversary presence in the network and categorize attacks. AI can understand the user and network behavior to incorporate more intelligence for effective decision-making in calculating the likelihood of fatalities for any dangers.
2. **Faster Response Time**: AI can help you solve any security problem faster with high precision than the other traditional techniques. Numerous available alert systems and protected data logs are used for speeding up the threat analysis process.
3. **Massive data management**: Securing such exponential data with increased complexity is a difficult challenge because of the enormous creation of data from various log files, system-generated warnings and so on. AI could help with data selection and security, which is difficult to do manually [13, 26].

7 Conclusion and Future Work

Mobile Ad hoc networks are vulnerable to network threats because of the lack of wire links and the fixed architecture. This paper discussed the state of existing security-based solutions in MANET. It has recently been discovered that the MANET concept's intrinsic features are the underlying cause of various issues. After evaluating the existing system, we found that the majority of current security approaches have limitations that fill gaps in network security. Indeed, there is still a need for a complete security plan for MANET that ensures resource efficiencies. We have explained some optimization-based approaches of AI that receive little attention in resolving the main problems of MANET security. Our paper's goal is to encourage other researchers to develop the technology of AI in the security of wireless networks. In our future work, we will create a mechanism that integrates AI with blockchain technology to strengthen the protection of MANET.

References

1. Van Assen M, Lee SJ, De Cecco CN (2020) Artificial intelligence from A to Z: from neural network to legal framework. Eur J Radiol 129:109083. https://doi.org/10.1016/j.ejrad.2020.109083
2. Tong W, Hussain A, Bo WX, Maharjan S (2019) Artificial intelligence for vehicle-to-everything: a survey. IEEE Access 7:10823–10843. https://doi.org/10.1109/ACCESS.2019.2891073
3. Hoebeke J, Moerman I, Dhoedt B, Demeester P (2004) An overview of mobile ad hoc networks: applications and challenges. J Commun Netw 3(3):60–66
4. Burchfiel J, Tomlinson R, Beeler M (1975) Functions and structure of a packet radio station. In: IEEE Computer Society, Los Alamitos, CA, USA, pp 245–251 (1975). https://doi.org/10.1145/1499949.1499989

5. Robinson YH, Julie EG (2019) MTPKM: multipart trust based public key management technique to reduce security vulnerability in mobile ad-hoc networks. Wirel Pers Commun 109(2):739–760. https://doi.org/10.1007/s11277-019-06588-4

6. Rani P, Verma S, Nguyen GN (2020) Mitigation of black hole and gray hole attack using swarm inspired algorithm with artificial neural network. IEEE Access 8:121755–121764. https://doi.org/10.1109/ACCESS.2020.3004692

7. Basagni S, Conti M, Giordano S, Stojmenovic I (2004) Mobile ad hoc networking. Wiley, New York

8. Agalawe SU, Chopde NR (2014) Security issues: the big challenge in MANET. Int J Comput Sci Mob Comput 3(3):417–424

9. Dorri A, Kamel SR, Kheirkhah, E.: Security challenges in mobile ad hoc networks: a survey. IJCSES 6(1). https://doi.org/10.48550/arXiv.1503.03233

10. Verma S, Sharma J, Sima (2016) A study of active and passive attacks in Manet. Int J Sci Res Dev 4:2321–0613 (2016)

11. Prasanna S (2018) Complete analysis of various attacks in Manet. Int J Pure Appl Math 119(15):1721–1727

12. Meddeb R, Triki B, Jemili F, Korbaa O (2017) A survey of attacks in mobile ad hoc networks. In: 2017 international conference on engineering & MIS (ICEMIS). IEEE, pp 1–7 (2017)

13. Shivashankar TM, Shivakumar SB (2019) Insights on security improvements and implications of artificial intelligence in MANET. Communications 7:1–9. https://doi.org/10.1109/ICEMIS.2017.8273007

14. Syed SA, Shahzad A (2022) Enhanced dynamic source routing for verifying trust in mobile ad hoc network for secure routing. Int J Elect Comp Eng 12(1):425–430. https://doi.org/10.11591/ijece.v12i1.pp425-430

15. Jawhar I, Trabelsi Z, Al-Jaroodi J (2014) Towards more reliable and secure source routing in mobile ad hoc and sensor networks. Telecommun Syst 55(1):81–91. https://doi.org/10.1007/s11235-013-9753-7

16. Ullah K, Das R, Das P, Roy A (2015) Trusted and secured routing in MANET: an improved approach. In: 2015 international symposium on advanced computing and communication (ISACC). IEEE, pp 297–302. https://doi.org/10.1109/ISACC.2015.7377359

17. Abolhasan M, Wysocki T, Dutkiewicz E (2004) A review of routing protocols for mobile ad hoc networks. Ad Hoc Netw 2(1):1–22. https://doi.org/10.1016/S1570-8705(03)00043-X

18. Khalfaoui H, Farchane A, Safi S (2022) Review in authentication for mobile ad hoc network. In: Advances on smart and soft computing. Springer, Singapore, pp 379–386. https://doi.org/10.1007/978-981-16-5559-3_31

19. Amin U, Shah MA (2018) A novel authentication and security protocol for wireless ad hoc networks. In: 24th international conference on automation and computing (ICAC), Newcastle Upon Tyne, UK, pp 1–5 (2018). https://doi.org/10.23919/IConAC.2018.8748982

20. Lv X, Li H (2013) Secure group communication with both confidentiality and non-repudiation for mobile ad-hoc networks. IET Inform Sec 7(2):61–66. https://doi.org/10.1049/iet-ifs.2010.0314

21. Mandal B, Sarkar S, Bhattacharya S, Dasgupta U, Ghosh P, Sanki D (2020) A review on cooperative bait based intrusion detection in MANET. In: Proceedings of industry interactive innovations in science, engineering & technology (I3SET2K19). https://doi.org/10.2139/ssrn.3515151

22. Borkar GM, Mahajan AR (2020) A review on propagation of secure data, prevention of attacks and routing in mobile ad-hoc networks. Int J Commun Netw Distrib Syst 24(1):23–57. https://doi.org/10.1504/IJCNDS.2020.103858

23. Yamini KAP, Suthendran K, Arivoli T (2019) Enhancement of energy efficiency using a transition state mac protocol for MANET. Comput Netw 155:110–118. https://doi.org/10.1016/j.comnet.2019.03.013

24. Haenlein M, Kaplan A (2019) A brief history of artificial intelligence: on the past, present, and future of artificial intelligence. California Manag Rev 61(4):5–14. https://doi.org/10.1177/0008125619864925

25. Park WJ, Park JB (2018) History and application of artificial neural networks in dentistry. Eur J Dent 12(04):594–601. https://doi.org/10.4103/ejd.ejd_325_18

26. Madanan M, Venugopal A (2020) Designing an artificial intelligent MANET to reduce and detect security threats and concerns. In: Proceedings of the 2006 14th IEEE international conference on network protocols (ICNP'06), pp 75–84

27. Luong NT, Vo TT, Hoang D (2019) FAPRP: a machine learning approach to flooding attacks prevention routing protocol in mobile ad hoc networks. Wirel Commun Mob Comput 2019:1–17. https://doi.org/10.1155/2019/6869307

28. Laqtib S, Yassini KE, Hasnaoui ML (2019) A deep learning methods for intrusion detection systems based machine learning in MANET. In: Proceedings of the 4th international conference on smart city applications, pp 1–8. https://doi.org/10.1145/3368756.3369021

29. Sbai O, El boukhari M (2020) Data flooding intrusion detection system for MANETs using deep learning approach. In: Proceedings of the 13th international conference on intelligent systems: theories and applications, pp 1–5. https://doi.org/10.1145/3419604.3419777

30. Chatterjee B, Saha HN (2019) Parameter training in MANET using artificial neural network. Int J Inform Sec 11(9):1–8. https://doi.org/10.5815/ijcnis.2019.09.01

31. Bisen D, Mishra S, Saurabh P (2021) K-means based cluster formation and head selection through artificial neural network in MANET. https://doi.org/10.21203/rs.3.rs-667651/v1

32. Kaushik P (2021) An improved Black hole detection and prevention mechanism in MANET using firefly and neural network. Int J Innov Sci Res Technol 6(10)

33. Hiremath PS, Anuradha T, Pattan P (2016) Adaptive fuzzy inference system for detection and prevention of cooperative black hole attack in MANETs. In: 2016 international conference on information science (ICIS). IEEE, pp 245–251. https://doi.org/10.1109/INFOSCI.2016.7845335

34. Kalimuthu VK, Samydurai G (2021) Fuzzy logic based DSR trust estimation routing protocol for MANET using evolutionary algorithms. Tehnicki vjesnik/Technical Gazette 28(6):2006–2014. https://doi.org/10.17559/TV-20200612102818

35. Ghasemnezhad S, Ghaffari A (2018) Fuzzy logic based reliable and real-time routing protocol for mobile ad hoc networks. Wirel Pers Commun 98(1):593–611. https://doi.org/10.1007/s11277-017-4885-9

36. Praveen T, Anurag S, Vijay K, Ashwani K (2019) Artificial intelligence supported Manet and Dumbo Net: reliable and better controlled networks. Int J Eng Res Technol (IJERT) 7(12)

Cloud Virtualization Attacks and Mitigation Techniques

Syed Ahmed Ali, Shahzad Memon, and Nisar Memon

Abstract The exponentially growing cloud technology equip organizations which they need to scale up their business without spending much on the IT infrastructure indeed. Services-based Cloud computing is a Leading-edge platform that allures the organization by its pervasiveness, elasticity, on-demand availability of the services, pays per usage model. All these features make the cloud an ideal environment for organizations and individuals. Virtualization is a key component of cloud computing that makes the virtual instances of cloud computing resources, including Network, Storage, Servers so that these instances could run on multiple machines at the same time to reduce the workload and the financial cost. The virtualization is implemented by using a special kind of software built on a large complex code is known as hypervisor or Virtual Machine Monitor (VMM). the hypervisor is responsible to control the overall functioning of the virtualized infrastructure including the virtual machines running on top of the virtualized layer by abstracting the host hardware. The vulnerabilities found in the hypervisor may be exploited by the cybercriminals resulting in taking full control of the environment. This study highlights the cyber-attack on virtualization and how the malicious actor can use virtualization, as an attack surface to aggravate the cloud services. This paper presents various vulnerabilities found in different hypervisors, gives an overview of various cyber-attacks on the virtualization platform. These attacks are further categorized as hypervisor-based and Virtual machine-based with explanation and pinpointing the solutions by mining the literature.

Keywords Virtualization · Virtual machine · Hypervisor · Side-channel · Security breach

S. A. Ali (✉) · S. Memon · N. Memon
Faculty of Engineering and Technology, A. H. S. Bukhari Post Graduate Centre of ICT, University of Sindh, Jamshoro, Pakistan
e-mail: Syed.ahmedali@scholars.usindh.edu.pk

S. Memon
e-mail: Shahzad.memon@usindh.edu.pk

N. Memon
e-mail: nisar.memon@usindh.edu.pk

A. A. Abd El-Latif et al. (eds.), *Advances in Cybersecurity, Cybercrimes, and Smart Emerging Technologies*, Engineering Cyber-Physical Systems and Critical Infrastructures 4,
https://doi.org/10.1007/978-3-031-21101-0_12

1 Introduction

Cloud computing can be introduced as the most transformative technology ever [1]. Cloud computing is a modern-day approach for creating, delivering, accessing, and managing IT services. According to the National Institute of Science and Technology (NIST) which poses cloud computing as "*cloud computing is a model for enabling ubiquitous, convenient, on-demand network access to a shared pool of configurable computing resources (e.g., networks, servers, storage, applications, and services) that can be rapidly provisioned and released with minimal management effort or service provider interaction*" [2].

On-demand service, Virtualization, instant service, pay as you go are the key features that transform any data center into the cloud [3]. It has three deployment models including, a private cloud in which the computing resources like network, hardware, and storage are managed completely under the sole control of a company's premises or allocation of dedicated servers at the provider's end. In the public cloud, the resources at Cloud Service Provider (CSP) are shared among the various users whereas the combining features of public and private cloud form the hybrid cloud. The cloud has three service models which are Infrastructure as a Service (IaaS), Platform as a Service (PaaS), and Software as a Service (SaaS). SaaS clients are allowed to use the cloud service provider's application in a network. PaaS allows the client to develop, maintain, deploy, and run their applications at the provider's end. In IaaS clients are provided the hardware structure, like processing, storage, and network capacity on a rental basis. Since the inception of cloud technology, cyber-criminals have always been strayed to find loopholes in the technology to breach. Exploiting the cloud gives a high scale of destruction as compared to the traditional tactics. Virtualization is the key component in the cloud environment and is susceptible to cyber-Breach. The rest of the paper is categorized into four different sections. Section 2 poses the virtualization and its architecture, Sect. 3 describes the virtualization attacks and their categories and sub-categories, Sect. 4 describes some remediation techniques against these breaches and Sect. 5 concludes the paper.

2 Virtualization and Its Architecture

Virtualization gives the optimal utilization of the hardware resources. It facilitates running multiple instances of operating systems on a single machine. Virtualization is referred to as the abstraction of hardware and software that allow the application to run on top of the virtual environment known as Virtual Machine (VM), regardless of knowing underlying available resource [4]. The main objective of virtualization is load balancing which is acquired by converting customary computing to provide cost-effective security, operation, isolation, and fast recovery. The application of virtualization can be made on Servers, operating systems, and hardware levels. The benefit of virtualization is the reduction in energy and hardware cost [5]. To attain virtualization, the software called a hypervisor, which is also known as the virtual machine monitor (VMM) is installed on top of the hardware server machine. Virtualization is a cutting-edge technique in the information technology sector that is rapidly gaining acceptance. It provides a large number of logical resources from a single server [6]. Simply the technology that lets to run multiple VMs by sharing the resources of the host machine [7]. It creates a clear-cut boundary among hardware, host machine, and the virtual machines running on the virtualized platform. The hypervisor allows installing multiple instances of guest operating systems that share the hardware resource of a single-host server machine. The virtual machine is an instance of the guest operating system which runs multiple applications. VM is a replica of the real machine [8]. Virtualization is classified as Native and Hosted. In native virtualization, the hypervisor software such as (VMware, Microsoft Hyper-V, Citrix XenServer, Oracle VM) is directly installed on the host hardware. Figure 1 illustrates the architecture of the Native or Bare-metal hypervisor, which install directly on the hardware without any host operating system. The Native hypervisor is managed by the software which is installed separately on another machine to manage its overall operation.

To manage the native virtualization, a different machine is required to configure the management software such as CIRBA, Correlsense, Embotics, Netuitive, etc., and connect with the machine where the hypervisor is running on. Native virtualization is hardware-based implementation and not the scope of this paper. This paper encircles the hosted virtualization, the hypervisor is installed on the host operating system which manages the hardware resources. In this type of virtualization, the hypervisor is assumed as an ordinary application like other host applications. Figure 2 illustrates the hosted virtualization architecture which employs the host operating system instead of installing directly on the hardware. This kind of virtualization is susceptible to various kinds of virtualization attacks.

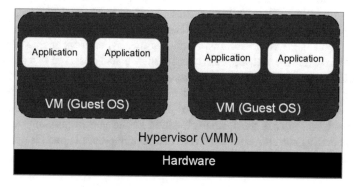

Fig. 1 Native virtualization architecture

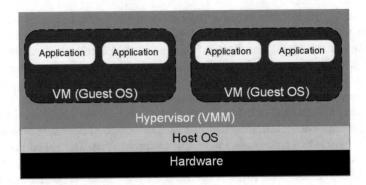

Fig. 2 Hosted virtualization architecture

3   Virtualization Breach

Virtualization technology contains several vulnerabilities and hazards that an attacker can exploit to breach cloud computing security and privacy systems [6]. A hypervisor vulnerability could lead to data loss, performance degradation, service stoppage, and hijacking of the hypervisor to take control of its functioning [9]. Virtualization can be exploited in two different ways, breach through the hypervisor and the virtual machine. The attacker may breach the virtualization environment by hijacking the hypervisor which works like the command and control for all VMs. Common Vulnerabilities and Exposure (CVE) is a database system that maintains all the publicly known security vulnerabilities. Information collected from the CVE database that the various kind of vulnerabilities in the different hypervisors have been found in the year 2021, and 2021.

Figure 3 shows the number of vender-based hypervisor vulnerabilities in the year 2020, and 2021. These vulnerabilities are related to Denial of Service (DoS), arbitrary code execution, information leakage, privilege escalation, memory corruption. We found these vulnerabilities in the various hypervisors such as Microsoft Hyper-V, Linux KVM, Vmware ESXi, and Xen. We also observed that some of the vulnerabilities found in the Xen are not found in the other hypervisors.

Figure 4 shows various kinds of attacks on virtualization technology. These attacks are mainly classified as hypervisor-based and virtual machine-based attacks. Hypervisor-based attacks are further classified as the guest operating system and the host operating system. Whereas the virtual machine-based attack is classified as side-channel, VM creation, VM migration, VM schedular attacks.

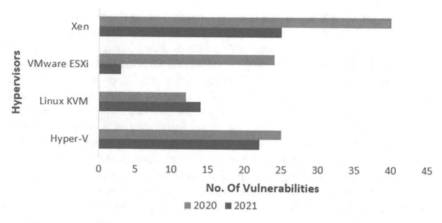

Fig. 3 Vender-based vulnerabilities

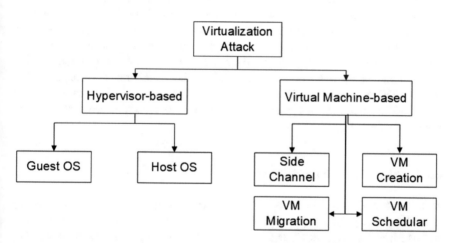

Fig. 4 Virtualization attacks

3.1 Hypervisor Based Attack

The hypervisor plays a vital role to manage virtualization and enable multiple virtual machines to run on a single host while maintaining isolation between them and sharing the underlying resources. Vulnerabilities in the hypervisor may exploit by the attackers [10] which allows them to compromise each of the virtual machines running on the underline host. Hypervisor vulnerabilities also empower the attacker to take control of the underlying host machine and the application running on it. Effective exploitation of the existing hypervisor may provide a way for criminals to compromise the entire virtualized infrastructure [11].

3.1.1 Attack Through Host OS

Virtualized infrastructure can be exploited through the host machine if the attacker finds vulnerabilities in the host operating system, as most of the implementation of virtualization is done on the hosted architecture as shown in Fig. 4. In this kind of exploit, the attacker takes the advantage of vulnerabilities in the host operating system which enable them to take control of the application running on it for instance the hypervisor. Aalam et al. [12] has explained three levels of privilege for CPU. Level 0 is referred to as kernel-level which is considered the heart of the operating system. All the major functions of OS are done by level 0, such as memory and task management. The software executes on level 0 has maximum control over the hardware of the host machine. The hypervisor is a kind of software that runs on level 0 and provides accessibility to the underlying hardware that can be exploited resulting in a security breach to the entire cloud environment. Level 1, and level 2 are responsible to run OS services. Different kinds of attacks can compromise the host OS.

One most common OS-level attacks are a buffer overflow attack which is a result of a coding error while writing a program. Adversaries may exploit this vulnerability to gain unauthorized access to the host machine. DoS/DDoS attack is another kind of attack that is launched with the intent to discontinue the services of the server to its legitimate users. The attack floods network traffic to the host machine to make its resources unavailable to the authorized users. In the cloud environment, the multiple VMs running on top of the host may also suffer if the underlying host is experiencing the DoS/DDoS attack. The DDoS is a most lethal version of DoS in which the host is targeted from the various intermediary compromised machines controlled by the remote attacker.

3.1.2 Attack Through Guest OS

In the world of Cloud virtualization, it is comparatively easy to exploit the hypervisor from the guest operating system as compared to the inverse direction coined as "Virtual Machine Extrospection". The naïve user or any malicious user having control of the virtual machine can breach the underlying hypervisor [13]. Such Breach has been demonstrated by Kortchinsky [14, 15]. The virtual machine or guest machine can be breached by deploying the rootkit that compromises the security of the hypervisor and provides full control to the system [12].

3.2 Virtual Machine Based Attack

In a cloud environment, such attacks exploit the vulnerabilities in the virtual machine that leads to affect the services of the cloud and violation of the data protection. Multiple VM sharing the same physical resources may cause security for various breaches.

3.2.1 Cross VM Side-Channel Attack

The Breach between two virtual machines is known as a cross virtual machine [16] this scenario attacker's virtual machine can abuse the neighboring VM and collect what resources are being utilized by the target VM. Resources information like CPU shared memory and the cache may be gathered by this attack. The attack may cause the change of structure of virtual machines while deploying on the cloud [17]. Resource sharing using virtualization between the customer and the services is the major cause of cross virtual machines [17].

3.2.2 VM Creation Attack

In this kind of attack, malicious code can be injected into a virtual machine resultantly the newly created virtual machine has the same malicious code which causes security issues in the cloud. This malicious VM can bypass the hypervisor and give access to the attacker to the underlying host [10].

3.2.3 VM Migration Attack

The migration of the virtual machine is done to achieve multiple resource management tasks including power management, load balancing, fault tolerance, and maintenance of the system [18]. Migration of the virtual machine has been done into different ways live or non-live. In live migration, all the data such as memory state, CPU, and storage information is shifted from the source host to the destination host. The non-live migration is achieved by pausing, copy, and resuming a state of VM [19]. The log file exists on the VM which maintain all the execution record is used for the rollback process. The chances of Breach on VM may increase if the log file is accessed by the intruder during migration from one physical server to another physical server.

3.2.4 VM Scheduler-Based Attack

Scheduling is the process of creating the policies that regulate the order of task execution in a process. Different forms of scheduling algorithms exist for cloud computing. The vital characteristics of the scheduling algorithm comprise distribution and the planning of resources [20]. VM Scheduling is the heart of cloud computing resource management and a vital technique for increasing system performance [21]. A poor scheduling app may affect the performance of virtual machines [22]. Susceptibilities found in the scheduler may cause of stealing of resources. For instance, a VM might be booked to pursue a particular time while holding the credit adjust of the VM execution time slice.

4   Prevention of Virtualization Attacks

Virtualization may be exploited through the vulnerabilities found in the hypervisor or virtual machines. To provide a secure cloud environment, several techniques have been presented by the researcher.

4.1   Prevention of Host OS Attack

All the virtual instances in cloud computing execute on top of the host operating system. Any malicious act performed by the adversaries on the host OS may affect the VM's performance on the cloud [23]. Buffer is a temporary storage space in the memory for holding the data. Buffer overflow attack is a result of bad programming or an insecure coding approach. The term is used for a specific situation when the buffer crosses its limit for holding the reproach incoming data into the adjacent memory location. Journal et al. [24] served the several preventive techniques against the buffer overflow attack these include Address Space Layout Randomization (ASLR), Address Space Layout Permutation (ASLP), Address Space Randomization (ASR). More important preventive measure is to write flawless secure code, validate the user input, implement some checks on input such as a bound check. To mitigate the DoS/DDoS attack various researchers presented their techniques. In [25] presented various types of DDoS attacks on the cloud servers and the algorithm to mitigate such attacking techniques. The study presented various cloud providers who offer their services to tackle the DDoS attack and presented their algorithm to handle the attack.

4.2   Prevention of Guest OS Attack

The probability exists to breach the guest virtual machine by the hacker to control its host. To find the vulnerability in VMWare, which lets the attacker breach into the host operating system [26]. However, the vulnerability has been fixed by VMWare later. Hypervisor escape is a kind of vulnerability in which the administrative privilege is granted to the attacker over the hypervisor itself, which enables the attacker to execute haphazard code to access the private data. This happens when an aggressor escapes from a virtual machine to assault other virtual machines or the hypervisor itself. In [27] poses the universal idea of blind hypervisor architecture which applies the VMs privacy in case of any hypervisor exploitation. The dependency of the methodology is based on the extension of hardware and the software in terms of a dedicated memory management unit including a reliable module to shift the images of VM in memory which are unreachable to the hypervisor alleviating it of such tasks. The Blind hypervisor reduces the privilege of the hypervisor concerning memory access. Consequently, any malicious actor having hypervisor privilege cannot reach the data of VMs. Jiang Wu and his colleagues [28] propose a model for the prevention of virtual machine escape by implementing the Prevention Virtual machine escape model over the full virtualization architecture.

4.3   Prevention of Cross VM Side Channel

The VM-based side-channel is a kind of attack which collects the resource usage information, including cryptographic keys, from the victim VM running on the same physical machine. The modern hypervisor including Xen and VMWare implements logical isolation between virtual machines with the help of a sandboxing technique. This logical isolation using side channels can be exploited by the attacker [29]. The cross VM Side channel is categorized as cache-based shared memory and CPU load. The category is based on the hardware component exploited by the attacker because of highly rated interaction and sharing between processes. In [29] explain the flushing and injecting noise prevention techniques to prevent cache side-channel attack that exploits the CPU cache. Fangfei with his partners [30] proposed a defeating mechanism called CATalyst which compartmentalizes the Low-Level Cache (LLC) into hardware and software-managed cache. They remediate the side channel by running a Xen hypervisor and Linux operating system on the server with the Intel processor. They stick a pool of page outlines forever in the store, overseen by the VMM and the visitor OS. The shared memory in a virtual machine is also a threat in which the data can be theft by the administrator of the virtual machine at the cloud end as he has the maximum privileged in the cloud. To mitigate such an attack [31] presented the algorithm for dynamic algorithm placement where the move is made based on previously selected server policy.

4.4   Prevention of VM Creation Attack

To control the maximum, virtualize architecture, the malicious user compromises the VM machine by injecting the malicious code which causes the creation of an already compromised new VM controlled by the intruder. For the protection of the user cloud, the researchers suggest the introspection cloud model [32] which is based on the technology of cloud introspection. In their proposed model they deployed the component of the cloud probe to collect the information of the various state of the cloud including network communication, memory system, and disk storage. Afterward, the collected information has been synchronized on the independently deployed cloud introspection. By reconstructing and analyzing the data in the introspection cloud the model can enhance the ability to trace the security actions state of the user cloud. In [33] propose Intrusion Prevention System for the cloud platform. The system is based on Software Designed Network (SDN) and utilizing its programming features the system sends the malicious information to the controller on detection of any intrusion detected by the Intrusion Detection System. The controller passes the security parameters to the virtual switch that discards the packets causing intrusion by filtering them. In such a way blocking intrusion behavior is achieved dynamically.

4.5   Prevention of VM Migration Attack

The migration of virtual machines is considered a prevailing approach that empowers the data center operator with the capability to conform to the placement of virtual machines for the enhancement of performance, minimization of power consumption, load balancing, and fault tolerance. During the migration process of the virtual machine from one physical machine to another, the contents become the cause of numerous attacks. The executing log which is kept for the implementation of roll-back can be accessed during the process of migration. An effective configuration of security policy or resume and suspend action may secure the migration process. Chandrakala and his colleague in [34] improve the secure migration by designing the placement strategy of the VM to the other physical server. The policy permits the VM which is created in the data center to be located on the servers with matching policies. By the arrangement of such secured placement, the migration of gust virtual machine migrates before the successful attack. The algorithm is based on the survivability and Discrete-Time Markov Chain (DTMC). Ahmed M and his colleagues proposed [35] a secured model for live migration of VM. The model covers the requirements needed for secure migration. The model is comprised of various runtime monitoring agents that inform any malicious action to the control manager, the heart of the proposed system. The control manager can terminate any of the attacks.

4.6   Prevention of Scheduler-Based Attack

Vulnerabilities found in the scheduler may cause stealing the services in the cloud environment. A virtualized environment should prevent the secret channel to the usage of shared resources. a study was done by [36] which examines the effect of hypervisor scheduling on the feasibility of secure channel exploits in a virtual environment. The consequences of their study show that load balancing has the highest effect on the covert channel because it controls its accessibility. Automatic load balancing makes the VE considerably more secure in terms of Covert Channel Attacks. In [36] based on Xen hypervisor assessed empirically the influence of various scheduling related factors including Time slot, Running Time, Number of Virtual CPUs, and Physical CPUs, Workloads, Slice Length, Slice Arrangements on the usefulness of side-channel attacks. Their study shows all the factors impact more or less but the slice arrangement has a higher impact (Table 1).

Table 1 Attack and proposed securing methods for cloud virtualization

Attack surface	Attack type	Implication	Proposed mitigation techniques
Hypervisor based attacks	Host OS	An intruder can get control due to the vulnerabilities found in Guest OS	Need to fill the security holes in host OS [23–25]
	Guest OS	An attacker can breach the physical resources by using collected information	Access control, intrusion detection, updating the system [26–28]
Virtual machine based attacks	VM side channel	Resource Information like CPU, Memory, Cache may be gathered by the attacker	The development of a Strong mechanism requires that protects the information leakage [29–33]
	VM creation	A malicious VM can be created by injecting malicious code into the target VM	Intrusion prevention system may detect injection of malicious code and prevent harm [32, 33]
	VM migration	The log file containing all information of VM may be captured during VM migration	Controlled management plane to place VM migration [34, 35]
	VM scheduler	VM might be booked to pursue a particular time while holding the credit adjust of the VM execution time slice	A strong scheduling policy prevents various attacks on scheduling attacks [36, 39]

5 Conclusion

Cloud computing is a state-of-the-art computing technology that compels the organization to adopt cutting-edge technology cost-effectively. It provides the opportunity for the organizations to concentrate on their business instead of spending a lot, on maintaining computing resources to run their business. Virtualization is the backbone of cloud technology for hardware resources abstraction. The virtualization layer can be compromised by the attacker to take control of cloud infrastructure. This paper presents a basic understanding of the virtualization architecture and the vulnerabilities in hypervisor software. We explore various kinds of attacks on the virtualized architecture and categorize them into hypervisor-based and virtual machine-based attacks. To protect the cloud virtualized environment various mitigating techniques have also been presented to cope with such types of attacks by digging into the literature.

References

1. Ruan K, Carthy J, Kechadi T, Crosbie M (2011) Cloud forensics, pp 35–46. doi: https://doi.org/10.1007/978-3-642-24212-0_3
2. Mell P, Grance T (2011) The NIST definition of cloud computing recommendations of the national institute of standards and technology. Natl Inst Stand Technol Inf Technol Lab 145:7. https://doi.org/10.1136/emj.2010.096966
3. Khorshed T, Ali ABMS, Wasimi SA (2012) A survey on gaps, threat remediation challenges and some thoughts for proactive attack detection in cloud computing. Futur Gener Comput Syst 28(6):833–851. https://doi.org/10.1016/j.future.2012.01.006
4. Jain R. Virtualization security in data centers and clouds, pp 1–12
5. Malhotra L, Agarwal D, Jaiswal A (2014) Information technology & software engineering virtualization in cloud computing, vol 4, no 2, pp 2–4. https://doi.org/10.4172/2165-7866.1000136
6. Almutairy NM, Al-Shqeerat KHA, Al Hamad HA (2019) A taxonomy of virtualization security issues in cloud computing environments. Indian J Sci Technol 12(3):1–19. https://doi.org/10.17485/ijst/2019/v12i3/139557
7. Shukur H, Zeebaree S, Zebari R, Zeebaree D, Ahmed O, Salih A (2020) Cloud computing virtualization of resources allocation for distributed systems. J Appl Sci Technol Trends 1(3):98–105. https://doi.org/10.38094/jastt1331
8. Saleem M (2017) Cloud computing virtualization. Int J Comput Appl Technol Res 6(7):290–292. https://doi.org/10.7753/ijcatr0607.1004
9. Zhu G, Yin Y, Cai R, Li K (2017) Detecting virtualization specific vulnerabilities in cloud computing environment. In: IEEE international conference cloud computing CLOUD, vol 2017, June 2017, pp 743–748. https://doi.org/10.1109/CLOUD.2017.105
10. Tank D, Aggarwal A, Chaubey N (2019) Virtualization vulnerabilities, security issues, and solutions: a critical study and comparison. Int J Inf Technol. https://doi.org/10.1007/s41870-019-00294-x
11. Daneshkhah A, Hosseinian Far A, Carroll F, Montasari R, Macdonald S (2021) Network and hypervisor-based attacks in cloud computing environments. Int J Electron Secur Digit Forensics 1(1):1. https://doi.org/10.1504/ijesdf.2021.10036493
12. Aalam Z, Kumar V, Gour S (1950) A review paper on hypervisor and virtual machine security. J Phys Conf Ser 1:2021. https://doi.org/10.1088/1742-6596/1950/1/012027
13. Xiao J, Lu L, Huang H, Wang H (2018) Virtual machine extrospection: a reverse information retrieval in cloud. IEEE Trans Cloud Comput 9:401–403. https://doi.org/10.1109/TCC.2018.2855143
14. Kortchinsky K. Immunity, Inc. 05/26/10 1
15. Elhage N (2011) Virtunoid: breaking out of KVM KVM: architecture overview attack surface bypassing ASLR
16. Ren XY, Zhou YQ (2016) A review of virtual machine attack based on Xen. In: MATEC Web of Conferences, vol 61, p 03003. EDP Sciences
17. Narayana KE, Jayashree K (2020) Survey on cross virtual machine side channel attack detection and properties of cloud computing as sustainable material. Mater Today Proc 45:6465–6470. https://doi.org/10.1016/j.matpr.2020.11.283
18. Ahmad RW, Gani A, Hamid SHA, Shiraz M, Yousafzai A, Xia F (2015) A survey on virtual machine migration and server consolidation techniques for cloud data centers. J Netw Comput Appl 52:11–25. https://doi.org/10.1016/j.jnca.2015.02.002
19. Malhotra D (2018) A critical survey of virtual machine migration techniques in cloud computing. In: 2018 first international conference on secure cyber computing and communication (ICSCCC), pp 328–332. IEEE
20. Tohidirad Y, Abdezadeh S, Aliabadi ZS, Azizi A, Moradi M (2015) Virtual machine scheduling in cloud computing environment. Int J Manag Pub Sect Inf Commun Technol 6(4):1–6. https://doi.org/10.5121/ijmpict.2015.6401

21. Liu L, Qiu Z (2017) A survey on virtual machine scheduling in cloud computing. In: 2016 2nd IEEE International Conference Computing Communication, ICCC 2016—Proceedings, pp 2717–2721. https://doi.org/10.1109/CompComm.2016.7925192
22. Supreeth S, Patil KK (2019) Virtual machine scheduling strategies in cloud computing—a review. Int J Emerg Technol 10(3):181–188
23. Kumar V, Rathore RS (2018) Security issues with virtualization in cloud computing. In: Proceedings—2018 international conference on advances in computing, communication control and networking, ICACCCN 2018, pp 487–491. https://doi.org/10.1109/ICACCCN. 2018.8748405
24. Cowan C, Wagle F, Pu C, Beattie S, Walpole J (2000). Buffer overflows: attacks and defenses for the vulnerability of the decade. In: Proceedings DARPA information survivability conference and exposition, DISCEX'00. IEEE, vol 2, pp 119–129
25. Sahu SK, Khare DRK (2020) DDOS attacks & mitigation techniques in cloud computing environments. Gedrag Organ Rev 33(2). https://doi.org/10.37896/gor33.02/246
26. Higgins KJ (2009) Hacking tool lets a vm break out and attack its host. Darkreading. [Online], 4. https://www.darkreading.com/risk/hacking-tool-lets-a-vm-break-out-and-attack-its-host
27. Dubrulle P, Sirdey R, Doré P, Aichouch M, Ohayon E (2015) Blind hypervision to protect virtual machine privacy against hypervisor escape vulnerabilities. In: Proceeding—2015 IEEE international conference on industrial informatics, INDIN 2015, pp 1394–1399. https://doi.org/10.1109/INDIN.2015.7281938
28. Wu J, Lei Z, Chen S, Shen W (2017) An access control model for preventing virtual machine escape attack. Futur Internet 9(2). https://doi.org/10.3390/fi9020020
29. Khan MA (2016) A survey of security issues for cloud computing. J Netw Comput Appl 71:11–29. https://doi.org/10.1016/j.jnca.2016.05.010
30. Liu F, Ge Q, Yarom Y, Mckeen F, Rozas C, Heiser G, Lee RB (2016) Catalyst: defeating last-level cache side channel attacks in cloud computing. In: 2016 IEEE international symposium on high performance computer architecture (HPCA), pp 406–418. IEEE
31. Maheswara Reddy Gali A, Koduganti VR (2021) Dynamic and scalable virtual machine placement algorithm for mitigating side channel attacks in cloud computing. Mater Today Proc. https://doi.org/10.1016/j.matpr.2020.12.1136
32. Zhang J, Zheng L, Gong L, Gu Z (2018) A survey on security of cloud environment: threats, solutions, and innovation. In: 2018 IEEE third international conference on data science in cyberspace, pp 910–916. https://doi.org/10.1109/DSC.2018.00145
33. Chi Y (2017) Design and implementation of cloud platform intrusion prevention system based on SDN, pp 847–852
34. Chandrakala N, Rao BT (2018) Migration of virtual machine to improve the security in cloud computing, vol 8, no 1, pp 210–219. https://doi.org/10.11591/ijece.v8i1.pp210-219
35. Mahfouz AM, Rahman L, Shiva SG. Secure live virtual machine migration through runtime monitors
36. Vateva-Gurova T, Suri N, Mendelson A (2015) The impact of hypervisor scheduling on compromising virtualized environments. In: Proceedings—15th IEEE International Conference on Computing Information Technology, CIT 2015, 14th IEEE International Conference Ubiquitous Computing Communication, IUCC 2015, IEEE international conference on automated software engineering, pp 1910–1917. https://doi.org/10.1109/CIT/IUCC/DASC/PICOM.201 5.283
37. Modi C, Patel D, Borisaniya B, Patel H (2013) A survey of intrusion detection techniques in Cloud. J Netw Comput Appl 36(1):42–57. https://doi.org/10.1016/j.jnca.2012.05.003
38. Tadokoro H, Kourai K, Chiba S (2012) Preventing information leakage from virtual machines. Memory IaaS Clouds 7:1421–1431
39. Zhou F, Desnoyers P (2011) Scheduler vulnerabilities and coordinated attacks in cloud computing. https://doi.org/10.1109/NCA.2011.24

Advances in Smart Emerging Technologies

Short Survey on Using Blockchain Technology in Modern Wireless Networks, IoT and Smart Grids

Moez Krichen⑩, Meryem Ammi⑩, Alaeddine Mihoub⑩, and Qasem Abu Al-Haija⑩

Abstract Blockchain is a cutting-edge technology that has changed the way people communicate and trade. It's a chain of blocks in a distributed and decentralized peer-to-peer (P2P) network that stores information with digital signatures. This method was first used to develop digital currencies such as bitcoin and Ethereum. However, some recent research and industrial studies have focused on the prospects that blockchain presents in a variety of other application fields in order to take advantage of the technology's fundamental qualities, such as decentralization, persistency, anonymity, and audibility. In this study, We give a thorough evaluation of blockchain's use in Wireless Networks, the Internet of Things (IoT), and Smart Grids (SGs). We also present the main obstacles of Blockchain in order to allow researchers to solve them and improve the technology's use.

Keywords Blockchain · Wireless Networks (WNs) · Internet of Things (IoT) · Smart Grids (SGs)

M. Krichen (✉)
FCSIT, Al-Baha University, Al-Baha, Saudi Arabia
e-mail: mkreishan@bu.edu.sa

ReDCAD Laboratory, University of Sfax, Sfax, Tunisia

M. Ammi
Naif Arab University for Security Sciences, Riyadh, Saudi Arabia
e-mail: moez.krichen@redcad.org

A. Mihoub
Department of Management Information Systems and Production Management, College of Business and Economics, Qassim University, P.O. Box: 6640, Buraidah 51452, Saudi Arabia
e-mail: mammi@nauss.edu.sa

Q. A. Al-Haija
Department of Computer Science/Cybersecurity, Princess Sumaya University for Technology (PSUT), Amman 11941, Jordan
e-mail: q.abualhaija@psut.edu.jo

© The Author(s), under exclusive license to Springer Nature Switzerland AG 2023
A. A. Abd El-Latif et al. (eds.), *Advances in Cybersecurity, Cybercrimes, and Smart Emerging Technologies*, Engineering Cyber-Physical Systems and Critical Infrastructures 4,
https://doi.org/10.1007/978-3-031-21101-0_13

1   Introduction

Blockchain is a breakthrough paradigm that has introduced new notions for securely sharing data and information. This current technology comprises of a chain of blocks that allows for the secure storage of all committed transactions over shared networks [1]. Cryptographic hashes, distributed consensus techniques, and digital signatures are among the basic technologies used to achieve this goal. All transactions are decentralized, eliminating the need for any intermediaries to authenticate or verify them. Decentralization, persistence, anonymity, and auditability are some of the key characteristics of blockchain [2] (Table 1).

One revolutionary paradigm that has brought new concepts on securely sharing data and information is Blockchain. This modern technology consists of a chain of blocks that allow storing all committed transactions using shared securely, and distributed networks [1]. Several basic technologies are adopted to fulfill this goal, like cryptographic hash, distributed consensus algorithms, and digital signatures. All transactions are carried out in a decentralized way, removing the need for any mediators to confirm and verify them. Blockchain has some key characteristics [2] such as: Decentralization, Persistency, Anonymity, and Auditability (Table 1).

Our major contributions of this work concentrate on the recent research activities related to including blockchain technology in modern applications by emphasizing the corresponding challenges, advantages, and limitations, and challenges. In particular, we review the blockchain architecture in Sect. 2 and then we present a short review on the use of blockchain in Wireless Networks (Sect. 3); Internet of Things (IoT) (Sect. 4); Smart Grids (Sect. 5). Section 6 lists the main open challenges related to the use of Blockchain technology. Section 7 is a general conclusion of the paper.

Table 1   Blockchain key characteristics

Decentralization	In blockchain a transection can be performed between any two nodes without central authentication. As a result, the use of blockchain can dramatically cut server expenses while also alleviating performance constraints at the central server
Persistency	It is nearly impossible to tamper with the system because each transection must be validated and recorded in blocks depressed across the whole network
Anonymity	Each user can communicate with the blockchain network with a created address. Furthermore, a user could generate many addresses to protect his identity
Auditability	Users can easily check and trace prior records by accessing any node in the distributed network because each transection is confirmed and stored with a timestamp

2 Blockchain Architecture

A blockchain is a continuously expanding collection of data blocks linked together to form a long chain [3] as described in Fig. 1. This network of connected data blocks represents a distributed ledger that is disseminated over a peer-to-peer network [4]. A distributed ledger can be seen as a collection of digital data synced, replicated, distributed, and shared through a peer-to-peer network. Each device linked to the network maintains the latest version of the common ledger, i.e., each peer in the network has a copy of the ledger that is identical to the other.

The ledger is mainly characterized by its safety, and the database can be expanded only by adding new blocks to the chain. Changes to records already registered to the chain are computationally impossible [5]. As a result, a primary benefit of the described distributed ledger is its decentralized nature. Indeed, there is no central authority that controls the ledger. However, each node updates its ledger when a new block is added to the blockchain, using a joint consensus mechanism [6].

Moreover, in the blockchain, data security is enhanced by its encryption using asymmetric algorithms [7]. In this sort of encryption, both the transmitter and receiver have a pair of keys consisting of a public key and a private one [8]. The private key is exclusively accessible to the nodes that created it, whereas the public key is freely spread throughout the network. The sender encrypts the data using the receiver's public key. Since data is encrypted using the receiver's public key, it can only be decrypted using the receiver's private key.

Furthermore, in the case of sending transactions on a blockchain network, a transaction is deemed complete only after it is digitally signed. The transaction is signed by the sender using his private key. For the receiver, the transaction's authenticity, i.e., the sender identity, can be checked using the associated public key (belonging to the sender). All transactions are automatically checked and authenticated by nodes, and the network rejects any unauthenticated transactions. Please note that on a blockchain network, an original, mined transaction is irreversible [9].

It is difficult to alter the data contained in blocks thanks to the cryptographic qualities of the Blockchain. Practically, the blocks are connected via a Hash reference since each subsequent block carries the previous block's hash value in addition to the actual block's hash value (Fig. 1). Generating a hash value is feasible using a mathematical and sophisticated cryptographic hash algorithm, which accepts any input type and outputs a fixed-length number termed the hash value. The primary characteristic of a hashing function is that if a single fraction in the input is changed, the entire value in the output will be altered [10].

Consequently, if an attacker attempts to edit data in Block 1 (B1), for instance, the hash value of that block (B1) will be modified in the following block (B2), and so the intruder will have to modify the hash value of that block. Moreover, because B2 carries the hash of B1, any modification in the hash will alter the hash value of B2 in B3. As a result, if someone wants to modify a block, he or she must modify the data for all subsequent blocks on the Blockchain. Additionally, even if the hash

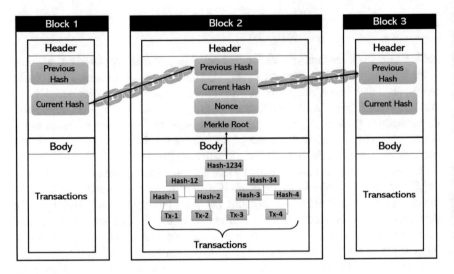

Fig. 1 Blockchain general architecture

value of a block is known, calculating the hash function's input is difficult due to the hash function's non-invertible feature [8].

3   Blockchain for Wireless Networks

Wireless applications such as broadband internet connections, mobile smartphones, and internet of vehicles [11, 12] all require radio spectrum [13]. Wireless networks, like cellular and Wi-Fi, are the most cost-effective way to provide broadband internet access. As a result, diverse spectrum management regimes are needed to optimize advantages from the utilization of available spectrum by mandating efficient spectrum usage while minimizing interference between consumers [14]. The traditional spectrum management regime has two major drawbacks. First, large portions of the licensed spectrum are underused. Second, this command-and-control spectrum management regime is slow to respond to market changes [15]. Spectrum sensing [16], supporting secondary spectrum trading marketplaces [17], spectrum sharing [18], and policy enforcement [19] are all possible uses for the blockchain technology in spectrum management [20].

Spectrum sensing and geo-location databases are the two leading technologies for providing dynamic spectrum access. Previously, these approaches were viewed as separate strategies in previous research. Because blockchain is a database technology, it may be used to create a unified method in which spectrum sensing techniques and geo-location database technology work in tandem. A more robust dynamic spectrum management framework will arise from combining these two spectrum access strategies. It is also necessary to investigate the integration of blockchains with the communication networks. The blockchain network could be set up as an overlay on top of the communication network, allowing communication network nodes to operate as complete nodes on the blockchain network. However, this network structure is energy-intensive and necessitates a specialized control channel for transferring blocks and transactions over blockchain networks [21].

4 Blockchain for Internet of Things

The Internet of Things (IoT) [22, 23] is the linking of smart devices for data collection and intelligent decision-making. Yet, IoT is prone to privacy and security risks due to the absence of inherent security measures. The dispersed and centralized architecture of the Internet of Things is a significant challenge [24]. Every node in an IoT infrastructure is typically a potential point of weakness that could be used to start cyber assaults. Data confidentiality and authentication are continuous and possibly one of the most serious threats. IoT data could be hacked and misused if data security is not established [25]. Data integrity is another issue for IoT. Decision support systems are one of the most important IoT applications. As a result, protecting the system from injection attacks, which attempt to insert bogus measures and so impact decision-making, is critical. For automated systems such as manufacturing sectors and vehicular networks [26] that handle real-time data, availability is crucial. Including a publicly verifiable audit trail that is not reliant on a trusted third party is essential, as it addresses all of these issues. Blockchain may assist solve major security concerns in IoT with its "security by construction" feature [27].

Blockchain is the final puzzle for resolving IoT privacy and dependability issues. The blockchain's inherent trustless, autonomous and decentralized characteristics make it suitable for various scenarios. Blockchain technology, for example, may store a permanent record of smart gadgets [28]. Furthermore, smart contracts may allow smart devices to perform autonomously, avoiding the need for human control or centralized authority. In addition, blockchain can establish a secure means for smart devices to communicate with one another [29].

The contribution in [30] can be viewed as a generic solution that can be used in any field of the IoT environment. Indeed, the authors of this paper developed a mechanism that would allow sensors to trade Bitcoins for data. Every node has a unique address that corresponds to the Bitcoin pub-key. When a user needs data from a sensor after locating it in a sensor repository, he sends a transaction directed to that sensor's public key. The sensor will reply by sending a transaction containing data to the client. This strategy is an extension of the solution provided in [31].

To summarize, the usage of blockchain for IoT applications provides excellent levels of security that prevent unwanted data access. Yet, scalability [32] is still an open question since the blockchain can grow in size over time, making it difficult to acquire and save the ledger.

5 Blockchain for Smart Grids

A smart grid [33] is a digital communications-based electrical network that provides the two-way flow of electricity and data in addition to the identification, reaction, and avoidance of changes in usage and other difficulties. Current smart grids integrate communication and control techniques into power networks, allowing considerable gains in energy efficiency and system safety.

Traditional centralized techniques of managing smart grids pose significant hurdles [34]. For instance, the centralized control method creates a dangerous single point of failure for the whole grid. In addition, many security issues have been growing, and external security assaults could result in significant financial losses. To overcome these limitations, the use of blockchain technologies is considered a good choice in several research and industrial projects [35]. Indeed the use of blockchain for smart grids may have the following advantages:

- The blockchain can potentially turn centralized grid administration into distributed intelligent administration.
- In terms of energy trading, a smart grid with blockchain technology can achieve optimum data flow and cash flow.
- Because of its decentralization and fault tolerance, blockchain can dramatically improve the privacy and security of power grids to improve risk management [36].

Electric vehicles can be thought of as mobile power grid terminals that perform key services. This is known as V2G technology, and it has the potential to increase the power grid's reliability, efficiency, and stability. On the other hand, electric vehicles are not properly linked with smart grids, and there are several issues such as energy shortages, security hazards, and data leakage. In this context, excessive charging load and unsteady voltage in electric vehicles can be addressed with blockchain technology as shown in [37]. In addition, the use of blockchain for connecting smart grids and electric vehicles can realize cost optimization through smart contracts. Furthermore, using blockchain technology to connect smart grids and electric vehicles might reduce costs using smart contracts as proposed in [38].

Although blockchain technology for smart grids appears to be promising, there are still challenges to fully converting to this new technology, as previously proved. Implementing blockchain in the smart grid, for example, entails huge infrastructural costs for re-architecting existing grid networks, making grid operators unwilling to include blockchain into their grid structure.

6 Open Challenges

In order to produce more useable and successful blockchain-based applications, several unsolved industrial difficulties [39–44] must be tackled and investigated further (Table 2).

Table 2 Open challenges

Solution benefits	When applied to replace existing solutions, blockchain is a new technology that has the potential to destabilize the market by introducing revolutionary ways that may transform the society. As a result, it is critical to establish whether a blockchain is truly required for a given application [39]
Proper implementation	Because blockchain is a general-purpose method to data manipulation that may be used in various reasons, its implementation has some degree of comprehension of maturity regarding its importance and the tradeoffs. Hence, its incorporation in different applications requires as in-depth and comprehensive study [40]
Standard testing mechanism	It is essential to adopt blockchain-based applications for diverse domains
Resilience to security risks	The resilience to security risks needs to be formally proved since blockchain may face malfunctions due to systems design or cyberattacks intended to compromise its security with large-scale applications
scalability	This issue is raised because blockchain-based transactions are very slow to be processed and verified. Processing the transections depend on the performance of the processing system. In [41], limitations of the proposed scaling methods are pointed out
Integration with other systems	This issue directly impacts organisations willing to adopt blockchain-enabled solutions. Indeed, the integration process will imply costs related to infrastructure change, trained staff, specialized developers, and management expectation [41]
Energy challenges	The use of blockchain will undoubtedly require an energy consumption much higher than the usual one. This challenge turns to an environmental issue when the energy used exceeds the load power, and the equipment is fully utilized [42]
Regulatory issues	Regulations are extremely important to generalize, regulate and accept the use of blockchain-enabled solutions
Storage	The integration of blockchain with data-intensive applications such as those based on IoT-raises the problem of data storage. Indeed, blockchain stores data into blocks that cannot support a large volume of date [43, 44]

Table 3 Summary about the main findings concerning the use of Blockchain in different fields

Wireless networks	[11, 12, 14–21]	1. Increasing spectrum access and utilization efficiency 2. Creating a secure spectrum sensing system 3. Improving the accuracy of spectrum sensing data 4. storing unoccupied spectrum bands and users geolocations 5. providing dynamic spectrum access 6. enabling collaborative sensing	1. Energy-intensive 2. Necessitates a specialized control channel for transferring blocks and transections over blockchain networks
Internet of Things (IoT)	[25–31]	1. Storing and Processing date at the Same Time while Maintaining Privacy 2. Establishing a Secure Means for Smart Devices to Communicate With one Another 3. Allowing Smart Devices to Perform Autonomously 4. Avoiding the Need for Human Control or Centralized Authority	1. Scalability still an open question since the blockchain can grow in size over time, making it difficult to acquire and save the ledger
Smart grids	[35, 37, 38]	1. Distributed Intelligent Administration 2. Improve Privacy and Security 3. Optimum Dataflow and Cash Flow	1. Large Infrastructural expenses needed

7   Conclusion

This research aimed to shed light on some recent studies involving blockchain technology integration in three important and modern domains: wireless networks, the Internet of Things (IoT), and smart grids. We offered some related blockchain technology examples for each domain, emphasizing on the benefits, limitations, and issues associated with each. Table 3 summarizes the solutions that have been reviewed.

For example, it would be interesting to investigate how Machine Learning (ML) Techniques [45–47] may be used in the context of Blockchain Technology for increasing security levels and for increasing the performance of the considered blockchain-based systems. It will be also important to apply some formal testing approaches for blockchain-based systems for improving their quality and increasing their robustness [48–50].

References

1. Salah K, Rehman MHU, Nizamuddin N, Al-Fuqaha A (2019) Blockchain for ai: review and open research challenges. IEEE Access 7:10127–10149
2. Kouhizadeh M, Sarkis J (2018) Blockchain practices, potentials, and perspectives in greening supply chains. Sustainability 10(10):3652
3. Srivastava G, Dhar S, Dwivedi AD, Crichigno J (2019) Blockchain education. In: 2019 IEEE Canadian conference of electrical and computer engineering (CCECE). IEEE, pp 1–5
4. Lemieux VL (2016) Trusting records: is blockchain technology the answer? Records Manag J
5. Abu Al-Haija Q, Alsulami AA (2021) High performance classification model to identify ransomware payments for heterogeneous bitcoin networks. Electronics
6. Johar S, Ahmad N, Asher W, Cruickshank H, Durrani A (2021) Research and applied perspective to blockchain technology: a comprehensive survey. Appl Sci 11(14):6252
7. Li X, Jiang P, Chen T, Luo X, Wen Q (2020) A survey on the security of blockchain systems. Futur Gener Comput Syst 107:841–853
8. Yan B, Yang Z, Ren Y, Tan X, Liu E (2017) Microblog sentiment classification using parallel svm in apache spark. In: 2017 IEEE international congress on big data (BigData congress). IEEE, pp 282–288
9. Nakamoto S (2008) Bitcoin: a peer-to-peer electronic cash system. Decentral Busin Rev 21260
10. Zheng Z, Xie S, Dai H-N, Chen X, Wang H (2018) Blockchain challenges and opportunities: a survey. Int J Web Grid Serv 14(4):352–375
11. Jabbar R, Dhib E, Ben Said A, Krichen M, Fetais N, Zaidan E, Barkaoui K (2022) Blockchain technology for intelligent transportation systems: a systematic literature review. IEEE Access
12. Jabbar R, Fetais N, Kharbeche M, Krichen M, Barkaoui K, Shinoy M (2021) Blockchain for the internet of vehicles: how to use blockchain to secure vehicle-to-everything (v2x) communication and payment? IEEE Sens J
13. Leyton-Brown K, Milgrom P, Segal I (2017) Economics and computer science of a radio spectrum reallocation. Proc Natl Acad Sci 114(28):7202–7209
14. Weiss MB, Werbach K, Sicker DC, Bastidas CEC (2019) On the application of blockchains to spectrum management. IEEE Transactions on Cognitive Communications and Networking 5(2):193–205
15. Anker P (2017) From spectrum management to spectrum governance. Telecommunications Policy 41(5–6):486–497
16. Ariyarathna T, Harankahadeniya P, Isthikar S, Pathirana N, Bandara HD, Madanayake A (2019) Dynamic spectrum access via smart contracts on blockchain. In: 2019 IEEE wireless communications and networking conference (WCNC). IEEE, pp 1–6
17. Qiu J, Grace D, Ding G, Yao J, Wu Q (2019) Blockchain-based secure spectrum trading for unmanned-aerial-vehicle-assisted cellular networks: An operator's perspective. IEEE Internet Things J 7(1):451–466
18. Han S, Zhu X (2019) Blockchain based spectrum sharing algorithm. In: 2019 IEEE 19th international conference on communication technology (ICCT). IEEE, pp 936–940
19. Careem MAA, Dutta A (2019) Sensechain: blockchain based reputation system for distributed spectrum enforcement. In: 2019 IEEE international symposium on dynamic spectrum access networks (DySPAN). IEEE, pp 1–10

20. Pei Y, Hu S, Zhong F, Niyato D, Liang Y-C (2019) Blockchain-enabled dynamic spectrum access: cooperative spectrum sensing, access and mining. In: 2019 IEEE global communications conference (GLOBECOM). IEEE, pp 1–6
21. Liang Y-C (2020) Dynamic spectrum management: from cognitive radio to blockchain and artificial intelligence. Springer Nature
22. Krichen M, Alroobaea R (2019) A new model-based framework for testing security of iot systems in smart cities using attack trees and price timed automata. In: 14th international conference on evaluation of novel approaches to software engineering - ENASE 2019
23. Jabbar R, Shinoy M, Kharbeche M, Al-Khalifa K, Krichen M, Barkaoui K (2019) Urban traffic monitoring and modeling system: an iot solution for enhancing road safety. In: 2019 international conference on internet of things, embedded systems and communications (IINTEC). IEEE, pp 13–18
24. Abu Al-Haija Q, Al-Badawi A (2022) Attack-aware iot network traffic routing leveraging ensemble learning. Sensors
25. Krichen M, Lahami M, Cheikhrouhou O, Alroobaea R, Maâlej AJ (2020) Security testing of internet of things for smart city applications: a formal approach. Smart infrastructure and applications. Springer, Cham, pp 629–653
26. Jabbar R, Kharbeche M, Al-Khalifa K, Krichen M, Barkaoui K (2020) Blockchain for the internet of vehicles: A decentralized iot solution for vehicles communication using ethereum. Sensors 20(14):3928
27. Li D, Deng L, Cai Z, Souri A (2020) Blockchain as a service models in the internet of things management: systematic review. Trans Emerg Telecommun Technol e4139
28. Kshetri N (2017) Can blockchain strengthen the internet of things? IT Profess 19(4):68–72
29. Suliman A, Husain Z, Abououf M, Alblooshi M, Salah K (2019) Monetization of iot data using smart contracts. IET Networks 8(1):32–37
30. Wörner D, von Bomhard T (2014) When your sensor earns money: exchanging data for cash with bitcoin. In: Proceedings of the 2014 ACM international joint conference on pervasive and ubiquitous computing: adjunct publication, pp 295–298
31. Zhang Y, Wen J (2015) An iot electric business model based on the protocol of bitcoin. In: 2015 18th international conference on intelligence in next generation networks. IEEE, pp 184–191
32. Maâlej AJ, Krichen M (2016) A model based approach to combine load and functional tests for service oriented architectures. In: VECoS, pp 123–140
33. Fang X, Misra S, Xue G, Yang D (2011) Smart grid-the new and improved power grid: A survey. IEEE Commun Surv Tutor 14(4):944–980
34. Smadi AA, Tobi Ajao B, Johnson BK, Lei H, Chakhchoukh Y, Abu Al-Haija Q (2021) A comprehensive survey on cyber-physical smart grid testbed architectures: Requirements and challenges. Electronics
35. Andoni M, Robu V, Flynn D, Abram S, Geach D, Jenkins D, McCallum P, Peacock A (2019) Blockchain technology in the energy sector: a systematic review of challenges and opportunities. Renew Sustain Energy Rev 100:143–174
36. Abu Al-Haija Q, Smadi AA, Allehyani MF (2021) Meticulously intelligent identification system for smart grid network stability to optimize risk management. Energies
37. Li Y, Hu B (2020) A consortium blockchain-enabled secure and privacy-preserving optimized charging and discharging trading scheme for electric vehicles. IEEE Trans Industr Inf 17(3):1968–1977
38. Liu C, Chai KK, Zhang X, Lau ET, Chen Y (2018) Adaptive blockchain-based electric vehicle participation scheme in smart grid platform. IEEE Access 6:25657–25665
39. Wüst K, Gervais A (2018) Do you need a blockchain? In: 2018 crypto valley conference on blockchain technology (CVCBT). IEEE, pp 45–54
40. Monrat AA, Schelén O, Andersson K (2019) A survey of blockchain from the perspectives of applications, challenges, and opportunities. IEEE Access
41. Islam MR, Rahman MM, Mahmud M, Rahman MA, Mohamad MHS, et al. (2021) A review on blockchain security issues and challenges. In: 2021 IEEE 12th control and system graduate research colloquium. IEEE, pp 227–232

42. Dwivedi SK, Roy P, Karda C, Agrawal S, Amin R (2021) Blockchain-based internet of things and industrial iot: a comprehensive survey. Secur Commun Netw 2021
43. Zaabar B, Cheikhrouhou O, Jamil F, Ammi M, Abid M (2021) Healthblock: a secure blockchain-based healthcare data management system. Comput Netw 200:108500
44. Zaabar B, Cheikhrouhou O, Ammi M, Awad AI, Abid M (2021) Secure and privacy-aware blockchain-based remote patient monitoring system for internet of healthcare things. In: 2021 17th international conference on wireless and mobile computing, networking and communications (WiMob). IEEE, pp 200–205
45. Abu Al-Haija Q, Krichen M, Abu Elhaija W (2022) Machine-learning-based darknet traffic detection system for iot applications. Electronics 11(4):556
46. Mihoub A, Fredj OB, Cheikhrouhou O, Derhab A, Krichen M (2022) Denial of service attack detection and mitigation for internet of things using looking-back-enabled machine learning techniques. Comput Electr Eng 98:107716
47. Ben Fredj O, Mihoub A, Krichen M, Cheikhrouhou O, Derhab A (2020) Cybersecurity attack prediction: a deep learning approach. In: 13th international conference on security of information and networks, pp 1–6
48. Lahami M, Krichen M (2021) A survey on runtime testing of dynamically adaptable and distributed systems. Software Qual J 29(2):555–593
49. Lahami M, Krichen M, Jmaïel M (2015) Runtime testing approach of structural adaptations for dynamic and distributed systems. Int J Comput Appl Technol 51(4):259–272
50. Lahami M, Krichen M, Barhoumi H, Jmaiel M (2015) Selective test generation approach for testing dynamic behavioral adaptations. IFIP international conference on testing software and systems. Springer, Cham, pp 224–239

Predicting Sleeping Quality Using Convolutional Neural Networks

Vidya Rohini Konanur Sathish, Wai Lok Woo, and Edmond S. L. Ho

Abstract Identifying sleep stages and patterns is an essential part of diagnosing and treating sleep disorders. With the advancement of smart technologies, sensor data related to sleeping patterns can be captured easily. In this paper, we propose a Convolution Neural Network (CNN) architecture that improves the classification performance. In particular, we benchmark the classification performance from different methods, including traditional machine learning methods such as Logistic Regression (LR), Decision Trees (DT), k-Nearest Neighbour (k-NN), Naïve Bayes (NB) and Support Vector Machine (SVM), on 3 publicly available sleep datasets. The accuracy, sensitivity, specificity, precision, recall, and F-score are reported and will serve as a baseline to simulate the research in this direction in the future.

Keywords Machine learning · Convolutional neural networks · Sleep stage classification · Deep learning · Machine learning classifiers

1 Introduction

The collection of data by numerous healthcare sensors assisted the Artificial Intelligence (AI) systems in predicting and analysing various types of health-related issues [1]. Such input data can be used as input for Machine Learning (ML) algorithms to further analyze the patients' health status automatically [2]. Deep Learning (DL) is a trending expansion of the classical neural network method where a greater number

V. R. K. Sathish · W. L. Woo · E. S. L. Ho (✉)
Department of Computer and Information Sciences, Northumbria University, 2 Ellison Place, Newcastle Upon Tyne NE1 8ST, United Kingdom
e-mail: e.ho@northumbria.ac.uk

V. R. K. Sathish
e-mail: vidya.sathish@northumbria.ac.uk

W. L. Woo
e-mail: wailok.woo@northumbria.ac.uk

© The Author(s), under exclusive license to Springer Nature Switzerland AG 2023
A. A. Abd El-Latif et al. (eds.), *Advances in Cybersecurity, Cybercrimes, and Smart Emerging Technologies*, Engineering Cyber-Physical Systems and Critical Infrastructures 4, https://doi.org/10.1007/978-3-031-21101-0_14

of complex non-linear data patterns can be explored. Another side of the popularity of DL is that complex operations and computations on data can be executed easily [3]. Deep learning algorithms like multi-layer perceptron, Recurrent Neural Network (RNN) and Convolution Neural Network (CNN) are applied successfully in various domains to solve challenging tasks. The elementary calculation unit in neural networks is a perceptron that achieves a linear combination of input features accompanied by a nonlinear transformation [4].

Recently, the understanding of health and well-being has improved in the society including quality of sleep, eating habits and physical activities. It is important to understand the relationship between health and sleep quality as they are heavily linked to each other. Sleep is an essential physiological activity of the human body and sleep quality will affect significantly our health [5]. The reduced sleep causes numerous sleep disorders such as insomnia, narcolepsy, sleep apnea, etc., affecting the overall health [6]. Sleep disorder can be diagnosed using Polysomnography (PSG) which records the Electroencephalography (EEG) signals at various locations over the head, electromyography (EMG), electrooculography (EOG) signals and many more. There are numerous times series data that are recorded over the night and every 30-second time segment will be allocated to a human sleep expert for a sleep stage analysis using reference nomenclature like those suggested by the American Academy of Sleep Medicine (AASM) [7]. Sleep apnea is a decreased or complete disturbance of breathing for a minimum of 10 seconds and it is a regularly observed issue among sleep disorders. Sleep apnea can be categorised into three types: obstructive, central and mixed. The breathing interruption of airflow causes the human body not to generate the basic necessary hormones and affects the life of an individual in unrestful, unhealthy and unbearable conditions [8].

An evaluation of sleep quality is therefore needed so that sleep disorders can be detected at the initial stages. The reason behind poor sleep quality is an accomplice with anxiety, physical activity, financial pressure, working hours, smoking, alcohol consumption, and stress. Numerous researchers have predicted that the changes in physical activity are connected to the changes in the rigidity of sleep disorder by causing breathing difficulties followed by disturbed or poor sleep [9]. The diagnosis and treatment of sleep-related diseases should be effective and prominently rely on the accurate classification of sleep stages. Therefore, sleep stage classification plays an important role in sleep analysis. Some of the existing work focuses on measuring how young adults convey their sleep habits on social media, as well as how the social media lifestyle is linked to the quality of sleep. Garett et al. [10] have associated the usage of electronic media and the sleep time and quality that has affected young adults.

The accessibility and scope of digital technologies for sleep measurement have drastically grown in recent years. Both medical and consumer-grade smart devices (remote sensing, mobile health, wearable gadgets-fitness tracker) across a range of areas are becoming more advanced and affordable. After the sleep data is preprocessed, the data modelling will be initiated for further analyzing the data. With the advancement of Deep Learning in the last decade, the implementation of Artificial

Neural Network (ANN) has had a positive impact on the health-related industry. A well-recognised algorithm among several deep learning models is CNN which is a leading technique in computer vision tasks [11].

In this paper, we propose a Convolution Neural Network (CNN) architecture that improves the classification performance. In particular, we benchmark the classification performance from different methods, including traditional machine learning methods such as Decision Trees (DT), k-Nearest Neighbour (k-NN), Naïve Bayes (NB) and Support Vector Machine (SVM), on 3 publicly available sleep datasets. The accuracy, sensitivity, specificity, precision, recall, and F-score are reported and will serve as a baseline to simulate the research in this direction in the future.

The contributions of this research can be summarized as:

- We propose 2 new CNN architectures for predicting sleeping quality from a wide range of data captured from smart sensors and surveys.
- We conducted experiments to evaluate the performance of the proposed CNNs and compared them with traditional machine learning algorithms on 3 public datasets.

2 Related Work

Mental stress is one of the major problems that lead to many other diseases, such as sleep disorders [12]. Analyzing sleeping patterns and the reasons that are leading to sleep disorders or insomnia is an active research area. For example, Garett et al. [10] analyze the relationship between the trending technologies (such as social media) and the quality of sleep. The sleep duration and quality are also dependent upon the physical activity performed that affects the production of the hormones in the body [13].

The identification of sleep stages plays a vital role in discovering sleep quality, which is estimated by analyzing the polysomnography (PSG) reports. A PSG study performed in laboratories consists of the analysis of electroencephalogram (EEG), electromyogram (EMG), and electrooculogram (EOG). These physiological signals are calculated by sensors that are attached to the patient's body [14]. As per the American Academy Of Sleep Medicine (AASM) rules, five different sleep stages are identified—Wake (W), Rapid Eye Movements (REM), Non-REM1 (N1), Non-REM2 (N2) and Non-REM (N3) as well as slow-wave sleep or even deep sleep.

Traditional sleep scoring algorithms either from actigraphy signals or PSG are likely to be created from an experimental perspective. These heuristic methods are built on prior experience/knowledge of sleep physiology and detecting modality [9]. Whereas, the estimation of traditional and ML algorithms is presented by applying the standard quality metrics like accuracy, recall, and precision for each dataset. By improving clinical metrics, ML techniques facilitate the doctors to be better informed and will be able to make appropriate clinical decisions [9].

There are wide range of models and methods that are suggested for sleep pattern recognition with self-supervised learning, for example, Zhao et al. [15] proposed a framework consisting of the recognition tasks including upstream and downstream modules. The upstream task is composed of pre-training and feature depiction phases. The extraction of frequency-domain and rotation feature sets to develop new labels simultaneously with the original data. The downstream task is a dynamic Bidirectional Long-Short Term Memory (BiLSTM) module for modelling the transient sleep data [15]. In the recent decade, unobtrusive sleep monitoring is one of the popular topics for sleep pattern recognition. Most of the surveys are interested in the sleep stage classification that is dependent on the study of cardiorespiratory features [16].

A recent study has shown that it is possible to estimate the sleep structure based on the respiratory parameters or interpretation of heart rate variability. The sleep stage classification is based on the respiratory pattern, as it is recorded and dependent upon the body surface displacement caused by respiratory motions. The body surface displacement is recorded in a non-contact approach by Bioradiolocation (BRL) which is a remote detection method of limb and organ motions by using a radar [17].

Obstructive Sleep Apnea (OSA) is the prevalent kind of sleep breathing disorder and it is represented by repetitive incidents of partial or complete barriers or obstructions while sleeping, generally linked with a decreased blood oxygen saturation. Rodrigues et al. [18] proposed to focus on the data pre-processing approach with an exhaustive feature selection, and evaluated the method using over 60 regression and classification algorithms in the experiments. Predictive models are needed to assist the clinicians in the diagnosis of OSA using Home Sleep Apnea Tests (HSATs). While there are two types of etiologies of non-diagnostic HSATs, both require a referral for PSG. In [19], machine learning technique is used for predicting non-diagnostic HSATs. Compared to traditional statistical models like logistic regression, machine learning algorithms have stronger predictive power whereas by incurring a decrement in the capability to draw interpretations regarding the relationships between the variables [19].

For classifying sleep stages, various approaches have been proposed in the literature. For example, features can be obtained using the Time-Frequency Image (TFI) representation of EEG signals and the sleep stages are then classified by a Multi-Class Least Squares Support Vector Machine (MC-LSSVM) classifier. The statistical measures of EEG epochs are recorded onto a dense network for feature extraction and a k-means classifier is applied for sleep stage classification. By using a phase encoding algorithm, the complex-valued non-linear features are obtained and used as the input to Complex-Valued Neural Network (CVNN) for classifying sleep stages. A feature extraction method identified as the statistical behaviour of local extrema is indicated for the sleep stage classification. The statistical features calculated from the segmented EEG-epochs are plotted onto the graph, then the structural comparison properties of the graph are classified based on the sleep stages with k-means clustering [20].

3 Datasets

To evaluate the classification performance of the different machine and deep learning methods, 3 publicly available sleep datasets are downloaded from Kaggle (https://www.kaggle.com/). The details of the datasets are explained in the rest of this section.

3.1 Dataset 1: Sleep-Study

This is a survey-based study of the sleeping habits of individuals within the United States of America [21]. The dataset consists of 104 sleeping habits records of individuals. Each record contains six attributes, namely *Enough (yes/no), Hours, Phone Reach (yes/no), Phone Time (yes/no), Tired (1 being not tired, 5 being very tired), and Breakfast (yes/no)*. The task is to predict whether the participant has *enough* sleep based on the rest of the attributes (i.e. 5) as input features.

3.2 Dataset 2: Sleep Deprivation

This dataset [22] is composed of 86 sleep records and each record has 80 attributes that are related to the demographic background of the participant and the response to the Karolinska Sleep Questionnaire. Only 7 key variables are selected as the input feature set, including *Age group, Anxiety rate, Depression rate, Panic, Worry, Health problems, and Nap duration*, to predict the *Overall sleep quality* and whether the participant has *enough* sleep.

3.3 Dataset 3: Sleep Cycle Data

This dataset [23] consists of 50 sleep records with 8 attributes, namely *Start, End, Sleep Quality, Time in Bed, Wake up, Sleep notes, Heart rate, and Activity (Number of Steps per day)*. Since the *Sleep note* is a text-based field for the participant to provide a textual description of the activities on that day, this attribute is not included in the classification and the rest of the attributes are used as input features to predict the *Sleep Quality*.

3.4 Data Pre-processing

After the data collection process, each sleep dataset is divided into training and testing sets on a 50–50 data split. We further performed data normalization on the data to facilitate the classifier training process.

4 Methodology

In this section, the proposed CNN architectures will be introduced. Inspired by the encouraging results in classifying infant movements using 1D and 2D CNNs [24], we propose a general CNN framework for classifying the sleep data and the network architecture is illustrated in Fig. 1.

Specifically, the initial input layer is followed by a convolution layer and then the ReLU activation layer. A Max Pooling layer is then added for obtaining a more abstract representation from the input. Such a Conv-ReLU-MaxPool structure is repeated before feeding the deep representation into a Fully-connected layer which is then followed by a dropout layer before the classification. Finally, the output layer comprises of softmax layer and the predicted class label can be obtained.

Overfitting is a known challenge when training machine learning models with small datasets such as those being used in this research. At the initial stage of experimenting the training, validation and testing performance were very poor. By adding dropout layers, the impact of overfitting was alleviated.

4.1 1D Convolutional Neural Networks

Here, we propose 2 1D CNN architectures based on the general structure illustrated in Fig. 1. The 2 new networks, namely $CONV-1D_1$ and $CONV-1D_2$, share the

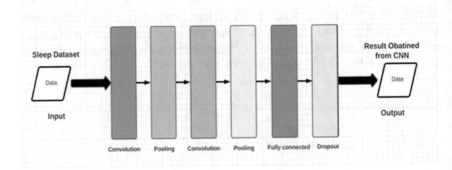

Fig. 1 The general network architecture of the proposed CNNs

same network architecture but with different dimensional in the intermediate layers. Specifically, the Max Pooling layers in $CONV-1D_1$ reduce the dimensionality of the intermediate representation while $CONV-1D_2$ will preserve the size of the representation as in the output of the convolutional layer. Such a design enables us to evaluate how the Max Pooling affects the performance of the CNNs.

5 Experimental Results

5.1 Evaluation Metrics

Several metrics are used for evaluating the classification performance in this research, including accuracy (AC), sensitivity (SE) and specificity (SP) which are calculated from true positive (TP), true negative (TN), false positive (FP) and false negative (FN). We also calculate the precision (PR), recall (RE), and F1 Score (F1). The equations for the calculation are as follows:

$$SE = \frac{TP}{TP + FN} \tag{1}$$

$$SP = \frac{TN}{TN + FP} \tag{2}$$

$$AC = \frac{TP + TN}{TP + FN + TN + FP} \tag{3}$$

$$PR = \frac{TP}{TP + FP} \tag{4}$$

$$RE = \frac{TP}{TP + FN} \tag{5}$$

$$F1 = 2 \cdot \frac{PR \cdot RE}{PR + RE} \tag{6}$$

5.2 Classification Results

In the first experiment, we report the performance of the Logistic Regression Classifier on the 3 datasets as an overview. Several evaluation metrics were used as stated in Sect. 5.1. The results are presented in Table 1. It can be seen that the datasets

Table 1 The performance of the logistic regression classifier on the three datasets

	Sleep-study	Sleep deprivation (%)	Sleep cycle data (%)
AC	63.46	58.02	55.11
SE	64.52	62.41	57.02
SP	61.90	59.85	56.45
F1	59.01	51.32	53.33
PR	66.67	56.13	52.23
RE	52.97	65.44	49.36

are challenging in general, with a classification accuracy ranging from 55.11% to 63.46%. With the F1 scores, the best performance (59.01%) was obtained on the Sleep-Study dataset which is the largest one among the 3 datasets. In contrast, the lowest performance (51.32%) was obtained in the Sleep Deprivation dataset.

In the second experiment, we further evaluate the classification performance using different classifiers [25, 26]. Table 2 reports the accuracy obtained using different methods on the 3 datasets for further evaluating the classification performance. For decision trees, 56.22, 61.26% and 51.33% were obtained from the 3 datasets. For k-NN classifier with k = 1 is being used, the accuracies were 55.33%, 59.50% and 55.94%, respectively. When the k=10 is used with k-NN, 61.83, 64.70 and 58.11% were obtained. It can be seen that performance was improved when k=10 is used instead of k = 1. In addition, Naïve Bayes achieved 59.65, 53.69 and 56.03% on the datasets. The SVM classifier achieved 59.21, 62.66 and 59.26% on the three datasets respectively. Compared to decision trees, k-NN and Gaussian Naïve Bayes classification models of machine learning, the SVM classifier performed better on all three sleep datasets.

For the proposed CNN-based classifiers, $CONV-1D_1$ obtained the accuracies of 59.18, 64.54 and 59.00% on the 3 datasets. $CONV-1D_2$ performed slightly better than $CONV-1D_1$ with 61.22, 65.23 and 58.73% obtained from the datasets, which shows the intermediate representations have to be in higher dimensionality in order to better model the data. It can be seen that $CONV-1D_1$ and $CONV-1D_2$ have shown better performances, more robust and more consistent than the traditional machine learning approaches. Although Logical Regression and SVM performed the best on Sleep-Study and Sleep Cycle Data datasets, the proposed CNN-based methods are more consistent on all 3 datasets and obtained comparable accuracy with those two methods.

6   Conclusions

In this paper, we proposed new CNN classification frameworks for predicting sleeping quality from a wide range of data captured from smart sensors and surveys. To show the effectiveness of the proposed CNN architectures, we evaluate the pro-

Table 2 The performance of different classifiers on the three datasets

	Sleep-study (%)	Sleep deprivation (%)	Sleep cycle data (%)
Logical regression	**63.46**	58.02	55.11
Decision Tree	56.22	61.26	51.33
k-NN (k=1)	55.33	59.40	55.94
k-NN (k=10)	61.83	64.70	58.11
Naïve Bayes	53.69	59.65	56.03
SVM	59.21	62.66	**59.26**
$CONV-1D_1$	59.19	64.54	59.00
$CONV-1D_2$	61.22	**65.23**	58.73

posed classification framework on 3 publicly available datasets. It is a challenging task since the datasets are small in general. We further conducted the experiments using traditional classifiers including Logistic Regression, Decision Trees, k-Nearest Neighbour, Naïve Bayes and Support Vector Machine as baselines for comparison. Experimental results highlighted the robustness of the proposed CNN architectures with highly consistent results obtained from different datasets.

References

1. Lam PS, Dinh Son N, Chi HP, Phuoc Van NT, Duc Minh N (2019) Novel algorithm to classify sleep stages. In: 2019 13th international conference on sensing technology (ICST), pp 1–6. https://doi.org/10.1109/ICST46873.2019.9047717
2. Ho ESL (2022) In: Abd El-Latif AA, Abd-El-Atty B, Venegas-Andraca SE, Mazurczyk, W, Gupta BB (eds.) Data security challenges in deep neural network for healthcare IoT systems. Springer, Cham, pp 19–37. https://doi.org/10.1007/978-3-030-85428-7_2
3. Jiang F, Jiang Y, Zhi H, Dong Y, Li H, Ma S, Wang Y, Dong Q, Shen H, Wang Y (2017) Artificial intelligence in healthcare: past, present and future. Stroke Vascul Neurol 2(4):230–243. https://doi.org/10.1136/svn-2017-000101
4. Biswal S, Sun H, Goparaju B, Brandon Westover M, Sun J, Bianchi M (2018) Expert-level sleep scoring with deep neural networks. J Am Med Inf Assoc 25(12):1643–1650. https://doi.org/10.1093/jamia/ocy131
5. Zhang J, Wu Y (2018) Complex-valued unsupervised convolutional neural networks for sleep stage classification. Comput Methods Progr Biomed 164:181–191. https://doi.org/10.1016/j.cmpb.2018.07.015
6. Anishchenko L, Zhuravlev A, Razevig V, Chizh M, Evteeva K, Korostovtseva L, Bochkarev M, Sviryaev Y (2019) Non-contact sleep disorders detection framework for smart home. In: 2019 PhotonIcs electromagnetics research symposium - Spring (PIERS-Spring), pp 3553–3557. https://doi.org/10.1109/PIERS-Spring46901.2019.9017470
7. Chambon S, Galtier MN, Arnal PJ, Wainrib G, Gramfort A (2018) A deep learning architecture for temporal sleep stage classification using multivariate and multimodal time series. IEEE Trans Neural Syst Rehab Eng 26(4):758–769. https://doi.org/10.1109/TNSRE.2018.2813138
8. Timuş OH, Bolat ED (2017) K-nn-based classification of sleep apnea types using ecg. Turkish J Electr Eng Comput Sci 25(4):3008–3023. https://doi.org/10.3906/elk-1511-99

9. Perez-Pozuelo I, Zhai B, Palotti J, Mall R, Aupetit M, Garcia-Gomez JM, Taheri S, Guan Y, Fernandez-Luque L (2020) The future of sleep health: a data-driven revolution in sleep science and medicine. NPJ Digital Med 3(1):42. https://doi.org/10.1038/s41746-020-0244-4
10. Garett R, Liu S, Young SD (2018) The relationship between social media use and sleep quality among undergraduate students. Inf, Commun Soc 21(2):163–173. https://doi.org/10.1080/1369118X.2016.1266374
11. Yamashita R, Nishio M, Do RKG, Togashi K (2018) Convolutional neural networks: an overview and application in radiology. Insights Imaging 9(4):611–629. https://doi.org/10.1007/s13244-018-0639-9
12. Machado Fernández JR, Anishchenko L (2018) Mental stress detection using bioradar respiratory signals. Biomed Signal Process Control 43:244–249. https://doi.org/10.1016/j.bspc.2018.03.006
13. Seixas AA, Henclewood DA, Williams SK, Jagannathan R, Ramos A, Zizi F, Jean-Louis G (2018) Sleep duration and physical activity profiles associated with self-reported stroke in the united states: application of bayesian belief network modeling techniques. Front Neurol 9:534–534. https://doi.org/10.3389/fneur.2018.00534
14. Dafna E, Tarasiuk A, Zigel Y (2018) Sleep staging using nocturnal sound analysis. Sci Rep 8(1):13474. https://doi.org/10.1038/s41598-018-31748-0
15. Zhao A, Dong J, Zhou H (2020) Self-supervised learning from multi-sensor data for sleep recognition. IEEE Access 8:93907–93921. https://doi.org/10.1109/ACCESS.2020.2994593
16. Tataraidze A, Anishchenko, L, Korostovtseva, L, Bochkarev, M, Sviryaev, Y, Ivashov, S (2017) Estimation of a priori probabilities of sleep stages: a cycle-based approach. In: 2017 39th annual international conference of the IEEE engineering in medicine and biology society (EMBC), pp 3745–3748. https://doi.org/10.1109/EMBC.2017.8037671
17. Tataraidze AB, Anishchenko LN, Korostovtseva LS, Bochkarev MV, Sviryaev YV (2018) Non-contact respiratory monitoring of subjects with sleep-disordered breathing. In: 2018 IEEE international conference "quality management, transport and information security, information technologies" (IT QM IS), pp 736–738. https://doi.org/10.1109/ITMQIS.2018.8525001
18. Rodrigues JF, Pepin J-L, Goeuriot L, Amer-Yahia S (2020) An extensive investigation of machine learning techniques for sleep apnea screening. Association for Computing Machinery, New York, pp 2709–2716
19. Stretch R, Ryden A, Fung CH, Martires J, Liu S, Balasubramanian V, Saedi B, Hwang D, Martin JL, Della Penna N, Zeidler MR (2019) Predicting nondiagnostic home sleep apnea tests using machine learning. J Clin Sleep Med?: JCSM?: Offic Publ Am Acad Sleep Med 15(11):1599–1608. https://doi.org/10.5664/jcsm.8020
20. Taran S, Sharma PC, Bajaj V (2020) Automatic sleep stages classification using optimize flexible analytic wavelet transform. Knowl-Based Syst 192:105367. https://doi.org/10.1016/j.knosys.2019.105367
21. Lomuscio M (2019) Sleep study - a survey-based study of the sleeping habits of individuals within the US. https://www.kaggle.com/mlomuscio/sleepstudypilot
22. Feraco, F (2018) Sleep deprivation. https://www.kaggle.com/feraco/sleep-deprivation
23. Işen K (2020) Sleep cycle data. https://www.kaggle.com/robottesla/sleepcycledata
24. McCay KD, Ho ESL, Shum HPH, Fehringer G, Marcroft C, Embleton ND (2020) Abnormal infant movements classification with deep learning on pose-based features. IEEE Access 8:51582–51592. https://doi.org/10.1109/ACCESS.2020.2980269
25. McCay KD, Ho ESL, Marcroft C, Embleton ND (2019) Establishing pose based features using histograms for the detection of abnormal infant movements. In: 2019 41st annual international conference of the IEEE engineering in medicine and biology society (EMBC), pp 5469–5472. https://doi.org/10.1109/EMBC.2019.8857680
26. McCay KD, Hu P, Shum HPH, Woo WL, Marcroft C, Embleton ND, Munteanu A, Ho ESL (2022) A pose-based feature fusion and classification framework for the early prediction of cerebral palsy in infants. IEEE Trans Neural Syst Rehab Eng 30:8–19. https://doi.org/10.1109/TNSRE.2021.3138185

A Dynamic Routing for External Communication in Self-driving Vehicles

Khattab M. Ali Alheeti and Duaa Al Dosary

Abstract Vehicular ad hoc networks (VANETs) are considered as a modification of mobile ad hoc networks (MANETs). They provide wireless ad hoc communication between vehicles and vehicle to roadside equipment. VANET provides an automated, safe, and secure traffic system. Several types of routing protocols have been modified for automated traffic. Many previous protocols such as Ad Hoc On-demand Distance Vector routing (AODV) have shown some weakness in performance. In this paper, a Dynamic AODV (DAODV) routing protocol is proposed that can adapt to various types of mobility and traffic scenarios, such as highway, rural and urban. AODV routing protocol is enhanced by designing a cooperative virtual bridge between vehicles and the nearest Road Side Unit (RSUs) to provide more than one path between the source and destination cars. This work was tested for thirty rounds. In each round, the cooperative routing protocol generated ten various paths between two nodes. In these cases, each RSU finds all available paths between nodes simultaneously. This protocol will have the ability to provide more than one path and reduce the possibility of packet loss between nodes. Simulation results prove that DAODV can improve the communication performance between self-driving and semi self-driving vehicles.

Keywords Vehicular ad hoc networks · AODV · Routing protocols · Road side units (RSUs) · DAODV

1 Introduction

Recently, Vehicular Ad hoc Network (VANET) technology was introduced to provide safe and secure traffic control systems in response to heavy congestion caused by the high density of vehicles, in the city environment. There are three main types of communications in VANET, as presented below [1]:

K. M. Ali Alheeti (✉) · D. Al Dosary
University of Anbar, Anbar, Iraq
e-mail: co.khattab.alheeti@uoanbar.edu.iq

© The Author(s), under exclusive license to Springer Nature Switzerland AG 2023
A. A. Abd El-Latif et al. (eds.), *Advances in Cybersecurity, Cybercrimes, and Smart Emerging Technologies*, Engineering Cyber-Physical Systems and Critical Infrastructures 4,
https://doi.org/10.1007/978-3-031-21101-0_15

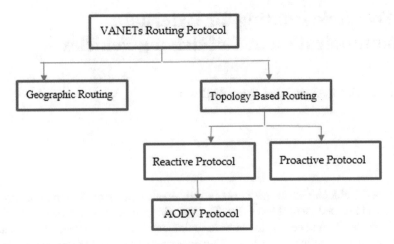

Fig. 1 Classification of routing protocols in VANET

- Vehicle-to-Vehicle communications (V2V)
- Vehicle-to-Infrastructure communications (V2I)
- Hybrid

The Ad hoc Mobile Network (MANET), is modified to provide autonomous and self-organizing wireless communication network technology. In order to provide autonomous wireless communication between vehicles, the safety, unique identity, and traffic conditions of every vehicle must be provided. For providing autonomous wireless communication, the client or server must be created to deliver the information between every vehicle/node. There are various types of routing protocols that can be employed to implement VANET in traffic systems, as presented in Fig. 1.

With the proactive routing protocols, routing information is exchanged periodically by nodes, and routes are computed. This presents a waste of bandwidth and energy in the network. Furthermore, the routing information did not compute periodically, for the on-demand routing protocols. Instead, these protocols calculate a route only when it is requested for communication between two nodes [2]. The connection between the nodes is not constant because of the dynamic change in these networks. In some cases, packets cannot transmit from one node to the desired node, and the outcome packets may be wasted. The performance of the route may be affected when packets are wasted in various ways. The proposed routing plays an important role for self-driving cars which are considered one of the most vital applications in the latest research area [3]. If there is no communication, the reactive routing protocol does not preserve routing information at the nodes. This protocol establishes the connection and searches for the route during transmission and receiving packets between nodes. There are many types of reactive routing protocols, such as TORA, AODV, and DSR, as presented below [4].

Ad hoc On-demand Distance Vector (AODV) is an old and simple routing protocol in the VANET system. The AODV routing protocol has become a more reliable and effective routing protocol. A drawback of AODV is that the mechanisms for the detection of the route are not well invested in the reply message of route loss. A significant waste in performance resulted from the route reply message loss [5]. In AODV, only one path from the source node to the destination node is maintained. In the case of communication path disconnection, the performance metrics are decreased, such as end-to-end delay, power consumption, and the packet delivery ratio when the source node must restart again the process of route discovery [6]. In order to avoid AODV routing protocol problems, a reverse AODV routing protocol has to be dependent. In this paper, the AODV routing protocol is enhanced by designing a cooperative virtual bridge between vehicles and the nearest Road Side Unit (RSU). This new technique will provide more than one path between the source and destination cars. The Dynamic AODV (DAODV) routing protocol proposed in this paper has the ability to adapt to various types of mobility and traffic scenarios, such as highway, rural and urban. On the one hand, the proposed system is heavily based on an unrestricted number of hops between nodes. On the other hand, it tries to achieve its target to deliver packets between nodes without any loss.

The main contributions of this paper are summarised below:

- Enhancing the process on the current routing protocol, which is AODV by building a virtual line with all of the nearest RSUs.
- Reducing the number of packets being lost that are sent/transmitted between cars at the same radio coverage area.
- Building dynamic paths between all of the mobility cars.
- Generating dynamic routing paths that update every four seconds, to provide all available routing paths.

The remainder of this paper is organised as follows: In Sect. 2, the literature reviews are presented. Section 3 discusses some details of Ad hoc On-Demand Distance Vector (AODV). The methodology of the proposed routing protocol is presented in Sect. 4. The results are presented in Sect. 5, and finally, the conclusion is detailed in Sect. 6.

2 Literature Reviews

Many types of research are presented on this topic. In [1], a modified AODV routing protocol with the help of a queue for VANET, is proposed. The results show that the proposed protocol delivers a more outstanding performance and only needs minor overhead to compare to existing AODV protocols. The authors in [7], aimed to research the efficiency of the reactive routing approach in Vehicular Ad Hoc Networks (VANETs). The simulation results obtained by their proposal have shown the efficiency of this solution in reducing the incurred overhead, under different scenarios. An improved AODV routing protocol in VANET is proposed in [8], to decrease overhead and enhance route stability. The results show that the proposed protocol achieves better performances in packet delivery ratio and link stability. In [9], a novel algorithm was proposed to enhance AODV for the internet of vehicles, to achieve efficient and stable clustering to simplify routing and ensure a better quality of service. The results prove that the proposed protocol enhances the overall delivery ratio throughput with minor delay, and less routing load than the original AODV.

The performance of AODV improved in [10], by modifying the original protocol by providing stable clusters and implementing routing by gateway nodes and cluster heads. The results prove that the improved protocol reduces the overhead and the efficiency can be improved. In [11], the AODV routing protocol is modified to make it adaptive, and establishing more stable routes for VANET. Two parameters, direction and position, are used in the proposed protocol for next-hops selected during the route discovery phase. The results show that the proposed protocol presents well in different traffic situations and the overall overhead is less, compared to existing protocols. Another study to improve the AODV routing protocol for vehicular ad hoc networks is presented in [12] to reduce the packet overhead. The results prove that the proposed protocol was efficient in improving the performance of a VANET.

Routing Table Flag (RTF) was checked to present a good performance by AODV and improve neighbor's discovery, in [13]. According to the results, the proposed AODV routing protocol presents better compare to existing protocols. In order to improve VANET performance, the AODV routing protocol is modified in [14]. The results prove that the modified AODV can reduce the average end-to-end delay, compared to the original AODV. The study in [15], aimed to report on and evaluate Ad hoc On-Demand Distance Vector (AODV) and Dynamic Source Routing performance for VANETs. This research aimed to provide knowledge regarding the best possible routing protocol selection. The results prove that improving the link throughput of the VANET, a combination of proper channel model, and an efficient routing protocol could be used. According to the results, routing protocols and network size impact the packet delivery ratio, packet loss, overhead, and average end-to-end delay. In [16], the authors proposed a new AODV routing protocol by exploiting social relationship mining from the historical Global Position System (GPS) trajectory data of vehicle nodes in VANET. The results prove that the proposed AODV protocol presents a better performance in packet delivery ratio and average transmission delay. In [17], the authors developed and enhanced the number of service quality parameters like packets received productivity, delay, and node power consumption. In this study, VANETS multi-cast transmission technology is introduced and optimized to increase throughput, reduce delay, and reduce packet loss. A comprehensive study of the various routing protocols in VANETs is described and presented in [18] to provide designers and researchers with an effective comparison for better analysis.

By reviewing and analysing the previously proposed approaches, our study is distinguished from others by presenting a Dynamic AODV (DAODV) routing protocol that can adapt to various mobility and traffic scenarios, such as highway, rural and urban. Our study involves the enhancing of the existing AODV routing protocol by proposing a cooperative virtual bridge between vehicles and the nearest

RSUs, to provide more than one path between the source and destination cars. This protocol will have the ability to provide more than one path, and reduce the possibility of packet loss between nodes.

3   Ad Hoc On-Demand Distance Vector (AODV)

On-demand is dependent in AODV, when each node is needed within the network. When data is transmitted to the desired node by the source node, then the source node initially transmits a Route Request (RREQ) message, which is broadcast by other nodes to deliver the data to the desired node. A route reply (RREP) message is unicast to the sending node when the desired node has a valid route to the required address. AODV routing protocol is capable of multicasting and unicasting.

Figure 2 shows the process of route discovery when a route query packet is propagated by the source node to its neighbors. If a route to the destination is contained by any of the neighbors, then it accepts the request; otherwise, the neighbor's nodes resend the request packet. Some requests are delivered to the destination node. The maintenance of the route in this routing protocol is shown in Fig. 3. N2 generates and propagates a RERR message when the communication link between the source node and destination node breaks. Finally, the RERR message finishes up at the source node. The source node will create a new RREQ message when receiving the RERR message.

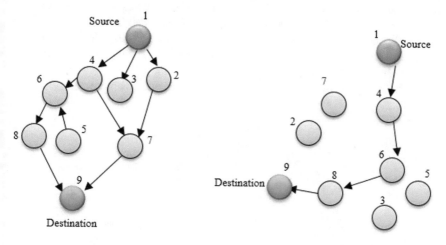

Fig. 2 Route discovery in AODV

Fig. 3 Route maintenance
in AODV

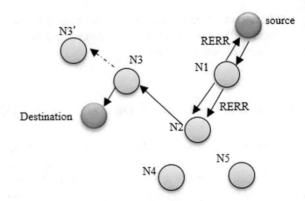

4   The Proposed Dynamic AODV Routing Protocol

In this paper, a Dynamic Ad hoc On-Demand Distance Vector (DAODV) protocol is proposed which is a development of the original AODV routing protocol. Moreover, the AODV routing protocol is enhanced by designing a cooperative virtual bridge between vehicles and the nearest Roadside unit (RSUs). This new technique will provide more than one path between the source and destination cars. This work is tested over thirty rounds. In each round the cooperative routing protocol generates ten various paths between two nodes. In this case, each RSU finds all available paths between nodes simultaneously. This protocol will have the ability to provide more than one path and reduce the possibility of packet loss between nodes. The communication system of autonomous vehicles is presented in Fig. 4. These paths are generated between the nodes which are heavily based on all of the closest RSUs. In other words, these infrastructures on-road enable the existing route to adapt with VANETs by creating a virtual bridge.

The procedure of the proposed DAODV routing protocol is presented below:

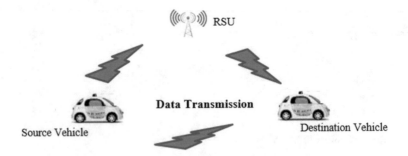

Fig. 4 Communication system of vehicles for proposed DAODV

- Building the paths of data transmitting to RSUs, by using the AODV routing protocol.
- Receiving the data by RSUs and forwarding it to RSUs.
- Transmitting the data by the RSUs to the destination vehicle by the AODV routing protocol.
- Collecting the real-time road traffic parameters, average speed, vehicle number, and calculating the connectivity probability of each road section.
- Assigning unique IDs for each road section and RSU.
- Loading the latest RSU coverage area division results by vehicles in the road network and updating them via the original high-power wireless communication systems.
- Creating a vehicle position table by each vehicle for storing the geographic positions of itself and the RSU, in the current coverage area.
- Creating a coverage area table for storing the road section information in the current coverage area.

The general structure of the proposed protocol is explained in Fig. 5.

Fig. 5 Basic structure of the proposed DAODV routing protocol

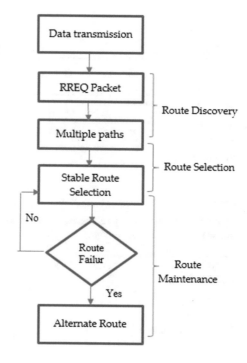

5 Experimental Results

The proposed DAODV routing protocol is tested by exploiting various performance metrics such as PDR, overhead rate, and throughput rate. In order to determine the efficiency of specific communication conditions dependent on measuring performance metrics of the DADOV, is compared with the original AODV for evaluation of its efficiency. In addition, the factors considered during the testing of the DAODV performance are broken links, the number of cars, and the speed of cars. The parameters used in this simulation are presented in Table 1.

Different metrics are calculated to evaluate the proposed DAODV routing protocol. Each of them represents an indication of the performance of the proposed protocol. Performance metrics for DAODV is presented in Tables 2 and 3. PDR and average throughput are calculated according to the following equations [19, 20]:

$$PDR = \frac{\Sigma(Total\ packets\ received)}{\Sigma(Total\ packets\ send)} \tag{1}$$

Table 1 Simulation results

Simulation parameters	Value
Routing protocol	AODV
Source node number	1
Destination node number	90
Radio coverage area	900 * 900 m
Number of nodes	90
Broadcast range of cars	250 m

Table 2 Performance metrics for AODV

Routing protocols	Packet deliver rate (%)	Throughput (kbps)	Time (s)	No. of vehicles	Overhead (%)	Broken links
AODV	94	11.23	4	500	81	14
DAODV	95.2	14.49			87	8

Table 3 Style performance for AODV

Routing protocols	Speed of cars (km)	Packet deliver rate (%)
AODV	30	92
DAODV		96
AODV	50	86
DAODV		91

$$Throughput = \left(\frac{Store\ received\ packet's\ size/}{(Simulation\ stop\ time - Simulation\ start\ time)} \right) * (8/1000) \quad (2)$$

Figures 6 and 7 show the packet delivery rate of the proposed DAODV, compared to the original AODV with respect to time in seconds.

Performance metrics for DAODV presented in Table 3, and Fig. 8, shows that packet delivery rate of the proposed DAODV compared to the original AODV with respect to the speed of the cars.

Table 3 reveals that the PDR rate is declining when the speed of cars increased. Table 4 shows the path from the source node (1) to the destination node (90). Hops represent the number of intermediate nodes the source node has taken to reach the destination node.

According to Table 4, we can quickly see that the routing protocol achieves so many paths between source and destination, to reduce any accident or break-in path. Thus, the source car can send sensitive and important information to the destination by various paths. The proposed routing protocol provides the VANETs with flexibility in the sending/receiving mechanisms between vehicles.

Fig. 6 Packet delivery rate (AODV vs. DAODV)

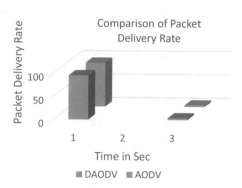

Fig. 7 Throughput rate (AODV vs. DAODV)

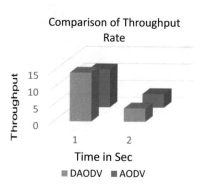

Fig. 8 Performance metrics
of DAODV versus AODV

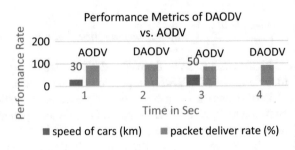

Table 4 Dynamic routing

Source node:1, Destination node: 90 for all routing					
Routing number	Number of hops	Waiting time	Routing number	Number of hops	Waiting time
1	14, 9	2	23	39, 17, 9	3
2	6, 21, 41, 28	4	24	15	1
3	11	1	25	5, 20, 2	3
4	40, 58	2	26	12, 62, 3	3
5	18	1	27	–	0
6	2	1	28	10, 41	2
7	9, 2	2	29	–	0
8	–	0	30	4, 30, 46, 67	4
9	36	1	31	42, 37, 17	3
10	–	0	32	74, 40	2
11	22, 68	2	33	29, 26	2
12	2	1	34	11, 87	2
13	–	0	35	41, 24, 22	3
14	32, 5	2	36	5, 84	2
15	8, 35	2	37	–	0
16	27	1	38	6, 20, 48	3
17	18	1	39	11, 48	2
18	13	1	40	7, 4, 45	3
19	31	1	41	2	1
20	17, 8, 31	3	42	4, 51	2
21	9, 24	2	43	2	1
22	49, 8	2	44	74, 80	2
45	9	1	70	8, 36, 33	3
46	3, 38, 20	3	71	–	0
47	84	1	72	14, 52, 45, 4	4
48	84	1	73	20, 3, 7	3
49	–	0	74	35, 27, 18	3

(continued)

Table 4 (continued)

Source node:1, Destination node: 90 for all routing

Routing number	Number of hops	Waiting time	Routing number	Number of hops	Waiting time
50	8	1	75	10, 2, 6	3
51	5, 80, 21	2	76	5, 2, 7	3
52	13, 37	2	77	11, 14	2
53	17, 5, 6	3	78	25, 20	2
54	–	0	79	35, 81, 70	3
55	21, 82	2	80	6, 17	2
56	–	0	81	11, 42	2
57	4, 7, 12, 27	4	82	67	1
58	3, 5, 31	3	83	25, 11	2
59	8, 3, 10	3	84	6, 13, 45, 3	4
60	19	1	85	11, 9, 6	3
61	26	1	86	65, 9	2
62	10, 20, 46, 3	4	87	21	1
63	9	1	88	18, 4	2
64	3, 38, 20	3	89	14	1
65	84	1	90	11, 56	2
66	6, 31, 20	3	91	84	1
67	21	1	92	19, 64	2
68	32	1	93	2, 26, 79	3
69	29, 73, 48	3	94	–	0
95	23, 63	2	120	5	1
96	63, 3, 40	3	121	3, 48	2
97	47, 18	2	122	7	1
98	3, 28	2	123	23, 18	2
99	14	1	124	11	1
100	47	1	125	41	1
101	68, 2	2	126	49	1
102	–	0	127	10, 2, 15	3
103	28, 11, 70	3	128	18, 12, 13	3
104	11, 41, 64, 37	4	129	33	1
105	63	1	130	–	0
106	10, 6, 26	3	131	16, 20, 57	3
107	–	0	132	35	1
108	6	1	133	10, 21	2
109	15	1	134	21	1

(continued)

Table 4 (continued)

Source node:1, Destination node: 90 for all routing

Routing number	Number of hops	Waiting time	Routing number	Number of hops	Waiting time
110	15, 72, 83	3	135	2	1
111	3, 4	2	136	–	0
112	13	1	137	–	0
113	–	0	138	21, 22, 82	3
114	12	1	139	10	1
115	33	1	140	–	0
116	10	1	141	22, 54	2 ara>
117	39, 14	2	142	53	1
118	32	1	143	7, 43	2
119	9	1	144	3, 81	2
145	25, 14, 8, 48	4	170	15	1
146	15, 13, 80, 31	4	171	–	0
147	77, 4, 11	3	172	81, 20	2
148	14	1	173	34, 75, 24	3
149	52, 6	2	174	37, 20, 46, 16	4
150	–	0	175	2, 64, 12	3
151	2, 4, 20, 38	4	176	15, 32	2
152	–	0	177	53	1
153	27	1	178	38	1
154	19	1	179	58, 68	2
155	29, 80	2	180	17	1
156	6	1	181	73, 16	2
157	–	0	182	6, 72	2
158	–	0	183	9, 68, 39	3
159	25	1	184	72, 20	2
160	34, 8, 9, 45	4	185	85, 11	2
161	3, 28	2	186	50	1
162	35, 6, 53	3	187	46, 18	2
163	22, 27	2	188	2	1
164	2	1	189	–	0
165	16	1	190	30	1
166	5	1	191	8, 13	2
167	9, 47	2	192	30, 7	2
168	33	1	193	15	1
169	45, 16	2	194	25, 8	2

6 Conclusions

In this paper, the DAODV routing protocol for VANETs is proposed. This protocol is proposed to enhance the original AODV routing protocol, and to improve the external communication system for vehicles by building a virtual line with all of the nearest RSUs. This new technique will provide more than one path between the source and destination cars. Different metrics are calculated to evaluate the proposed DAODV routing protocol, with each of them representing an indication for the performance of the proposed protocol. The experimental results demonstrate that the proposed protocol can enhance the external communication systems of vehicles. DAODV achieves better delivery of packets between nodes without losing any rate, compared to the original AODV in VANETs.

Conflicts of Interest The authors declare that they have no conflicts of interest to report regarding the present study.

References

1. Saha S, Roy U, Sinha D (2014) AODV routing protocol modification with dqueue (dqAODV) for VANET in city scenarios ICHPCA. ICHPCA, Bhubaneswar India, pp 1–6
2. Perkins CE, Royer EM (1999) Ad-hoc on-demand distance vector routing. In: Proc. WMCSA'99. NY, USA, pp 90–100
3. Abd N, Alheeti KMA, Al-Rawi SS (2020) Intelligent intrusion detection system in internal communication systems for driverless cars. Webol J 17(2):376–393
4. Goyal S (2015) A comparative performance analysis of AODV and DSR routing protocols for vehicular Ad-hoc networks (VANETs). Inter J Adv Res Comp Eng Technol (IJARCET) 4(4):1607–1613
5. Kim C, Talipov E, Ahn B (2006) A reverse AODV routing protocol in ad hoc mobile networks. In: Inter conf. on embedded and ubiquitous computing. Berlin, Heidelberg, pp 522–531
6. Chakeres ID, Belding-Royer EM (2004) AODV routing protocol implementation design. In: Proceedings of the International Conference on Distributed Computing Systems. Tokyo, Japan, pp 698–703
7. Moussaoui B, Djahel S, Khelifi H, Merniz S (2015) Towards enhanced reactive routing in urban vehicular Ad hoc networks. In: Inter. Conf. on Prot. Engineering (ICPE) and Intern. Conf. on New Technologies of Dist. Syst. (NTDS). Paris, France, pp 1–6
8. Ding B, Chen Z, Wang Y, Yu H (2011) An improved AODV routing protocol for VANETs. In: Inter. Conf. on Wireless Comm. and Signal Processing (WCSP). Nanjing, China, pp 1–5
9. Ebadinezhad S (2021) Design and analysis of an improved AODV protocol based on clustering approach for the internet of vehicles (AODV-CD). Int J Electron Telecommun 67(1):13–22
10. Aswathy MC, Tripti C (2012) An enhancement to AODV protocol for efficient routing in VANET–A cluster-based approach. Adv Intell Soft Comput 167(2):1027–1032
11. Abedi O, Fathy M, Taghiloo J (2008) Enhancing AODV routing protocol using mobility parameters in VANET. In: International conference on computer systems and applications. Doha, Qatar, pp 229 235
12. Mubarek FS, Aliesawi SA, Alheeti KMA, Alfahad NM (2018) Urban-AODV: an improved AODV protocol for vehicular ad-hoc networks in the urban environment. Int J Eng Technol 7(4):3030–3036

13. Keshavarz H, Noor RM, Mostajeran E (2013) Using routing table flag to improve performance of AODV routing protocol for VANETs environment. Adv Intell Syst Comput 209(3):73–82
14. Raju S (2015) Performance improvement in VANET by modifying AODV routing protocol. Comp Eng Intell Syst 6(5):92–99
15. Malik S, Sahu PK (2019) A comparative study on routing protocols for VANETs. ScienceDirect Heliyon 5(8):e02340
16. Zhang W, Xiao X, Wang J, Lu P (2018) An improved AODV routing protocol based on social relationship mining for VANET. In: International Conference on Communication and Information Processing. ACM Qingdao, China, pp 217–221
17. Al-Shabi MA (2020) Evaluation the performance of MAODV and AODV protocols in VANETs models. Int J Comput Sci Secur 14(1):1
18. Smiri S (2020) Performance analysis of routing protocols with roadside unit infrastructure in a vehicular Ad Hoc network. Int J Comput Netw Commun 12(4)
19. Khairnar M, Vaishali D, Kotecha D (2013) Simulation-based performance evaluation of routing protocols in vehicular Ad-hoc network. Int J Sci Res Publ 3(10)
20. Ali Alheeti KM, Al-ani MS, McDonald-Maier K (2018) A hierarchical detection method in external communication for self-driving vehicles based on TDMA. PLOS One 13(1):1–19

Analysis of the Air Inlet and Outlet Location Effect on Human Comfort Inside Typical Mosque Using CFD

Mohammed O. Alhazmi and Abdulaziz S. Alaboodi

Abstract Air distribution plays a significant role in the mosque design process because the building inhabitants anticipate acceptable indoor air quality. The temperature, speed, direction, and volume flow rate of indoor air, might make the worship uncomfortable if not designed well. It makes no difference how efficient the ventilation system is if the air is not evenly distributed. This study aims to study the air distribution for the best energy consumption by comparing three cases of variation in inlet and outlet locations. Furthermore, to avoid local temperature differences by extracting warmed and contaminated air before dispersing across the space. The impacts of inlet and outlet air position to room heat sources on air distribution and thermal comfort were investigated using Computational Fluid Dynamics CFD approaches. The result shows that inlet and outlet location significantly affect mosque air quality.

Keywords CFD · Mosque · Airflow · Human comfort

1 Introduction

Several studies have shown that increased ventilation rates in building spaces can be beneficial [1, 2]. The removal of harmful indoor air and the supply and dispersion of fresh (outside) air into the inside environment is part of ventilation. Buildings require fresh air to supply oxygen for breathing and prevent odors [3]. The physical variables that affect thermal sense (air temperature, air velocity, relative humidity means radiant temperature) were defined by Macpherson [1]. These criteria may not be enough to determine thermal comfort [2].

The position of the outlet and the parameters of the diffuser is critical for indoor air quality and thermal comfort. Because most room or space air is not adequately mixed, thermal comfort can vary significantly at any one time. Therefore, the mean

M. O. Alhazmi · A. S. Alaboodi (✉)
Mechanical Engineering Department, College of Engineering, Qassim University, Buraydah, Saudi Arabia
e-mail: alaboodi@qu.edu.sa

© The Author(s), under exclusive license to Springer Nature Switzerland AG 2023
A. A. Abd El-Latif et al. (eds.), *Advances in Cybersecurity, Cybercrimes, and Smart Emerging Technologies*, Engineering Cyber-Physical Systems and Critical Infrastructures 4, https://doi.org/10.1007/978-3-031-21101-0_16

value for thermal comfort is insufficient. When mixing ventilation, the air is used to mix room air thoroughly. As a result, contaminant concentrations are equally distributed throughout the space, and contaminants are neutralized by incoming ventilation air [4].

One sustainable solution to maintaining healthy and comfortable environmental conditions in buildings is natural ventilation. However, practical design, construction, and operation of naturally ventilated buildings require a good understanding of complex airflow patterns caused by buoyancy and wind effects. The 3D computational fluid dynamics (CFD) analysis could be employed to investigate the occupants' environmental conditions and thermal comfort of a highly-glazed naturally ventilated meeting room [5]. The RNG model was developed to renormalize the Navier–Stokes equations to account for the effects of more minor scales of motion. In the standard k-epsilon model, the eddy viscosity is determined from a single turbulence length scale, so the calculated turbulent diffusion occurs only at the specified scale, whereas in reality, all scales of motion will contribute to the turbulent diffusion. The RNG k-ε model is slightly better than the standard k-ε model and is therefore recommended for simulations of indoor airflow [6]. Effective wind-driven cross-ventilation requires a well-designed opening, such as a window, to allow sufficient exchange between indoor and outdoor air, especially when the building is surrounded by other buildings [7]. The study of mosque acoustics, concerning acoustical characteristics, sound quality for speech intelligibility, and other applicable acoustic criteria, has been neglected [8].

Many large confined spaces in tropical countries utilize a combination of power-driven fans and natural ventilation for space cooling purposes. However, due to low wind velocity and the inability of mechanical fans to remove warm air, this cooling method cannot provide good thermal comfort to the occupants. A computational fluid dynamic CFD method was utilized in one study research to predict airflow and temperature distributions and examine the effects of installing exhaust fans on the thermal comfort condition inside the mosque [9]. The same design of a building, if repeated irrationally from one place to another, even within an identical climatic region, the same design of a building gives rise to some grave problems that can compromise multiple dimensions of sustainability [10]. The air conditioning in mosques is considered critical HVAC applications in which indoor air quality depends on the airflow behavior, the distributions of temperatures and relative humidity, and the concentration of carbon dioxide due to the high occupancy load of people. One of the research studies was carried out in Al-Masa'a between Al-Safa and Al-Marwa hills inside Al-Haram mosque located at Makkah, Saudi Arabia. The work focuses on airflow, thermal behavior, and carbon dioxide dispersion by studying different cases by modifying ACH to reveal the impact of the variation on the whole people [11].

Large mosques used for Friday prayers are designed to accommodate many worshippers. However, only 15% of the total capacity is present at each of the remaining five-daily prayers. The indoor thermal comfort needs a considerable amount of energy for cooling. Hence, a rear zone idea is proposed in this study to alleviate this dilemma. Simulation methods and fieldwork were used to predict the reduction in energy consumption besides the payback of the incurred costs [12]. Air distribution plays an essential aspect in the design process of a building as building occupants expect good indoor air quality standards. Indoor air may feel uncomfortable because of its temperature, speed, direction, and volume flow rate. It does not matter how efficient the ventilation equipment is if the air is not distributed well [13].

The building process of a mosque is categorized by its unique spatial requirements and intermittent occupancy pattern to create an appropriate ambiance for prayers and maintain the thermal comfort of the interior. As mosques are skin-load-dominated buildings, the climatic conditions play a defining role in the thermal comfort of the users and the energy efficiency of the building. Therefore, an appropriate building envelope design that ensures optimal thermal performance is required to protect the interior from the harsh external climate and confirm a user's thermal comfort. Moreover, The improvement of the overall thermal performance of the building allows the reduction of energy consumption and confirms the improvement of energy efficiency [14]. The Temples represent a unique type of building owing to their ample inner space, purpose, and intermittent use. The review has revealed that indoor air quality, thermal comfort, and energy consumption are interconnected. However, the energy consumption in worship places is often high, which may not consist of a comfortable indoor environment. Furthermore, the air quality depreciates rapidly during prayers, and natural ventilation is usually insufficient to ventilate the space [15].

The CFD simulations approach is applied in a real scenario of a recently refurbished former church built in the fourteenth century converted to be used as a concert and conference hall. The restoration investor was concerned about temperature fluctuations caused by variable occupancy and their possible negative impact on the historical stucco decorations and the original wooden trusses of the former church. Moreover, only natural ventilation through window openings at the street level and roof windows is possible in order to preserve the original look of the building. The study elaborated in the paper is based on the results of CFD simulations with simplified models of visitors acting as heat sources under two different occupancy scenarios [16, 17]. The literature review provides a practical understanding of mosque energy usage patterns and identifies the essential features to be considered in reducing energy consumption in mosque buildings. Numerous research gaps have been identified through this literature review that may be pivotal in designing energy-efficient mosques [18].

The objective of the current paper is to study the effects of inlet and outlet air position in a mosque and their relationship to the mosque, which were explored numerically using computational fluid dynamic (CFD) software. Furthermore, the CFD analysis will better understand the ventilation input and exit location before achieving good air dispersion.

2 Modeling

Using ANSYS Software, the effect of several intakes and outlet air ventilation locations and their interaction with the air distribution in the indoor environment were
numerically examined. The model mosques in Fig. 1 are $(29.6 \times 24.6 \times 14.7) \, m^3$ in
size, $(1 \times 1) \, m^2$ four inlets and $(1 \times 1) \, m^2$ two outlets. The inlet and out a comparison
between three different cases is shown in Fig. 2.

All of the room's boundary conditions and parameters were established during
setup. For all room geometry, the domain must first be determined. The domain
now includes the fluid type and its attributes. The parameters were chosen to have
air as a continuous fluid and the wall as a solid. Additionally, the model must be
chosen. It is crucial to calculate the turbulence model. The (Re) for predicting the
laminar to turbulent flow transition was used before the modeling solver selected
the laminar model. For this flow scenario, the Reynolds number can be used. The
Reynold number equation is shown in Eq. (1).

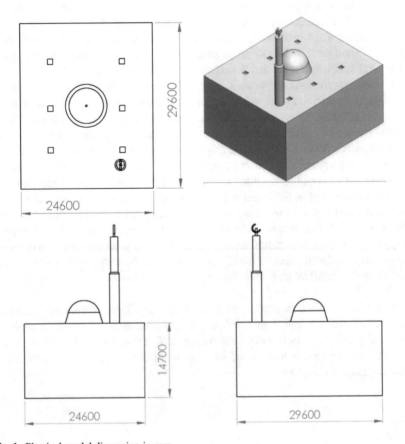

Fig. 1 Physical model dimension in mm

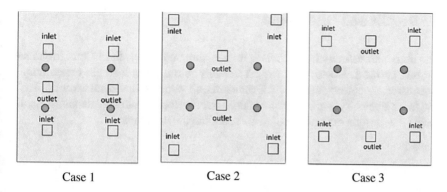

Fig. 2 The inlet and outlet locations for each case of the study

$$Re = \frac{\rho v D}{\mu} \tag{1}$$

Another modeling setup energy equation was that the air movement and detached mass were directed to momentum and energy equations using a finite volume method.
Momentum equation:

$$\frac{\partial(\rho u)}{\partial t} + \nabla.(\rho u V) = -\frac{\partial p}{\partial x} + \frac{\partial \tau xx}{\partial x} + \frac{\partial \tau yx}{\partial y} + \frac{\partial \tau zx}{\partial z} + \rho f x \tag{2}$$

$$\frac{\partial(\rho v)}{\partial t} + \nabla.(\rho v V) = -\frac{\partial p}{\partial y} + \frac{\partial \tau xy}{\partial x} + \frac{\partial \tau yy}{\partial y} + \frac{\partial \tau zy}{\partial z} + \rho f y \tag{3}$$

$$\frac{\partial(\rho w)}{\partial t} + \nabla.(\rho w V) = -\frac{\partial p}{\partial z} + \frac{\partial \tau xz}{\partial x} + \frac{\partial \tau yz}{\partial y} + \frac{\partial \tau zz}{\partial z} + \rho f z \tag{4}$$

Energy equation

$$\frac{\partial}{\partial t}\left[\rho\left(e + \frac{V2}{2}\right)\right] + \nabla \cdot \left[\rho\left(e + \frac{V2}{2}\right)V\right]$$
$$= \frac{\rho \dot{q} + \frac{\partial}{\partial x}\left(k\frac{\partial T}{\partial x}\right) + \frac{\partial}{\partial y}\left(k\frac{\partial T}{\partial y}\right) + \frac{\partial}{\partial z}\left(k\frac{\partial T}{\partial z}\right) - \frac{\partial(up)}{\partial x} - \frac{\partial(vp)}{\partial y} - \frac{\partial(wp)}{\partial z}}{\partial x}$$
$$+ \frac{\partial(u\tau yx)}{\partial y} + \frac{\partial(u\tau zx)}{\partial z} + \frac{\partial(v\tau xy)}{\partial x} + \frac{\partial(v\tau yy)}{\partial y} + \frac{\partial(v\tau zy)}{\partial z} + \frac{\partial(w\tau xz)}{\partial x}$$
$$+ \frac{\partial(w\tau yz)}{\partial x} + \frac{\partial(w\tau zz)}{\partial z} + \rho f \cdot V \tag{5}$$

3   Results and Discussions

The higher air velocity has a cooling impact, one of the essential factors for human thermal comfort. However, if the air velocity is too high, it might create human discomfort. According to ASHRAE standards, air velocity should not exceed 0.2 m/s, and low air velocity is not recommended. The air velocity was estimated numerically inside the mosque at various points surrounding the occupants.

Case 1

The four air inlets are located apart from the corners, and two outlets are located at the center in Fig. 2. The results of the 3D model illustrated in Fig. 3 represent the air velocity streamline inside the building. The airflow is concentrated under the inlet location, while the rest of the building has low airflow. The same results are represented by contour with two altitudes at 5, and 14 m, representing the airflow is concentrated only under the inlet locations in Fig. 4. Figure 5 shows the axial distribution of airflow velocity and pressure under the ceiling surface. The two humps represent the highest airflow velocity of 40 m/s while the pressure is at a lower value around 12,410 Pa. Figure 6 shows the velocity and pressure distribution at y-direction, representing the flow velocity increase from lower altitude toward higher altitude while the pressure was decreased.

Case 2

In this case, four air inlets are located near the corners, and two outlets are located at the center. The results of the 3D model illustrated in Fig. 7 represent the air velocity streamline inside the building and the airflow is distributed that fills all the buildings. The same results are represented by contour with two altitudes at 5, and 14 m, representing the airflow is concentrated under the inlet and outlet locations as in Fig. 8. Figure 9 shows the axial distribution of airflow velocity and pressure under the ceiling surface. It shows that the two humps away from each other represent the highest airflow velocity of 50 m/s. In contrast, the pressure is the opposite at a lower value around 12,245 Pa. Figure 10 shows the velocity and pressure distribution at

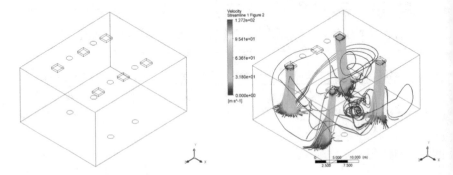

Fig. 3   3D model of the building showing the air velocity streamlines inside the building, case 1

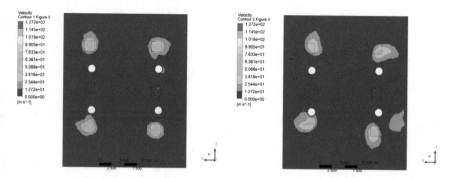

Fig. 4 The contour of the air velocity at two axial surfaces y = 5 m (left) and y = 14 m (right), case 1

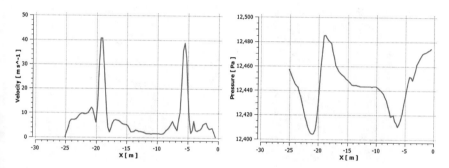

Fig. 5 Velocity and pressure distribution through x-direction passing through two inlets near the top surface, case 1

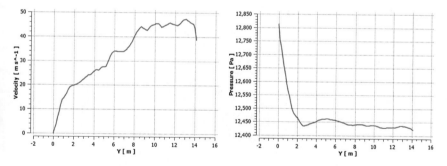

Fig. 6 Velocity and pressure distribution through y-direction at the center of one inlet, case 1

the y-direction, representing the sudden increase of flow velocity from lower altitude until it reaches 17 m/s decreases remains approximately content. In contrast, pressure take the opposite phenomenon.

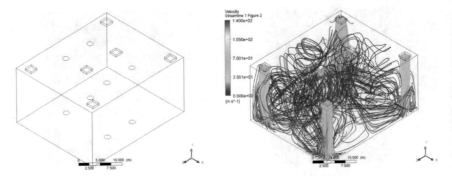

Fig. 7 3D model of the building showing the air velocity streamlines inside the building of case 2

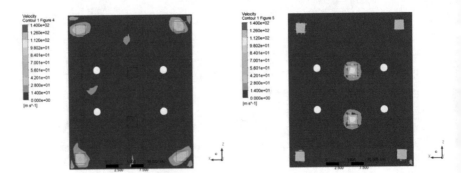

Fig. 8 The contour of the air velocity at two axial surfaces y = 5 m (left) and y = 14 m (right), case 2

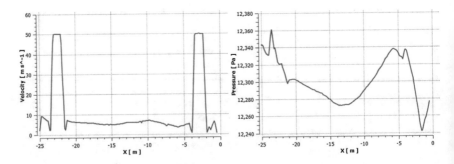

Fig. 9 Velocity and pressure distribution through x-direction passing through two inlets near the top surface, case 2

Case 3

Four inlets of air are located apart from the corners, and two outlets are located at the center. The results of the 3D model illustrated in Fig. 11 represent the air velocity

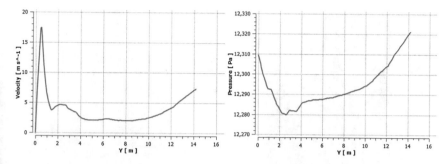

Fig. 10 Velocity and pressure distribution through y-direction at the center of one inlet, case 2

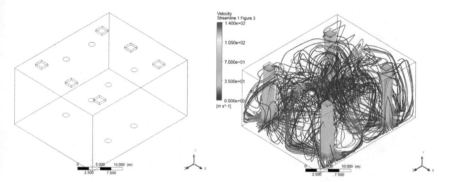

Fig. 11 3D model of the building showing the air velocity streamlines inside the building, case 3

streamline inside the building. The airflow is concentrated under the inlet location, while the rest of the building has low airflow. The same results are represented by contour with two altitudes at 5, and 14 m, representing the airflow is concentrated only under the inlet locations in Fig. 12. Figure 13 shows the axial distribution of airflow velocity and pressure just under the ceiling surface. It shows that the two humps represent the highest airflow velocity of 50 m/s while the pressure is at a lower value around 12,410 Pa. Figure 14 shows the velocity and pressure distribution at y-direction, representing the flow velocity increase from lower attitude toward higher altitude while the pressure was decreased.

The optimal inlet and outlet location was case 2, which gives a good distribution of velocity with an average of 0.15, which is comfortable for a mosque with more than 50 people praying simultaneously.

4 Conclusion

The current study uses a computational fluid dynamics CFD to simulate airflow in a mosque space. The method used was by arranging the air distribution systems

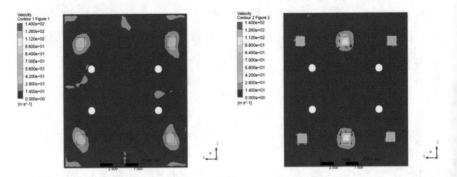

Fig. 12 The contour of the air velocity at two axial surfaces y = 5 m (left) and y = 14 m (right), case 3

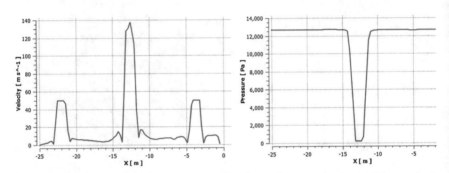

Fig. 13 Velocity and pressure distribution through x-direction passing through two inlets near the top surface, case 3

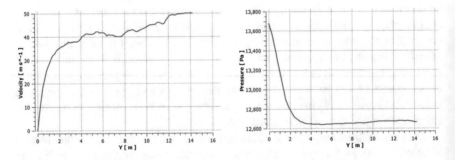

Fig. 14 Velocity and pressure distribution through y-direction at the center of one inlet, case 3

with varying inlet and outlet airflow locations inside the environment represented by the mosque. Three cases were used to compare the indoor environment ventilation systems. The analysis results revealed that placing the air inlet and outlet near the corners inside the mosque resulted in airflow distribution to cover the overall place, presented in case 2. In contrast, if the inlet and outlet are located on the same line arrangement, the outcome of the air circulation is less efficient and effective. Further studies are required to investigate the various arrangement of inlets and outlets on the celling and the building wall. Moreover, the inclusion of the people inside the mosque may give a more realistic simulation considering their breathing and heat generation from their bodies.

References

1. Al Horr Y, Arif M, Kaushik A, Mazroei A, Katafygiotou M, Elsarrag E (2016) Occupant productivity and office indoor environment quality: A review of the literature. Build Environ 105:369–389
2. Yao R, Li B, Liu J (2009) A theoretical adaptive model of thermal comfort–adaptive predicted mean vote (aPMV). Build Environ 44(10):2089–2096
3. Tarantini M, Pernigotto G, Gasparella A (2017) A co-citation analysis on thermal comfort and productivity aspects in production and office buildings. Buildings 7(2):36
4. Rupp RF, Vásquez NG, Lamberts R (2015) A review of human thermal comfort in the built environment. Energy Build 105:178–205
5. Hajdukiewicz M, Geron M, Keane MM (2013) Calibrated CFD simulation to evaluate thermal comfort in a highly-glazed naturally ventilated room. Build Environ 70:73–89
6. Chen Q (1995) Comparison of different k-ε models for indoor air flow computations. Numer Heat Transf Part B Fundam 28(3):353–369
7. Zaki SA, Kasim NFM, Ikegaya N, Hagishima A, Ali MSM (2018) Numerical simulation on wind-driven cross ventilation in square arrays of urban buildings with different opening positions. J Adv Res Fluid Mech Therm Sci 49(2):101–114
8. Abdou AA (2003) Measurement of acoustical characteristics of mosques in Saudi Arabia. J Acoust Soc Am 113(3):1505–1517
9. Kamar HM, Kamsah NB, Ghaleb FA, Alhamid MI (2019) Enhancement of thermal comfort in a large space building. Alexandria Eng J 58(1):49–65
10. Ahmad T, Thaheem MJ, Anwar A, ud Din Z (2016) Implications of stereotype mosque architecture on sustainability. Proc Eng 145:96–103
11. Jaafar RK, Khalil EE, Abou-Deif TM (2017) Numerical investigations of indoor air quality inside Al-Haram mosque in Makkah. Proc Eng 205:4179–4186
12. Al-Tamimi N, Qahtan A, Abuelzein O (2020) Rear zone for energy efficiency in large mosques in Saudi Arabia. Energy Build 223:110148
13. Hatif IH, Hariri A, Idris AFA (2020) CFD analysis on effect of air Inlet and outlet location on air distribution and thermal comfort in small office. CFD Lett 12(3):66–77
14. Azmi NA, Ibrahim SH (2020) A comprehensive review on thermal performance and envelope thermal design of mosque buildings. Build Environ 185:107305
15. Yüksel A, Arıcı M, Krajčík M, Civan M, Karabay H (2021) A review on thermal comfort, indoor air quality and energy consumption in temples. J Build Eng 35:102013
16. Zelenský P, Barták M, Hensen JL (2015) Simulation of indoor environment in the concert hall housed in a converted former Church. In: Proceedings of the BS2015 on 14th Conf. of Int. Building Performance Simulation Association. pp 1716–1721
17. Barták M, Drkal F, Hensen J, Lain M, Matuska T, Schwarzer J, Sourek B (2001) Simulation to support sustainable HVAC design for two historical buildings in Prague. In: Proc. 18th conference on passive and low energy architecture, PLEA. pp 903–908
18. Azmi NA, Arıcı M, Baharun A (2021) A review on the factors influencing energy efficiency of mosque buildings. J Clean Prod 292:126010

Hybrid Feature-Based Multi-label Text Classification—A Framework

Nancy Agarwal, Mudasir Ahmad Wani, and Mohammed ELAffendi

Abstract Multi-label Text Classification (MLTC) as a problem is a scenario in which a text document can belong to one or more classes simultaneously. Such classification tasks pose several general as well as specific research challenges. The general challenges include dependency among classes, imbalanced data, and scalability in the presence of an excessive number of labels. On the other hand, the MLTC-specific challenges include high dimensional feature space, obtaining contextual and semantic knowledge from the text, and understanding content diversity. This paper provides a brief description of the multi-label classification approaches such as problem transformation, algorithm adaptation, and ensemble learning along with their strengths and weaknesses. Furthermore, we proposed an MLTC framework referred to as HMTCS (Hybrid feature-based Multi-label Text Classification System) that handles both general multi-labeling issues and text categorization-specific issues. The proposed framework has three modules, namely, Labels Knowledge Base, Hybrid Feature Extraction, and Ensemble Learning.

Keywords Multi-label text classification · Natural language processing · Ensemble learning · Deep learning

This research work is supported and funded by the EIAS Laboratory, Prince Sultan University, Riyadh, Saudi Arabia.

N. Agarwal (✉)
Norwegian University of Science and Technology, Gjovik 2814, Norway
e-mail: nancya@ntnu.no

M. A. Wani · M. ELAffendi
College of Computer and Information Sciences (CCIS), Prince Sultan University (PSU), Riyadh 11586, Saudi Arabia
e-mail: mwani@psu.edu.sa

M. ELAffendi
e-mail: afffendi@psu.edu.sa

1   Introduction

Multi-label classification (MLC) problem is defined as a task in which an instance or an object can belong to multiple categories simultaneously, in contrast to traditional classification problems (i.e., binary or multi-class classification) where an instance belongs to a single class only [10] In other words we can also say that MLC refers to prediction of zero or more class labels. In traditional classification, the classes are mutually exclusive, i.e. the model aims to assign a single label to the unknown instance from the set of labels $L = l_1, l_2 \ldots l_m$. If the problem is associated with only two classes, say $|L| = 2$, it is termed as a binary classification problem, e.g. classification of emails into spam or normal email, and the problems with more than two classes, $|L| > 2$ come under the realm of multi-class classification, e.g. determining whether a news article reflects positive, negative, or neutral sentiment. However, in MLC, instead of assigning a single label to an instance, the model is designed to assign a subset of labels, $Y \in L$ to the unseen instances, e.g. identification of different objects in an image scenery.

Formally, on given a collection of multi-labeled training instances:

$$T = (x_1, y_1), (x_2, y_2) \ldots . (x_n, y_n) \tag{1}$$

where, $x_i \in X$ (set of instances) and $y_i \in Y = 2^L$, the goal of the multi-label classifier is to learn a function $h : X \to Y$.

MLC problem is transformed to Multi-label text classification (MLTC) when the data comprises text documents. In MLTC, a document may belong to more than two categories based on the degree of closeness to the conceptual background of the categories. For instance, the news categorization task where a news article can be classified into both political and economic domains, and the emotion detection task where a text can reveal multiple emotions such as fear, anger, and sadness. In another manner, we can say MLTC is a problem of mapping inputs (x) to a set of class labels (y), which are not mutually exclusive. For example, a movie can be mapped to one or more categories of art, music, or literature.

Since most of the conventional machine-learning algorithms are designed for single or multi-class data, treating multi-label problems is still a challenging task. The literature presents various techniques for building multi-label classifiers which can be broadly classified into three categories: problem transformation, algorithm adaptation, and ensemble methods [8, 10]. In the problem transformation approach, the MLC task is decomposed into several binary classification problems, and a decision is taken by combining the output of different classifiers using transformation methods such as binary relevance, classifier chaining, and stacking.The adaption method is associated with extending the learning algorithms such that they can handle multi-label data directly. For example, the study [12] proposed a new algorithm as an extended version of C4.5 decision tree to handle multi-label data. The last method comprises an ensemble of base classifiers to make multi-label predictions, and these classifiers can be designed using both problem transformation and algorithm adaptation approaches. Besides a plenty of benefits and application areas the MLC problems come with several challenges also dependencies in among non-exclusive classes, class-imbalance, scalability, etc. According to the literature [10], these inter-dependencies provide a promising source of information, and hence, must be modelled while building the classifier. Class imbalance is another common issue in multi-label task as the distribution of examples across multiple categories is mostly skewed. Classifiers built using problem transformation approaches are most likely expected to face this issue as the number of instances in a positive class can be way smaller than its negative counterpart. The scalability of algorithms is also one key issue in the MLTC domain. For example, the tasks related to product categorization for e-commerce and object identification in an image generally consist of hundreds of possible labels, and a computationally expensive algorithm might not be a practical solution in such domains.

The MLTC tasks pose some additional challenges besides the general MLC issues discussed above. The first extra overhead is raised due to the underlying latent, syntactical, semantic, and conceptual structure existing in the textual content. The MLTC model should be embedded with the understanding of several language constructs such as concealed relationships among words (e.g. poly, semy, synonymy) in order to effectively categorize a document. High dimensional feature space and class imbalance are other MLTC-specific issues. Mostly, in traditional text representation approaches such as bag-of-words and tf-idf, each word appearing in the training documents model edmonds to a feature, and thereby, a lengthy vector is produced depending on the size and tasksrsity in the document collection. Furthermore, the resultant vector is usually sparse as each document consists of only a small number of unique terms. The current study attempts to design an MLTC system that handles both general multi-labeling issues and text categorization-specific issues. The system involves the potential of several feature categories as explained in Sect. 4. Therefore, we referred to this new framework as a Hybrid feature-based Multi-label Text Classification System (HMTCS) The empirical focus of this research is to design the multi-labeling framework that leverages various sources of information such as labeled documents and conceptual background of the classes, comprises different categories of features for inducing language and problem domain expertise into the classifier, employs the suitable methods to represent the document in a dense vector format and utilizes an ensemble of different learning algorithms that include both conventional machine learning techniques such as Naïve Bayes and deep learning techniques such as the convolutional neural network.

The rest of the paper is structured as follows. Section 2 provides the literature for the multi-label classification problem. The characteristics of the proposed framework (HMTCS) are briefly mentioned in Sect. 3 framework and Sect. 4 provides the designing methodologies for the HMTCS. Finally, Sect. 5 concludes the overall work around Multi-label text classification.

2 Background Study

Text nowadays is considered as a rich source of information, but extracting insights from it can be hard and time-consuming, due to several shortcomings with it.

However, with the advancement in Natural Language Processing (NLP) and Deep Machine Learning (DML) technologies together with the power of artificial intelligence, sorting text data is getting easier. The literature has witnessed several popular text classification problems being solved with NLP and DML. For example, classification of reliable and unreliable textual content on social media [26], distinguishing users based on the emotions expressed by them on the social media platforms [1, 22, 27, 28], language error detection and correction [2], and sentiment, semantic, aspect, and opinion analysis, etc.

Here we explore the existing works around some inter-related research domains, such as multi-label classification and multi-label text classification in order to investigate the state-of-the-art. These research areas are further discussed as follows.

2.1 Multi-label Classification (MLC)

MLC approaches can be broadly classified into three categories, namely, problem transformation, algorithm adaptation, and ensemble methods [15] which are briefly described in the following subsections.

Problem Transformation Method These methods work by decomposing the MLC into several binary classification problems. Several transformation techniques have been proposed in literature which can be applied to make use of binary classifiers to predict multi-label data. Some are discussed below with their respective strengths and weaknesses.

Binary Relevance (BR)—In BR learning [4, 29], separate binary classifiers are trained for each class of the problem, and later used to predict whether the queried instance belongs to their corresponding label. Since the BR approach is simpler and computationally less expensive, it received considerable attention in MLC research. However, one of the fundamental issues in BR is that it is insufficient to capture label correlations in the dataset. Therefore, various studies proposed approaches that can model the relationships between labels and retain the simplicity of BR also. These include the classifier chaining approach that assumes partial order random label correlations by permutation technique and the stacking technique that models full-order label dependencies.

Classifier Chain (CC)—In the classifier chain method [20], a random order of the labels is selected from the permutation set, and binary classifiers are trained based on this order. Let $l_1 - l_2 - \ldots l_m$ be the selected order of labels, the training data of the h_j classifier would consist of instances $(x_i, y(i, 1), \ldots . y(i, j - 1))$ with $y(i, j)$ as assigned labels, i.e. original data is appended with the label information $l_1, l_2, \ldots l(j - 1)$ as features. During prediction, the classifiers are queried according to the order of the selected chain, and the predicted value is used as an input to the successive classifier as mentioned in the following equation:

$$y = (h_1(x), h_2(x, h_1(x)), h_3(x, h_2(x, h_1(x)))......)$$ (2)

The one limitation of the CC approach is that different orders of labels (chains) normally produce different results because of the difference in their training sets. Therefore, in order to reduce the influence of the order of the chain, the same study [20] proposes a method based on the ensemble of multiple CC classifiers. However, there is still one issue, the CC approach models the unconditional dependency only among the labels. In a study [6], the author presented the Probabilistic Classifier Chains (PCC) method that induces the concept of posterior probabilities in the CC method to model the conditional relationships among labels. In order to reduce the complexity, the two studies [19] presents Monte Carlo sampling and beam search respectively as an extension to PCC.

Stacking—The stacking method proposed in the study [9] is another approach that overcomes the limitation of BR by modeling the unconditional dependency among labels. In this approach, two groups of classifiers are built in the learning phase. The first group is based on the BR approach where for each label, an independent classifier is modeled. The second group also consists of m binary classifiers for m labels, however, the difference lies in the training set. They are trained using an augmented feature space which consists of the output vector of the first level binary classifiers along with existing feature values as given in the below equations.

First level classifiers:

$$h^1(x) = (h_1^1(x), ...h_m^1(x))$$ (3)

Second level classifiers:

$$h^2(x, \hat{y}) = (h_1^2(x, \hat{y}), ...h_m^2(x, \hat{y})), where \ \hat{y} = h_m(x)$$ (4)

During prediction, only the outputs of the second level classifiers are taken into consideration, the predictions at the first level are only used to supply the attribute values to the augmented feature set. The drawback of this approach is that the model complexity becomes higher for a problem having a large number of labels. The one way to deal with it is instead of appending all the labels in $h(j)$ meta classifier, supply only those labels information which are related to label l_j.

Label Powerset (LP)—This is another transformation method where each subset of labels is considered as a class value and correspondingly training sets are constructed for building the respective binary classifiers [23]. Given an unseen example, the classifier outputs the most probable class which is nothing but a set of predicted labels. However, this approach is also challenged by the problem with a high number of labels as the total number of label subsets gets increased which increases the computational cost.

Adaptation Method In this approach, the conventional binary algorithms are modified to assign the multi-labels to the data simultaneously. For example, the study [5] presents the adapted version of the C4.5 algorithm whose outcome of classification is a set of labels. In work, [21], two extensions of AdaBoost algorithms are developed where the former version is intended to predict the relevant labels and the latter is designed for first-levels based on their relevance. BP-MLL (Backpropagation of Multilabel Learning) based on the neural network architecture with a modified error function considers the labels correlations in the sense that relevant labels should receive higher scores than the non-relevant ones.

Ensemble Methods This method aims at designing multiple classifier systems using different subsets of feature second-level data and integrating these classifiers to make a final prediction. There exist numerous possible ways to construct an ensemble of classifiers because of four main reasons: (1) there is no limitation on the number of classifiers, (2) different models can be trained on either the same learning algorithm or different, (3) subsets of features or training data can be created in many ways, and (4) there are multiple ways to combine the decision of different classifiers. The ability to contribute classifiers of making diverse predictions is the one key property that makes an ensemble approach a viable solution. In the study [24], the author proposes RAkEL (RAndom k labELsets) which determines the labels based on the ensemble of the multilabel classifiers. For training these classifiers, the label set of the problem is broken into a number of subsets and the LP approach is adopted for learning. For predicting the unlabeled data, the output of all LP classifiers is taken into consideration via the voting scheme.

2.2 Multi-label Text Classification (MLTC)

In this section, the focus is on the studies which are dedicated to the MLTC problem. In [8], the authors proposed an ensemble-based method, Multi Label Rotation Forest (MLRF) where multiple classifiers are learned on the different subsets of features. The novelty in the work lies in the utilization of rotation forest as an ensemble technique with latent semantic indexing (LSI) as a dimensional reduction approach. The study [3] treats the MLTC problem as a ranking problem and employs the learning to rank (LTR) method for selecting the top-ranked labels of the documents. The authors designed a feature set that reflects the contextual similarity between the label terms and documents. Since label term is a single word, the knowledge representation for label l_i is obtained by the text content of the documents labeled with l_i. In studies [12, 18], the authors explore the neural network to directly predict the labels of a document. The proposed network comprise a hidden layer and feedforward output layer with a cross-entropy method as the loss function. They also modeled a threshold function to transform the probabilities of labels predicted by the network into their binary values. Since deep learning approaches have gained momentum in outperforming many NLP tasks, the MLTC research community has also been inclined towards adopting this methodology in solving their challenges. As an example the study [14], CNN as a deep learning model is tailored for the MLTC problem. The model consists of multiple filters followed by a dynamic max-pooling scheme for representing a document in a low-dimensional space. The model is trained using a binary cross-entropy loss function.

3 Characteristics of the HMTCS Framework

The primary goal of this paper is to design a Multi-label Document Classification System with following characteristics

- Understanding of language and semantic constructs of the textual content along with the problem domain.
- Considers label inter-dependencies while predictions.
- Applicable to hybrid category of features
- Holds continued Scalability.
- Outperforms state-of-the-art methods.

4 Hybrid Feature-Based Multi-label Text Classification System (HMTCS)

The proposed methodology pipeline for HMTCS consists of three modules as presented in Fig. 1. The description of each module is given as under.

Module 1—Knowledge Base of Labels: A small description of the background of the categories might help the model to find out the right set of labels for a textual document based on the assumption that the content of the document would be more similar to the description of relevant categories than the non-relevant ones. For example e, consider a task where a new article can belong to the following categories: politics, education, recruitment, and entertainment. Then, the conceptual knowledge of these textual classes would be useful for determining their similarity with a document, and therefore, would assist the model in making a more informed decision. The conceptual knowledge of classes can be built in two steps: Data Aggregation and Text Summarization.

Data Aggregation—It aims to collect the required content that is useful for building the knowledge base of the labels. It can be proceeded by simply aggregating all the instances in a training set which belong to a particular label [3], or by leveraging the existing knowledge resources such as Wikipedia and Probase [25].

Text Summarization—The data collected during aggregation will highly likely contain irrelevant information. Text summarization [13] method would help in reducing the

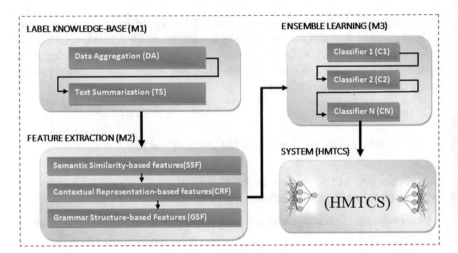

Fig. 1 Framework for Hybrid feature-based Multi-label Text Classification System (HMTCS)

information in the source text to the content which expresses only important points. Extractive summarization is one of the viable approaches which works by picking up the informational sentences based on a scoring function, and thereby, will produce the condensed conceptual knowledge of the textual labels.

Module 2—Feature Extraction: This module aims to automatically pre-process textual data and extract the different categories of features that would assist the model in learning the pattern from multiple aspects. Three categories of features are proposed in this methodology which is discussed below:

Semantic similarity-based features—The objective is to design the feature set that reflects the semantic relationships in document-label pairs. It can be calculated in two steps: *(1) Data Representation:* In order to evaluate the similarity, the labels, and document content need to be converted into some mathematical notation. n-gram, a bag of words, and tf-idf are some of the candidate solutions to design the mathematical vectors for both label and document. *(2) Similarity Estimation:* This step includes leveraging similarity metrics that would assist in computing the degree of closeness between labels and documents. Euclidean distance, cosine similarity, Jaccard coefficient, Pearson correlation coefficient, and kullback-leibler Divergence [11] are some of the similarity measures which are widely used by researchers. However, in particular to text, the study [16] mentions various similarity metrics that make use of WordNet to estimate the semantic closeness among textual content, e.g. shortest path-based measure, and palmer's measure, etc.

Contextual Representation-based features—Word embedding techniques are one of the popular approaches to represent a document contextually by converting a word into an array of continuous values such that the size of the vector is independent of the vocabulary size, and words with similar contexts are assigned similar values. Word embeddings are nowadays have become a standard for representing the text in deep learning models designed to deal with NLP problems. Various embedding models such as Word2Vec, GloVe, fastText, and ELMo [17] have been introduced by researchers from time to time to learn the representation of a word in a low dimensional semantic vector space. Sentence embedding models such as BERT (Bidirectional Encoder Representations from Transformer) [7] can also be used to transform a document into the contextual numerical format. Each embedding model offers its methodology to capture the context and semantic meaning of the words and therefore, yields a different vector.

Grammar Structure-based features—The vector that holds the grammar knowledge of the words in the text has the potential to enhance the MLTC performance as they are capable of handling the two classical problems in NLP: synonymy and polysemy. Synonymy refers to the scenario where two different words (e.g. car and automobile) hold the same meaning. It can be noticed that although they are two distinct terms but grammatically they are the same, i.e. a noun. Similarly, polysemy refers to the case when a word serves multiple meanings in different contexts. Named entity recognition (NER) and part-of-speech (POS) tagging are the two approaches that can be useful in capturing the syntactic knowledge of the data. NLTK and SpaCy are some of the libraries that provide functions to extract the NER-based and POS-based information from the text.

Module 3—Ensemble Learning The ensemble method has been proven successful to enhance the performance of the classification by utilizing the power of different classifiers. In this approach, instead of just one, multiple classifiers are trained, and during the prediction of unseen instances, the output of all the classifiers is taken into consideration via some scheme such as voting to form the final decision. These classifiers can be trained on different subsets of either features or data. Furthermore, these classifiers can be built using either the same machine learning (ML) algorithm or different algorithms. Since, the feature set in the proposed methodology belongs to three different categories, experimenting with different ML techniques might yield better performance. The base-level classifiers can belong to both traditional ML algorithms (e.g. SVM, NB, Random Forest, etc.) and deep learning algorithms (e.g. CNN, RNN, LSTM, GRU, Transformer, etc.). Semantic similarity-based and grammar structure-based features can be used as input to train both traditional ML and DL models. However, contextual representation-based features suit best to DL classifiers as their architecture is inherently designed to handle word-embedded vectors.

Upon combining all the three above-discussed modules we will be able to have a framework that will support handling the general as well as specific issues related to multi-label text document classification. The framework will be equipped with all the qualities mentioned in Sect. 3.

5 Conclusions

This paper discussed the three approaches to multi-label classification, namely, problem transformation, adaptation and ensemble training. And, based on the existing studies we proposed a framework that handles both general multi-labelling issues and text categorization-specific issues. The proposed framework has three modules, namely, *Labels Knowledge Base, Hybrid Feature Extraction* and *Ensemble Learning*. The knowledge about the classes have been built using Data Aggregation and Text Summarization techniques and the feature extraction module automatically preprocess textual data and extract the different categories of features and help the model in learning the pattern from multiple aspects. And finally the ensemble learning module utilises the power of multiple models for classification task. The proposed model will be able to understand the language and semantic knowledge of the textual content, consider label inter-dependencies while prediction, and can be applied to several category of features simultaneously.

The future work in this direction will focus on the implementation of proposed framework on real world textual datasets. We will also focus on the extension of proposed framework to solve the multi label classification related problems in audio signals and image contents.

Acknowledgements This research work is supported and funded by the EIAS Laboratory, Prince Sultan University, Riyadh, Saudi Arabia.

References

1. Agarwal N, Jabin S, Hussain SZ et al. (2019) Analyzing real and fake users in facebook network based on emotions. In: 2019 11th international conference on communication systems and networks (COMSNETS). IEEE (2019), pp 110–117
2. Agarwal N, Wani MA, Bours P (2020) Lex-pos feature-based grammar error detection system for the english language. Electronics 9(10):1686
3. Azarbonyad H, Dehghani M, Marx M, Kamps J (2021) Learning to rank for multi-label text classification: combining different sources of information. Natl Lang Eng 27(1):89–111
4. Boutell MR, Luo J, Shen X, Brown CM (2004) Learning multi-label scene classification. Pattern Recognit 37(9):1757–1771
5. Clare A, King RD (2001) Knowledge discovery in multi-label phenotype data. In: European conference on principles of data mining and knowledge discovery. Springer, pp 42–53
6. Dembczynski K, Cheng W, Hüllermeier E (2010) Bayes optimal multilabel classification via probabilistic classifier chains. In: ICML
7. Devlin J, Chang MW, Lee K, Toutanova K (2018) Bert: pre-training of deep bidirectional transformers for language understanding. arXiv preprint arXiv:1810.04805
8. Elghazel H, Aussem A, Gharroudi O, Saadaoui W (2016) Ensemble multi-label text categorization based on rotation forest and latent semantic indexing. Expert Syst Appl 57:1–11
9. Godbole S, Sarawagi S (2004) Discriminative methods for multi-labeled classification. In: Pacific-Asia conference on knowledge discovery and data mining. Springer, pp 22–30
10. Gunasekara I, Nejadgholi I (2018) A review of standard text classification practices for multi-label toxicity identification of online content. In: Proceedings of the 2nd workshop on abusive language online (ALW2), pp 21–25

11. Huang A et al. (2008) Similarity measures for text document clustering. In: Proceedings of the sixth new zealand computer science research student conference (NZCSRSC2008), Christchurch, New Zealand, vol 4, pp 9–56
12. Jiang M, Pan Z, Li N (2017) Multi-label text categorization using l21-norm minimization extreme learning machine. Neurocomputing 261:4–10
13. Kanapala A, Pal S, Pamula R (2019) Text summarization from legal documents: a survey. Artif Intell Rev 51(3):371–402
14. Liu J, Chang WC, Wu Y, Yang Y (2017) Deep learning for extreme multi-label text classification. In: Proceedings of the 40th international ACM SIGIR conference on research and development in information retrieval, pp 115–124
15. Madjarov G, Kocev D, Gjorgjevikj D, Džeroski S (2012) An extensive experimental comparison of methods for multi-label learning. Pattern Recognit 45(9):3084–3104
16. Meng L, Huang R, Gu J (2013) A review of semantic similarity measures in wordnet. Int J Hybrid Inf Technol 6(1):1–12
17. Mikolov T, Grave E, Bojanowski P, Puhrsch C, Joulin A (2017) Advances in pre-training distributed word representations. arXiv preprint arXiv:1712.09405
18. Nam J, Kim J, Loza Mencía E, Gurevych I, Fürnkranz J (2014) Large-scale multi-label text classification-revisiting neural networks. In: Joint European conference on machine learning and knowledge discovery in databases. Springer, pp 437–452
19. Read J, Martino L, Olmos PM, Luengo D (2015) Scalable multi-output label prediction: from classifier chains to classifier trellises. Pattern Recognit 48(6):2096–2109
20. Read J, Pfahringer B, Holmes G, Frank E (2011) Classifier chains for multi-label classification. Mach Learn 85(3):333–359
21. Schapire RE, Singer Y (2000) Boostexter: a boosting-based system for text categorization. Mach Learn 39(2):135–168
22. Shakil KA, Tabassum K, Alqahtani FS, Wani MA (2021) Analyzing user digital emotions from a holy versus non-pilgrimage city in saudi arabia on twitter platform. Appl Sci 11(15):6846
23. Sun L, Kudo M, Kimura K (2016) Multi-label classification with meta-label-specific features. In: 2016 23rd international conference on pattern recognition (ICPR). IEEE, pp 1612–1617
24. Tsoumakas G, Katakis I, Vlahavas I (2010) Random k-labelsets for multilabel classification. IEEE Trans Knowl Data Eng 23(7):1079–1089
25. Wang F, Wang Z, Li Z, Wen JR (2014) Concept-based short text classification and ranking. In: Proceedings of the 23rd ACM international conference on conference on information and knowledge management, pp 1069–1078
26. Wani M, Agarwal N (2021) Bours P (2020) Impact of unreliable content on social media users during covid-19 and stance detection system. Electronics 10:5
27. Wani MA, Agarwal N, Bours P (2021) Sexual-predator detection system based on social behavior biometric (SSB) features. Proc Comput Sci 189:116–127
28. Wani MA, Agarwal N, Jabin S, Hussain SZ (2018) User emotion analysis in conflicting versus non-conflicting regions using online social networks. Telemat Informat 35(8):2326–2336
29. Zhang ML, Li YK, Liu XY, Geng X (2018) Binary relevance for multi-label learning: an overview. Front Comput Sci 12(2):191–202

Voice Recognition and User Profiling

Bahaa Eddine Elbaghazaoui⊙, **Mohamed Amnai**⊙, **and Youssef Fakhri**⊙

Abstract Scientific exploration searches in different fields to define and outline an emergent area of research. One of the important challenges is that of profiling humans by their voice. In today's era of data technology, audio information plays an important role in increasing data volume. Voice recognition is one of the methodologies that aims to recognize the person speaking the words, rather than the words themselves. As technology has evolved, voice recognition has become increasingly embedded in our everyday lives, with voice-driven applications in digital appliances. Voice recognition systems enable consumers to interact with technology simply by speaking to them, enabling hands-free requests, reminders, and other simple tasks. This study could be viewed as a literature review on voice recognition technology with user profiling. The purpose of this research is to present voice recognition, to provide a global vision of human voice recognition over the big data era, and then highlight voice recognition techniques, use cases, and challenges. Finally, we conclude by discussing some future research directions.

Keywords Voice recognition · User profiling · Speaker recognition · Big data · Audio data

1 Introduction

Humans could now communicate with robots via voice commands and instructions due to speech recognition technologies [3]. As a result, this technology is now used in a variety of applications, including cellular systems, telephones, and other fields [4].

B. E. Elbaghazaoui (✉) · M. Amnai · Y. Fakhri
Laboratory of Computer Sciences, Faculty of Sciences Kenitra, Ibn Tofail University, Kenitra, Morocco
e-mail: elbaghazaoui.bahaa@gmail.com

The mismatch problem that emerges owing to the discrepancy between the testing and application settings that have been contaminated with noise is the primary cause of voice recognition system performance decline.

Digital data as audio or images are being taken, created, and stored in greater numbers as information technology progresses. In order to effectively and efficiently use the information held in these media forms, there has been a great interest in multimedia indexing and retrieval research and development [2]. The audio data is by far the most important signal that contains a wealth of information [1]. We can collect an endless amount of data and information in the speech signal.

The audio data are converted into a stream of words during the speech recognition process [5]. Several factors influence the performance of these systems, including the environment, vocabulary, speaker variability, and so on. Speech recognition systems work best in a clean environment with a limited vocabulary and few utterances per speech. In the presence of noise, the system's performance suffers.

All voice recognition systems have two primary stages that have a significant impact on the system's operation and recognition rate. One is the front-end stage, which turns voice samples into a stream of feature vector coefficients that only include the information needed to identify a specific utterance [6]. LPC (Linear Predictive Coding) and MFCC (Mel Frequency Cepstral Coefficient) are examples of feature extraction methods [7]. The classification or pattern matching stage identifies which category each pattern belongs to, such as SVM (Support Vector Machine) and HMM (Hidden Markov Model) [8].

This paper's structure, which summarizes all of the previous ideas, is as follows. We begin by outlining voice recognition and the strategies that have been offered. After that, we talk about the importance of voice recognition in the age of big data. Following that, we will show voice recognition techniques, use cases and challenges. Finally, we reach a broad conclusion with some recommendations for the future works.

2   Voice Recognition

A machine or program's capacity to accept and interpret dictation or recognize and carry out spoken commands is known as voice or speaker recognition [9]. With the emergence of AI and intelligent assistants like Amazon's Alexa, Apple's Siri, and Microsoft's Cortana, voice recognition has gained popularity and use. Voice recognition systems allow users to engage with technology merely by speaking to it, allowing them to make requests, set reminders, and perform other simple activities without having to use their hands.

Analog audio must be converted into digital signals by voice recognition software on computers, a process known as analog-to-digital conversion. A computer needs have a digital library, or vocabulary, of words or syllables, as well as a quick way to compare this data to signals in order to decode a signal [10]. When the application is started, the speech patterns are stored on the hard drive and loaded into memory. A comparator compares the recorded patterns to the A/D converter's output, a process known as pattern recognition.

In practice, the amount of a voice recognition program's effective vocabulary is exactly proportional to the computer's random access memory capacity [11] When compared to searching the hard drive for some of the matches, a voice recognition program operates many times faster if the complete vocabulary can be put into RAM. Processing speed is also important since it influences how quickly the computer searches the RAM for matches.

As the use of speech recognition technology expands and more users interact with it [12], companies using voice recognition software will have more data and information to send into the neural networks that fuel voice recognition systems, enhancing the products' capabilities and accuracy.

3 Profiling and Human Voice

Profiling from speech is the process of extracting personal characteristics as well as information about a speaker's circumstances and environment from their voice. Identifying data about the user's characteristics and interest domain is called user profile analysis or user profiling [20] and the process known by data profiling [21]. To begin, we should clarify the difference between the phrases voice and speech. The term "voice" refers to the sound made by the human vocal tract. The signal produced by modulating voice into meaningful patterns is referred to as "speech".

In computational voice profiling, there are numerous sorts of personal characteristics, often known as speaker parameters. Gender, weight, and height of the speaker are examples of physical parameters. Other characteristics, such as the speaker's age and heart rate, are used to capture the speaker's physiological state. Ethnicity, geographical origin, and native language are examples of demographic information about the speaker. The speaker's emotional state and sociological features are examples of psychological qualities. Medical criteria assess the speaker's overall health, drunkenness level, and presence of neurological problems, among other things.

Applied voice profiling has many application in law enforcement, security, and healthcare. Artificial intelligence and machine learning advancements have sparked a surge in research interest in the topic. The key assumption behind computational voice profiling is that if any parameter influences the production of voice, the relationship may be established. Voice creation, on the other hand, is a complex biomechanical process with a lot of variation. When a speaker makes the same sounds under the same controlled settings, the spectral and temporal representations of the signal can change. The significant degree of variability in the human vocalization process poses a number of obstacles for computational voice profiling, any system must be tolerant of the signal's inherent changes.

Variations in the produced signal can be caused directly (or indirectly) by speaker characteristics. These effects can be seen and investigated in order to figure out the relationship between speaker parameters and the change in the produced speech signal. Knowledge-driven and data-driven techniques to identifying these linkages can be broadly divided into two types.

The organs used in each subsystem of human voice production are affected by speaker characteristics. The created sound wave has distinct characteristics as a result of these effects, and these differences can be seen in the resulting voice signal. The basic premise of speech profiling is that speaker characteristics cause distinct observable patterns in the voice signal, which may be determined by analyzing these patterns [17].

4 Voice Recognition Techniques

Hearing, understanding, and then acting with spoken information are the most important aspects of speech recognition systems. The ASR or Automatic Speech Recognition system is broadly classified as shown in following Fig. 1.

The initial stage of the ASR system is analysis [13]. Speech data includes the speaking tract, the source of excitation, and the behavior attribute to reveal the speaker's identity. The speech analysis step is divided into three stages: segment analysis, subsegmental analysis, and suprasegmental analysis.

Feature extraction is the most important phase of an ASR system [14]. It is quite crucial in the system. By extracting features from the input sound, feature extraction aids the system in identifying the speaker.

In ASR systems, two modeling strategies are used: speaker identification and speaker recognition [16]. The information extracted by the speech signal aids in identifying the speaker. In the voice recognition process, acoustic-phonetic method and dynamic temporal warping (DTW) are two common modeling approaches.

The recognition of words is the emphasis of the Matching pattern approach. The voice recognition engine uses the identified word and then compares it to a previously known word [15]. This strategy works by using either a sub-word matching or a whole-word matching method.

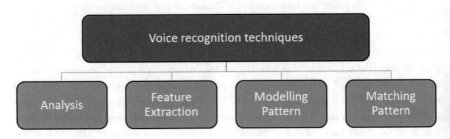

Fig. 1 Voice recognition techniques

5 Voice Recognition Over Big Data Era

Speech recognition is increasingly becoming a part of our daily life, whether it's for searching on small handheld devices, controlling home entertainment, or inputting, entering a destination address into a navigation system. Many other speech analysis jobs, particularly the 'paralinguistic' ones, are still in the works. The most common application in everyday life is speaker identification, such as in various telephone banking scenarios. Following that, most probable and frequently ignored are gender and age group in dialogue systems or just to customize the voice recognizer.

However, as different resources such as the internet, broadcast, voice communication, and greater use of speech-services such as self-monitoring allow access to 'large' amounts, it is becoming less and less the real speech data that is lacking in the big data era. Instead, it's the labels that aren't there. Fortunately, with the growing availability of crowdsourcing services and recent breakthroughs in weakly supervised, contextual, and reinforcement learning, new opportunities to mitigate this reality have emerged.

When we talk about 'big data,' we usually mean that the amount of data is so large that 'traditional' processing methods can't handle it [18]. This may necessitate data division and effort distribution [19]. While there is a large body of work on how to best distribute processing in the field of 'core' Machine Learning, it will remain to adjust these approaches to the needs of speech analysis. Distributed processing has been heavily focused on voice recognition, but not so much on paralinguistic problems, where only preliminary results have been published, such as optimal compression of feature vectors [22].

6 Application Profiling Use Cases

Due to the large number of applications and the development of interfaces or computing devices that can enable speech processing, Automatic Speech Recognition (ASR) is a thriving topic of research. Profiling has a wide range of applications [23]. The science of profiling is still in its early stages. As it improves in accuracy, additional applications will arise. The most common fields in which voice recognition can be used are depicted in the diagram Fig. 2.

6.1 Law Enforcement

In cases when voice is utilized as evidence, profiling can be used to provide additional and supporting evidence. Some crimes are performed purely by voice, and the criminal's voice is frequently the only concrete evidence of their identity in these cases. In these circumstances, profile information may be particularly useful in locating and

Fig. 2 Application areas
and uses of voice profiling

identifying the criminal [25]. The technology may be used to analyze voiceprints from a variety of different sources. Voice evidence from crimes recorded on a variety of security and surveillance devices, including wearable smart devices, includes radio-transmitted voice, wiretaps, telephone conversations, and voice evidence from crimes recorded on a variety of security and surveillance devices, including wearable smart devices. Its application in law enforcement, on the other hand, presents more obstacles than in other fields.

6.2   Security

A speaker's voice is compared to those in a database in most contemporary security systems. The system is said to do speaker identification when the purpose is to identify the speaker from a list of possible speakers. In this example, the voiceprint is compared to all of the recordings in the database, with the closest match determining the speaker's identity [26]. When the purpose is to verify that a speaker is who they say they are, on the other hand, the voiceprint is only matched to the specific speaker's voiceprint in the database if it exists. The system is said to perform speaker verification, or speaker authentication, in this situation. The basic technique for voiceprint comparison is the same in both scenarios.

6.3 Health Services

Voice is non-intrusive and may be easily transferred for examination to faraway sites. It can be used to assess, track, and monitor a person's physical and mental health status over time [27]. It can be used to track medical drug compliance, intoxication, substance misuse, suicidal thoughts, and a variety of other things. Telemonitoring or remote monitoring of poor populations, veterans, old people, disabled persons, children, and others is expected to have the most impact.

6.4 Social and Commercial Services

Using the power of human speech to better comprehend people on a deeper level can have a huge impact on society [28]. The technology can lead application servers to make automated decisions that are more appropriate for the human at any given time, such as speech interactive systems or voice-driven human-machine interfaces. Their responses can be adjusted to their current situation.

6.5 Gaming and Entertainment

Even in its early stages, when reconstructions aren't perfect in every way, recreation of the human in 2D and 3D from speech can offer up new gaming and entertainment possibilities [24]. For entertainment, profiling-centric attractions such as holographic reconstructions, themed creations of characters based on determined personality features of a speaker, and so on can be set up. Gaming servers can customize game content to the player's characteristics depending on the situation. For example, age and height appropriate characters in animated children's games.

7 Challenges

Approaches to speaker recognition confront a number of difficulties. Intra-speaker variability is the most typical obstacle in data-driven tactics [29]. Both text-dependent and text-independent speaker recognition tasks face similar difficulties. The general issues of speaker recognition were initially discussed in this part, followed by the technological and deployment challenges.

7.1   General Challenges

For both text-dependent and text-independent models, the following are the general challenges of speaker recognition tasks:

Data driven dependency: To teach the background procedures, a vast amount of knowledge is required. The database had to be formatted and organized in a very controlled manner, which necessitated significant human effort.

Intra speaker variability: Speaker recognition tasks are made more difficult by a variety of unpredictable sorts.

Speaker based variability: It shows how a speaker speaks differently and has an impact on ASR accuracy.

Conversation based variability: It depicts various circumstances involving voice contact with another person or system, as well as disagreements about a specific language or accent.

Technology based variability: It includes issues such as electromechanical, environmental, data quality duration, sampling rate, recording quality, and audio compression as well as the time and place of audio recording.

Low resource languages: When the concept of few resources is applied to these approaches, the outcome is less than ideal. As a result, there is a significant amount of effort required in the field of low-resource language ASR systems.

7.2   Technological Challenges

The key algorithms are inextricably linked to the technology challenges of speaker recognition frameworks. The following are a few technological challenges:

Limited data and constrained lexicon: In the most serious cases, the testing lexicon is selected because it closely matches the enrollment lexicon.

Channel usage: It's an important area where speech recognition architectures need to be improved.

Aging of speaker models: Behavioral changes have an impact on the models and scores, which is reflected in the efficiency.

7.3   Deployment Challenges

When translating the speaker recognition architecture into an actual application, the following deployment issues of speaker recognition systems are encountered:

Cost of deployment: Most developers want a better grasp of how accurate their protective layer is. They have a strong need to calculate original data for the traditional false accept, false reject, and prompt rates. As a result, the deployment costs are significant.

Forward compatibility: The speaker model's description of lexical features is an essential component, and any changes would reduce efficiency. It also governs what research can be used to improve an existing algorithm.

Today's speaker recognition algorithms are complicated by difficult audio datasets, diverse phonation styles, speech under stress in the record, channel mismatch, and speech modality. Solving these problems, on the other hand, necessitates ASR's vast advanced research capabilities.

8 Conclusion

Voice recognition is a well-studied study area that has been integrated into a variety of systems for identifying and verifying people. However, only a small amount of research has been done in this huge field, and most of it is now outdated. As a result, the chapter concentrates on a well-known research subject and investigates numerous aspects of such a fascinating research field.

The paper targets the voice recognition in multiple aspects, including state of art, use cases and challenges. However, according to this work, we could take it as a survey on profiling speaker based on voice recognition.

This work bolsters the existing defaults and variations of ASR systems, which will aid new researchers in fast adapting to the research domain concepts. It would also aid in the exploration of a broader view of speaker identification systems, such as architectural and feature extraction terminologies.

References

1. Singh AP et al (2018) A survey: speech recognition approaches and techniques. In: 2018 5th IEEE Uttar Pradesh section international conference on electrical, electronics and computer engineering (UPCON). IEEE, p 14. DOI.org (Crossref), https://doi.org/10.1109/UPCON.2018. 8596954
2. Lu G (2001) Indexing and retrieval of audio: a survey. Multimed Tools Appl 15:269–290. https://doi.org/10.1023/A:1012491016871
3. Gupta D et al (2018) The state of the art of feature extraction techniques in speech recognition. In: Agrawal SS et al (eds) Speech and language processing for human-machine communications, dit par, vol 664. Springer Singapore, p. 195207. DOI.org (Crossref), https://doi.org/10. 1007/978-981-10-6626-9_22
4. Bhabad SS, Kharate GK (2013) An overview of technical progress in speech recognition. Int J Adv Res Comput Sci Soft Eng 3(3)
5. Nehe NS, Holambe RS (2012) DWT and LPC based feature extraction methods for isolated word recognition. EURASIP J Audio, Speech, Music Process 2012(1):7. DOI.org (Crossref), https://doi.org/10.1186/1687-4722-2012-7. (decembre 2012)

6. Mishra AN et al (2010) Comparative wavelet, PLP, and LPC speech recognition techniques on the Hindi speech digits database, p 754634. DOI.org (Crossref), https://doi.org/10.1117/12. 856318
7. Mada Sanjaya WS et al (2018) Speech recognition using linear predictive coding (LPC) and adaptive neuro-fuzzy (ANFIS) to control 5 DoF Arm Robot. J Phys Conf Ser 1090:012046. DOI.org (Crossref), https://doi.org/10.1088/1742-6596/1090/1/012046. (septembre 2018)
8. Tiwari V (2010) MFCC and its applications in speaker recognition. Int J Emerg Technol 1(1):19–22
9. Chiu, C-C et al (2018) State-of-the-art speech recognition with sequence-to-sequence models. In: 2018 IEEE international conference on acoustics, speech and signal processing (ICASSP). IEEE, p 477478. DOI.org (Crossref), https://doi.org/10.1109/ICASSP.2018.8462105
10. Triwiyanto T et al (2021) Design of hand exoskeleton for paralysis with voice pattern recognition control. J Biomim Biomater Biomed Eng 50:5158. DOI.org (Crossref), https://doi.org/10.4028/www.scientific.net/JBBBE.50.51. (avril 2021)
11. Mantovani SCA, De Oliveira JR (2003) Evolvable hardware applied to voice recognition. In: Intelligent engineering systems through artificial neural networks, vol 13, pp 321–326
12. Meng J, Zhang J, Zhao H (2012) Overview of the speech recognition technology. In: Fourth international conference on computational and information sciences, vol 2012, pp 199–202. https://doi.org/10.1109/ICCIS.2012.202
13. Gaikwad SK et al (2010) A review on speech recognition technique. Int J Comput Appl 10(3):1624. DOI.org (Crossref), https://doi.org/10.5120/1462-1976. (novembre 2010)
14. Yadav KS, Mukhedkar MM (2013) Review on speech recognition. Int J Sci Eng 1(2):61–70
15. Anasuya MA, Katti SK (2009) Speech recognition by machine: a review. Int J Comput Sci Inf Secur 6(3):181–205.7
16. Luengo I et al (2010) Feature analysis and evaluation for automatic emotion identification in speech. IEEE Trans Multimed 12(6):490501. DOI.org (Crossref), https://doi.org/10.1109/TMM.2010.2051872. (octobre 2010)
17. Singh R (2019) Profiling humans from their voice. Open WorldCat, https://doi.org/10.1007/978-981-13-8403-5
18. Madden S (2012) From databases to big data. IEEE Internet Comput 16(3):46. DOI.org (Crossref), https://doi.org/10.1109/MIC.2012.50. (mai 2012)
19. Ghai W, Singh N (2012) Literature review on automatic speech recognition. Int J Comput Appl 41(8):4250. DOI.org (Crossref), https://doi.org/10.5120/5565-7646. (mars 2012)
20. Eke CI et al (2019) A survey of user profiling: state-of-the-art, challenges, and solutions. IEEE Access 7:14490724. DOI.org (Crossref), https://doi.org/10.1109/ACCESS.2019.2944243
21. Elbaghazaoui BE et al (2021) Data profiling over big data area: a survey of big data profiling: state-of-the-art, use cases and challenges. In: Intelligent systems in big data, semantic web and machine learning, dit par Noreddine Gherabi et Janusz Kacprzyk, vol 1344. Springer International Publishing, p 11123. DOI.org (Crossref), https://doi.org/10.1007/978-3-030-72588-4_8
22. Zhang Z et al (2014) Distributing recognition in computational paralinguistics. IEEE Trans Affect Comput 5(4):40617. DOI.org (Crossref), https://doi.org/10.1109/TAFFC.2014.2359655. (octobre 2014)
23. Fendji JLKE et al (2021) Automatic speech recognition using limited vocabulary: a survey. arXiv:2108.10254 [cs, eess]. (aot 2021)
24. Xia L (2018) China focus: technologies at summer Davos offer a glimpse into future. News Article, XinhuaNet, China, vol 20
25. Lally SJ (2003) What tests are acceptable for use in forensic evaluations? A survey of experts. In: Prof Psychol Res Pract 34(5):49198. DOI.org (Crossref), https://doi.org/10.1037/0735-7028.34.5.491. (octobre 2003)
26. Kreuk F et al (2018) Fooling end-to-end speaker verification with adversarial examples. In: 2018 IEEE international conference on acoustics, speech and signal processing (ICASSP). IEEE, p. 196266. DOI.org (Crossref), https://doi.org/10.1109/ICASSP.2018.8462693

27. Liu Z et al (2021) A novel bimodal fusion-based model for depression recognition. In: 2020 IEEE international conference on E-health networking, application & services (HEALTH-COM). IEEE, p. 14. DOI.org (Crossref), https://doi.org/10.1109/HEALTHCOM49281.2021.9399033

28. Nyagadza B (2020) Search engine marketing and social media marketing predictive trends. J Digit Media Policy 00(00):119. DOI.org (Crossref), https://doi.org/10.1386/jdmp_00036_1. (decembre 2020)

29. Kabir MM et al (2021) A survey of speaker recognition: fundamental theories, recognition methods and opportunities. IEEE Access 9:7923663. DOI.org (Crossref), https://doi.org/10.1109/ACCESS.2021.3084299

Compression-Based Data Augmentation for CNN Generalization

Tajeddine Benbarrad, Salaheddine Kably, Mounir Arioua, and Nabih Alaoui

Abstract Nowadays, deep learning is widely exploited in various fields due to its ability to solve complex problems. These networks have proven their efficiency compared to classical machine learning methods in several recent applications. Machine vision, as an innovative technology, represents a major element of the industrial transformation and is currently replacing human vision, which is outpaced by the speed and complexity of the tasks in most manufacturing processes. However, optimizing latency is an important challenge for integrating machine vision into real-world use cases. In this context, compression of collected images before they are transmitted and processed is crucial to save bandwidth and energy, and enhance latency in vision applications. Nevertheless, the degradation of image quality resulting from compression affects the performance of convolutional neural networks (CNNs) and reduces the accuracy of the results. In this paper, a compression-based data augmentation method is proposed to improve the classification performance of CNNs and generalize the models when tested on poor compression qualities. Three different models were trained and tested with images from the surface defect database. The obtained results in the performed experiments reveal that the compression-based data augmentation significantly increases the classification precision of CNNs, and improves the generalization of the models when tested on different compression qualities.

Keywords Machine vision · Deep learning · Classification · Data augmentation · Image compression

T. Benbarrad (✉) · S. Kably · M. Arioua
Laboratory of Information and Communication Technologies (LabTIC), National School of Applied Sciences, Abdelmalek Essaadi University, Tangier, Morocco
e-mail: tbenbarrad@uae.ac.ma

S. Kably · N. Alaoui
Ecole Supérieure d'Informatique et du Numérique, TICLab, Université Internationale de Rabat, Rabat, Morocco

© The Author(s), under exclusive license to Springer Nature Switzerland AG 2023
A. A. Abd El-Latif et al. (eds.), *Advances in Cybersecurity, Cybercrimes, and Smart Emerging Technologies*, Engineering Cyber-Physical Systems and Critical Infrastructures 4, https://doi.org/10.1007/978-3-031-21101-0_19

1 Introduction

Nowadays, deep learning is making machine vision technology for automated visual inspection more accessible and more powerful [1]. Deep learning mimics the way the human brain processes visual data, but accomplishes this task with the speed and robustness of a computerized system [2]. This technology helps ensure quality in manufacturing industries [3], control production costs and improve customer satisfaction [4].

The collected data by automated vision systems can be stored and processed on a remote server or on the edge of the network itself [5]. However, storage and bandwidth capacities are usually limited by available resources [6]. Indeed, the main challenge is to ensure that we get the right type of data at the right level of accuracy to ensure the optimal exploitation of our system [7]. Therefore, compression of collected images before they are transmitted and processed is crucial to save bandwidth and energy, and enhance latency in vision applications [8].

The aim of this compression is to minimize data redundancy in the images in order that they can be stored without taking up much space and transmitted quickly [9]. However, the problem with image compression is the degradation in quality that results in a negative impact on the classification performance of trained models [10]. Several methods have been proposed in the literature to mitigate the negative impact of this phenomenon on the classification performance of CNN models, including data augmentation and stability training [11]. Even though these techniques enhance the robustness of model classification against JPEG compression, there is still a significant degradation in classification accuracy when these trained models are applied to low quality JPEG compressed images.

In this paper, a compression-based data augmentation method is proposed to improve the classification performance of CNNs and generalize the models when tested on poor compression qualities. Each model was trained once with a dataset containing a mixture of different compression qualities and then tested on all compressed image datasets with different parameters.

The structure of this paper is as follows. Section 2 presents the related work. Section 3 describes the performed experiment in detail. Section 4 reports the results and analysis. Section 5 concludes this paper.

2 Related Work

CNNs are becoming extremely ubiquitous in the image classification task due to their ability to extract desired features from raw data [12–14]. The raw data introduced in the classification models are the pixel values of an image to be classified. Throughout the pipeline of data acquisition, encoding, transmission, and processing, the raw data fed into a CNN are generally lossy compressed. Since raw data is compressed with different compression ratios and qualities, many versions of compressed raw data can be produced [8]. This brings forth the following interesting question: Which version of the compressed raw data is the right one to optimize the performance of CNNs? In practice, JPEG is the most common encoder used to compress images [15]. Considering the image degradation induced by JPEG lossy compression algorithm affecting the performance of CNN models, various research activities have been carried out in recent years to study the impact of image compression on the performance of CNNs [16–18]. In addition, several methods have been proposed in the literature to mitigate the negative impact of JPEG compression on the classification performance of CNN models, including data augmentation and stability training [19, 20].

Generally, machine vision systems are trained and tested on high-quality image datasets, but in real-world deployments, it cannot be supposed that the input images are of high quality. Dodge et al. [16] evaluated four leading CNN models for image classification under quality distortions. They studied five types of quality distortions: blur, noise, contrast, JPEG compression and JPEG2000. The results showed that existing networks are sensitive to these quality distortions, especially blur and noise. Ghazvinian et al. [17] have studied the impact of JPEG 2000 compression on CNNs for metastatic cancer detection in histopathology images. They found that a trained model on lossy compressed images showed a classification performance that was significantly improved for the corresponding compression ratio. The authors also concluded that their CNN model was robust to compression ratios of up to 24:1 when trained on high quality uncompressed images. Grm et al. [18] studied the effects of different covariates on the verification performance of four recent CNN models using the Labelled Faces in the Wild dataset. They investigated the influence of covariates related to image quality and model characteristics, and analyzed their impact on the face verification performance of different CNN models. The results indicated that high levels of noise, blur, missing pixels, and brightness have a detrimental effect on the verification performance of all models, while the impact of contrast changes and compression artifacts is limited. Galteri et al. [19] studied the impact of compression artifacts on the localization and recognition of texts in nature. They also proposed a CNN that can remove text-specific compression artifacts and leads to improved text recognition. Experimental results on the ICDAR-Challenge4 dataset demonstrate that compression artifacts have a significant impact on localization and text recognition and that the proposed approach can improve both aspects, especially at high compression ratios. Zheng et al. [20] presented a general stability learning method for stabilizing deep networks against small input distortions that result from various types of common image processing, such as compression, rescaling and cropping. They demonstrated that their stabilized model provides robust state-of-the-art performance on large-scale near-duplicate detection, similar-image ranking, and classification on noisy datasets.

In our proposed study, a data augmentation technique based on image compression is presented. The main purpose of this approach is to generalize the surface defect classification of the experimented CNN models when they are tested on different compression qualities.

3   Experiment

Since data storage guidelines regarding image compression vary across different use cases, CNN models are trained and evaluated on a wider variety of image types with different compression ratios and qualities. This phenomenon results in a negative impact of compression on the classification performance of the trained models. In this experiment, we trained the classifiers using images compressed with a mixture of compression qualities in order to reduce the negative impact of compression on the classification performance of models to some extent. The results of these classifiers were compared first with trained models on the original dataset and second with trained models using the ImageDataGenerator class of the Tensorflow API to generate an additional training set by applying many geometric transformations for data augmentation (rescale, rotation, flip, zoom, etc.). The workflow for this experiment is shown in Fig. 1.

In practice, images are frequently compressed by JPEG encoders. The compression ratio (C_R) and compression quality of a JPEG image are controlled by a parameter called the quality factor (QF); the higher the QF, the lower the C_R and the better the compression quality. Therefore, the JPEG encoder compression parameters presented in Table 1 were used in this study. Also we noted the C_R, which evaluates the compression efficiency of an algorithm for an image. We used the formula given in Eq. (1) to calculate the C_R.

$$C_R = \frac{n_1}{n_2} \tag{1}$$

where n_1 is the original image size and n_2 is the compressed image size.

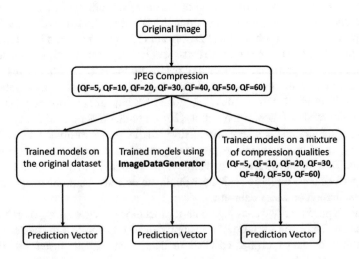

Fig. 1 The experiment's workflow

Table 1 JPEG compression parameters used in the experiment

Designation	Compression parameters		C_R
	Scale	Quality	
Q1	1/1	5	27.32
Q2	1/1	10	21.69
Q3	1/1	20	15.46
Q4	1/1	30	12.41
Q5	1/1	40	10.63
Q6	1/1	50	9.33
Q7	1/1	60	8.21

Three models with distinct architectures were chosen to study the impact of compression-based data augmentation on various types of CNN models with different characteristics:

MobileNets, a family of computer vision models for TensorFlow, designed for mobile devices and embedded vision systems. The architecture of these networks has witnessed the introduction of depthwise separable convolution, which reduces the size and complexity of the model to effectively optimize accuracy while considering the limited resources [21].

DenseNet is a CNN architecture based on dense connections between convolution layers. In which, each layer is connected to all other layers. For each layer, the feature maps of all previous layers are used as inputs, and its own feature maps are used as inputs for each subsequent layer [22].

VGG is a very deep CNN architecture developed by the Visual Geometry Group, which belongs to the Department of Science and Engineering at Oxford University. The main goal of the VGG's research on the depth of CNNs is to understand how the depth of these networks affects the precision and accuracy of large-scale image classification and recognition. The idea was to propose deep configurations (16 to 19 layers) using the structural stabilization technique. This technique allows to control the number of parameters in deep networks in order to avoid overfitting [23].

The Northeastern University (NEU) [24] surface defect database contains six types of surface defects. Therefore, we inserted a dense layer of six units at the end of all models with Softmax activation, since there were six classes to predict in output.

In this experiment, we used the Northeastern University (NEU) surface defect database to train and test the models. Six types of typical surface defects of hot rolled steel strip are contained in this dataset, i.e., Rolled-in Scale (RS), Patches (Pa), Crazing (Cr), Pitted Surface (PS), Inclusion (In) and Scratches (Sc). The database includes 1800 grayscale images: 300 samples each of six different types of typical surface defects with an original resolution of 200 × 200 pixels [24]. Sample images of six types of surface defects are shown in Fig. 2.

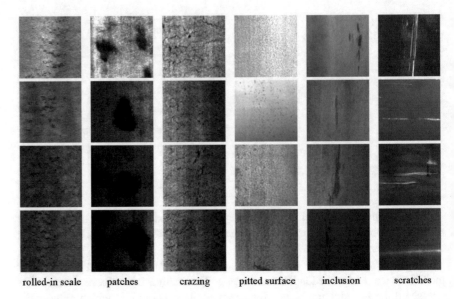

rolled-in scale   patches   crazing   pitted surface   inclusion   scratches

Fig. 2 Samples of six kinds of typical surface defects on NEU surface defect database. Each row shows one example image from each of 300 samples of a class [24]

The surface defect database has been divided into two subsets. Indeed, 80% of the data were used to train the models. While the remaining data which consists of 20% of the total dataset was kept for testing the trained models. The surface defect classification was written in Python using the KERAS package (Tensorflow backend).

In this experiment, we used a confusion matrix to evaluate the performance of the different models since we are dealing with a classification problem. The confusion matrix is a kind of summary of the prediction results for a particular classification problem. It compares the actual data for a target variable with those predicted by a model. The correct and false predictions are revealed and divided by class, allowing them to be compared with defined values. This matrix highlights four categories of outcomes:

- True Positives (*TP*) is the number of positive samples judged to be positive;
- True Negatives (*TN*) is the number of negative samples judged to be negative;
- False Negatives (*FN*) is the number of positive samples judged to be negative;
- False Positives (FP) is the number of negative samples judged to be positive.

The confusion matrix provides two main statistical measures that evaluate the efficiency of models. The first is recall, which represents the proportion of positives that have been correctly identified. The second is precision, which represents the ratio between TP and all positives. This is the percentage of true predictions out of the total predictions obtained for the models. We used formulas given in Eqs. (2) and (3) to calculate the recall and precision respectively. Recall and precision are to be used together for a complete vision of the performance. We try to have values as close as possible to 100% for both indicators. Depending on the case, we may choose to maximize recall rather than precision or vice versa.

$$\text{Recall} = \frac{TP}{TP + FN} \tag{2}$$

$$\text{Precision} = \frac{TP}{TP + FP} \tag{3}$$

4   Results and Analysis

Once training was completed, we evaluated the performance of models with new images that are not contained in the training dataset. First, we calculated the precision and recall of the trained models on the original dataset without data augmentation. We tested the constructed models using compressed images with the different parameters listed in Table 1. The obtained results are reported in Table 2 in the format Precision/Recall. Second, we noted the precision and recall for trained models on the original dataset, but this time using the ImageDataGenerator class of the Tensorflow API as the data augmentation technique. The obtained outcomes are shown in Table 3. Finally, the results of the trained models using compression-based data augmentation are displayed in Table 4.

The obtained results in this experiment clearly showed that the classification efficiency of the trained models on the original dataset was highly affected when tested on compressed image datasets with low quality, even when using ImageDataGenerator class of the Tensorflow API for data augmentation. This negative impact is more

Table 2 Classification performance of models when trained on the original dataset

Model/data	Q1	Q2	Q3	Q4	Q5	Q6	Q7
MobileNet	0.66/0.56	0.85/0.83	0.96/0.95	0.98/0.98	1.00/1.00	1.00/1.00	1.00/1.00
DenseNet 201	0.62/0.58	0.86/0.75	0.96/0.96	0.99/0.99	1.00/1.00	1.00/1.00	1.00/1.00
Vgg16	0.81/0.79	0.90/0.89	0.92/0.90	0.94/0.93	0.94/0.93	0.95/0.93	0.96/0.95

Table 3 Classification performance of trained models on the original dataset using ImageData-Generator for data augmentation

Model/data	Q1	Q2	Q3	Q4	Q5	Q6	Q7
MobileNet	0.61/0.23	0.88/0.81	0.96/0.96	0.98/0.98	0.99/0.99	1.00/1.00	1.00/1.00
DenseNet 201	0.53/0.63	0.89/0.74	0.96/0.95	0.98/0.98	0.98/0.98	0.98/0.98	0.98/0.98
Vgg16	0.71/0.66	0.91/0.90	0.92/0.91	0.93/0.92	0.93/0.92	0.95/0.93	0.96/0.94

Table 4 Classification performance of trained models using compression-based data augmentation

Model/data	Q1	Q2	Q3	Q4	Q5	Q6	Q7
MobileNet	0.94/0.92	0.98/0.97	0.99/0.99	1.00/1.00	0.99/0.99	0.99/0.99	1.00/1.00
DenseNet201	0.97/0.96	0.99/0.99	1.00/1.00	1.00/1.00	1.00/1.00	1.00/1.00	1.00/1.00
Vgg16	0.96/0.96	0.97/0.96	0.98/0.97	0.98/0.98	0.98/0.98	0.98/0.98	0.98/0.98

significant for higher compression ratios. Furthermore, it was found that this impact increases greatly for MobileNets and DenseNet201 that achieved perfect precision and recall scores when tested on original images.

In addition, the findings indicated that data augmentation based on compression considerably enhanced the classification efficiency of our models, even when tested with different compression qualities. This indicates that using compression for data augmentation improves the generalization of the models when tested with different compression qualities. In fact, augmenting the training dataset using image compression improved feature extraction by including the features of the compressed images. As a result, the constructed models were able to identify similar features and successfully classify poorly compressed images. Figure 3 shows the difference in precision between the trained models on the original dataset and the trained models with compression-based data augmentation. The solid curves in the graph clearly show that the trained models with the compression-based data augmentation achieved very high precision scores, even for compressed images with poor quality. While the dashed curves indicate a significant performance degradation of the trained models on the original dataset for poor compression qualities.

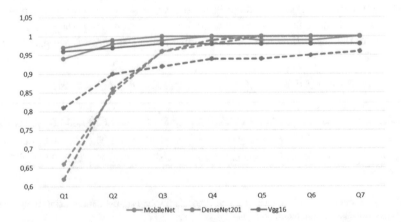

Fig. 3 Difference in precision between the trained models on the original dataset (dashed curves) and the trained models with compression-based data augmentation (solid curves)

5 Conclusion

This paper evaluated the impact of training CNN models with compression-based data augmentation on classification performance. The results of this study showed that using compression for data augmentation significantly improves the generalization of models when tested with different compression qualities. This method of data augmentation is extremely beneficial when the data storage guidelines for image compression vary in particular use cases. In the future, our aim is to exploit the results of this study to strike a balance between latency and accuracy in automated vision systems using image compression techniques.

References

1. Benbarrad T, Arioua M, Anas H (2022) Performance evaluation of transfer learning for surface defect classification. Lect Notes Netw Syst 393:977–988. https://doi.org/10.1007/978-3-030-94191-8_79
2. Nasir V, Sassani F (2021) A review on deep learning in machining and tool monitoring: methods, opportunities, and challenges. Int J Adv Manuf Technol 115(9):2683–2709. https://doi.org/10.1007/s00170-021-07325-7
3. Benbarrad T, Salhaoui M, Kenitar SB, Arioua M (2021) Intelligent machine vision model for defective product inspection based on machine learning. J Sens Actuator Netw 10(1):1. https://doi.org/10.3390/jsan10010007
4. Lee J, Ardakani HD, Yang S, Bagheri B (2015) Industrial big data analytics and cyber-physical systems for future maintenance & service innovation. Proc CIRP 38:3–7. https://doi.org/10.1016/j.procir.2015.08.026
5. Benbarrad T, Salhaoui M, Arioua M (2021) On the performance of deep learning in the full edge and the full cloud architectures. New York, NY USA. https://doi.org/10.1145/3454127.3457632
6. Nikoui TS, Rahmani AM, Balador A, Javadi HHS (2021) Internet of things architecture challenges: a systematic review. Int J Commun Syst 34(4):e4678. https://doi.org/10.1002/dac.4678
7. Benbarrad T, Salhaoui M, Kenitar SB, Arioua M (2020) Intelligent machine vision model for defective product inspection based on machine learning. https://doi.org/10.1109/ISAECT50560.2020.9523643.
8. Benbarrad T, Salhaoui M, Anas H, Arioua M (2022) Impact of standard image compression on the performance of image classification with deep learning. Lect Notes Netw Syst 393:901–911. https://doi.org/10.1007/978-3-030-94191-8_73
9. Marlapalli K, Bandlamudi RSBP, Busi R, Pranav V, Madhavrao B (2021) A review on image compression techniques. Commun Softw Netw 271–279. https://doi.org/10.1007/978-981-15-5397-4_29
10. Benbarrad T, Eloutouate L, Arioua M, Elouaai F, Laanaoui MD (2021) Impact of image compression on the performance of steel surface defect classification with a CNN. J Sens Actuator Netw 10(4). https://doi.org/10.3390/jsan10040073
11. Liu Z, et al (2018) DeepN-JPEG: a deep neural network favorable JPEG-based image compression framework. CoRR abs/1803.05788. http://arxiv.org/abs/1803.05788

12. Krizhevsky A, Sutskever I, Hinton GE (2012) ImageNet classification with deep convolutional neural networks. In: Pereira F, Burges CJC, Bottou L, Weinberger KQ (eds) Advances in neural information processing systems 25. Curran Associates, Inc., pp 1097–1105. http://papers.nips.cc/paper/4824-imagenet-classification-with-deep-convolutional-neural-networks.pdf. Accessed May 25 2020
13. Chatfield K, Simonyan K, Vedaldi A, Zisserman A (2014) Return of the devil in the details: Delving deep into convolutional nets. https://doi.org/10.5244/c.28.6
14. Chen K, Zeng Z, Yang J (2021) A deep region-based pyramid neural network for automatic detection and multi-classification of various surface defects of aluminum alloys. J. Build. Eng. 43:102523. https://doi.org/10.1016/j.jobe.2021.102523
15. Usage Statistics of JPEG for Websites (2022) https://w3techs.com/technologies/details/im-jpeg. Accessed 26 Mar 2022
16. Dodge S, Karam L (2016) Understanding how image quality affects deep neural networks. https://doi.org/10.1109/QoMEX.2016.7498955
17. Ghazvinian Zanjani F, Zinger S, Piepers B, Mahmoudpour S, Schelkens P, de With PHN (2019) Impact of JPEG 2000 compression on deep convolutional neural networks for metastatic cancer detection in histopathological images. J Med Imaging Bellingham Wash 6(2): 027501. https://doi.org/10.1117/1.JMI.6.2.027501
18. Grm K, Štruc V, Artiges A, Caron M, Ekenel HK (2018) Strengths and weaknesses of deep learning models for face recognition against image degradations. IET Biom 7(1):81–89. https://doi.org/10.1049/iet-bmt.2017.0083.
19. Galteri L, et al (2018) Reading text in the wild from compressed images. In: Proceedings–2017 IEEE International Conference on Computer Vision Workshops, ICCVW 2017, vol 2018, pp 2399–2407. https://doi.org/10.1109/ICCVW.2017.283
20. Zheng S, Song Y, Leung T, Goodfellow IJ (2016) Improving the robustness of deep neural networks via stability training. CoRR abs/1604.04326. http://arxiv.org/abs/1604.04326
21. Howard AG, et al (2017) MobileNets: efficient convolutional neural networks for mobile vision applications. ArXiv170404861 Cs. http://arxiv.org/abs/1704.04861. Accessed 06 Aug 2021
22. Huang G, Liu Z, Van Der Maaten L, Weinberger KQ (2017) Densely connected convolutional networks. In: IEEE conference on computer vision and pattern recognition (CVPR), pp 2261–2269. https://doi.org/10.1109/CVPR.2017.243
23. Simonyan K, Zisserman A (2015) Very deep convolutional networks for large-scale image recognition. In: 3rd International Conference on Learning Representations, ICLR 2015, San Diego, CA, USA, May 7–9, 2015, Conference Track Proceedings. http://arxiv.org/abs/1409.1556
24. Song K, Yan Y (2013) A noise robust method based on completed local binary patterns for hot-rolled steel strip surface defects. Appl Surf Sci 285:858–864. https://doi.org/10.1016/j.apsusc.2013.09.002

A Study of Scheduling Techniques in Ad Hoc Cloud Using Cloud Computing

Rajdip Das⬤ and Umesh Pal⬤

Abstract In a college lab, a large portion of the figuring gadgets like PCs and workstations lie underutilization. An impromptu cloud can be facilitated, and occupations can be designated on these gadgets. As part of this approach, the devices must be accessed sporadically and without commitment. The machine is unlikely to be run 24 h a day, seven days a week. As soon as the job is completed and before the machine can be accessed again, it becomes inoperable. Such inaccessibility can happen because of different reasons the most widely recognized being the client may simply turn the machine off. In this paper, we, subsequently, put forward a different technique to foresee the accessibility of machines and investigate them. Regression and classification models were built for the information collection. The methods were tested on real data collected from machines from multiple university labs. In the second part, a scheduling NSGA-II algorithm is proposed that has two objectives to optimize 1. Minimize the average job completion time and 2. Allocate jobs in such a way so that the machine remains available until the deadline elapses. We found them to perform extremely well.

Keywords Regression model · Classification model · Ad hoc · Resource availability · Scheduling · NSGA-II

1 Introduction

"Cloud computing is a model for enabling ubiquitous, convenient, on-demand networked access to a shared pool of configurable computing resources (e.g.,

R. Das
Shizuoka University, Hamamatsu, Japan
e-mail: das.rajdip.21@shizuoka.ac.jp

U. Pal (✉)
Jadavpur University, Kolkata, West Bengal 700032, India
e-mail: umeshp.cse.pg2@jadavpuruniversity.in

networks, servers, storage, applications, and services) that can be rapidly provisioned and released with minimal management effort or service provider interaction" [1]. Companies like Microsoft, Google, and Amazon offer cloud-based support. These are called public clouds. Some companies like Yahoo have their own dedicated infrastructure using which they set up a private cloud to serve the companies own internal needs. Gary McGilvary in [2] provides the only complete implementation of ad hoc cloud currently. The scheduler he used is naive. In a university environment, the machines can be expected to have regular usage patterns. Some of them are used to hold classes in labs. Classes are held according to schedule, so there will be regular usage patterns. Most of these machines can be expected to have regularity in their usage patterns. If these regular usage patterns are exploited to predict the future availability of the machines, then this information can be used to develop sophisticated schedulers. At present in the area of availability prediction, there is limited research. All the work found relevant are summarized. In the field of scheduling there too is limited past work in the context of ad hoc cloud so relevant scheduling methods from related areas have been summarized along with the schedulers used in past ad hoc cloud implementation. Karthick Ramachandran et al. propose a decentralized method of predicting resource availability in a desktop grid environment in [3]. Each machine in the desktop grid records its availability over some time. As it has been demonstrated earlier that the uptime of machines can be anticipated with adequate exactness, a strategy must be devised to use the prediction power for allocating jobs to machines. Such a strategy will only be feasible if the time required by a job can be predicted in advance. Let us assume that such prediction methods exist. In this work, a naive prediction method for a particular kind of job was used. One of the objectives of job allocation is to allocate jobs in such a way that the deadline can be met. Nevertheless, it is unlikely that a job allocation strategy of ad hoc cloud will have a single objective. So, a job allocation strategy should be flexible enough to work with multiple objectives.

2 Motivation

Some time ago to address the issues of organizations and exploration establishments in-house private groups and network registering were utilized. Lately, these arrangements have been progressively supplanted by distributed computing. Distributed computing makes it conceivable to scale as indicated by the interest, increment coordinated effort and offer data effectively, just as possibly diminish working costs, access limitless computational assets, and advantage from different benefits that distributed computing offers. Still, there can be circumstances where both private and public cloud arrangements are not appropriate. Like, the application or information can't be moved to the public cloud. The information might be excessively enormous for relocation. Hence the cost of migration to the cloud outweighs the advantages. The data may be sensitive and uploading it to a public cloud might be unsafe. The application may not be suitable for the public cloud model. This may happen when the execution cost of the application surpasses the value of the actual results or the application might be under development and a lot of trial run required can make it economically infeasible for executing it in the clouds. One solution to the above-mentioned problem is to set up an in-house private cloud. But that too is not generally feasible in a university environment the machines can be expected to have regular usage patterns. Some of them are used to hold classes in labs. Classes are held according to schedule, so there will be regular usage patterns. Some of the machines are used by research scholars and staff. Most of these machines can be expected to have regularity in their usage patterns. On the off chance that these customary utilization designs are abused to foresee the future accessibility of the machines, then, at that point, this data can be utilized to foster complex schedulers. As of now let us present the possibility of customary and sporadic execution follows. A standard execution follows repeats with a particular periodicity. It might rehash every day or week by week. An unpredictable execution follows is one that doesn't rehash the same thing. It occurs by some coincidence. For viable planning, an indicator should have the option to perceive whether the occasion of a machine being turned on is a piece of a customary execution follow or not.

3 Ad-Hoc Cloud Architecture

The architecture which forms the foundation of this paper was proposed by different researchers titled "Ad hoc Cloud Computing" [2–6]. The two main components of the proposed architecture are: The ad hoc server does the following tasks: (1). Acknowledges submitted work from specially appointed cloud users. (2). Timetables cloud occupations dependent on some booking algorithm. (3). Sends guidelines to specially appointed host for execution. (4). Screens and deals with the condition of the framework, and the specially appointed customer plays the accompanying role: (1). Gets directions from the specially appointed worker to be executed. (2). Intermittently takes designated spots of the virtual machine. (3). Timetables and sends designated spots to impromptu has. (4). Gets virtual machine designated spots from other impromptu has.

An ad hoc server is logically centralized. Physically it may run in one machine or multiple machines as a distributed system. Ad hoc clients run in sporadically available non-exclusive machines. These machines are called ad hoc hosts. Each ad hoc host executes an ad hoc client software. Ad hoc client software manages several virtual machines termed ad hoc guests. Cloud jobs are executed within these ad hoc guests A bunch of associated impromptu visitors that offers a specific execution climate is named a cloudlet. An ad hoc host can be part of multiple cloudlets. Dividing the underlying infrastructure into cloudlets helps to manage it better. Figure 1 shows the architecture of an ad hoc cloud platform.

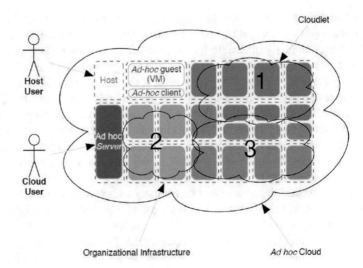

Fig. 1 The architecture of Ad Hoc cloud

3.1 Related Works

There have been many works on the ad-hoc cloud, availability predictions of various applications, and we will not repeat them here. A method was suggested by Gan et al. [7] for determining which workstations in an association are reliably turned-on design week by week. Utilizing this data, the creators guarantee that they can foresee the future accessibility of workstations and distinguish workstations to shape specially appointed PC groups. The proposed framework has three steps, such as data pre-processing, turned-on pattern analysis, and data aggregation report generation. The status collected from various machines through the existing monitoring system comes in the form of bit strings. Each bit in the string represents the status of the machine for a particular hour. A value of 1 means the machine was on at that hour and a value of 0 means the machine was on. The authors defined a function: $C (i, w, h) = 1$ if the ith workstation was found to be on at the hth hour else $C (i, w, h) = 0$. So, the final form of data representation is from Fig. 2 it is clear there are certain hours for which the workstations turn on consistently.

According to Artur Andrezak and colleagues [8], an assortment of committed and non-committed hosts may be used together to determine the handling degree of a web administration. They argued that future host accessibility might be a key factor when using non-dedicated hosts. PIL is a term coined by developers to describe the length of time that is expected. Each pil lasts for about an hour. Forecasts can be made for one or many pils in the future. In the comparison section, there is a 0 or 1 for each pil in the following information. The host had 0 methods and 1 method. Some classifiers were attempted, including innocent Bayes, support vector machines (SVM), K*, and multinomial computed regression models with edge assessors. According to their

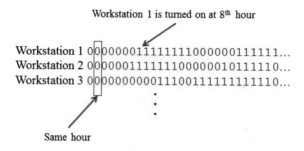

Fig. 2 Binary strings are generated for work stations

last expectation method, they relied on Gaussian models of access run. Based on all accessibility and non-accessibility runs, the normal deviations are calculated. To determine how many hours have passed since the last changeover, Gaussian arithmetic is used. $k + pil$ is more likely to be in the current accessibility run if the last-noticed state is accessibility. Let's call this likelihood p. It is concluded that the pil is accessible if the probability is at least [5].

Among the current works we evaluated, the Gan et al work most closely resembles ours. As a comparison, Gan's work has an accessibility expectation of 1 hour on the contrary ours is 10 mins, which may be easily changed to less or more time. Unavoidable circumstances outdoors might induce a change in usage design by 20 or 30 mins. Possibly, a client is stuck in rush hour traffic jams. Numerous variances were found in the data set we studied. A half-month period is examined in Gan's study, whereas ours looks at a few days' worths of data. Our work is more sensitive to accessibility design nuances. We hope that the work of Dr. Gan will complement our own. It is often used to identify consistently reliable computers.

4 Methodology

Ad hoc clouds are the primary focus of our suggested technique for predicting uptime, and job allocation. To determine the effectiveness of a method, it is important to know the specific environment in which it will be utilized.

4.1 Data Representation

The uptime of each machine is stored in the windows registry. The data from the registry is collected. This data is stored in the format of start time-end time. As shown in Fig. 3 this representation through space-efficient and easy to read is not suitable for learning algorithms. So, the data is modified to suit the needs. The data for each day is stored in a table. The table has 1440 rows and 1 column. There are a total of 1440 min in a day. Each of the 1440 rows represents the status of the machine

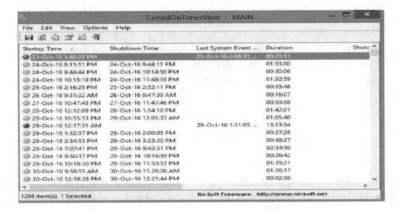

Fig. 3 Default format of record storage

Table 1 Data representation

Minute	Status
700	0
701	1
702	1
703	1
704	1
705	1
706	0

for that particular minute of the day. Table 1 shows a snapshot of the representation of the data. Using this representation two models are built the regression and the classification model. In a regression model, the amount of time that the machine will be available from a certain moment in time is estimated. The categorization model predicts whether or not the machine will be accessible at a certain moment in the future.

4.2 Regression Model

Table 2 has a five-column to represent the data needed to create the regression model. The explanatory variables are listed in the first four columns, while the response variables are listed in the last column. The five rows are proportional to the five ticks of the solar day that occur in sequential order throughout the day. This dataset is trained using neural networks (NN), support vector machines (SVMs), and random forest regressor methods (RFRM). A linear model was also used, but the results were too inadequate to include it, as because the explanatory and response variables are

Table 2 Information about the regression model in a table (actual data)

Mon	Tue	Wed	Thu	Fri
108	112	230	232	257
109	111	231	231	258
108	108	228	232	257
107	107	227	231	254
106	106	228	228	255
104	105	226	228	254
103	104	2325	227	251

not linearly related, there is no linear relationship in this data set. Without explicit feature engineering, the three algorithms utilized may represent such non-linearity. In the NN utilized, there are two unseen layers, each of which has 13 secrets units in it. There are four units in each day's input layer, and one unit in each day's output layer. When using the relu activation function, the following steps are followed. The maximum number of iterations is 2000. Learned Rate.001; Regularization Factor:01 The Gaussian kernel is used for support vector regression. It is set to 10000 for the C parameter and 0.1 for the gamma value. Trees are counted in a random forest regression model at 10. Maximum feature at each splitting point: 4.4 (maximum feature accessible at each splitting point). Nodes are separated until only one sample is left.

4.3 Classification Model

For the ordered dataset with just two qualities, 0 and 1 are utilized. We get ready three kinds of information portrayal for order. C1, C2, and C3 are their names. C1: consider Table 3 shown above. The four rows from the left represent the status for the four previous days and the fifth row represents the status of the fifth day. The learning algorithm won't be able to model such situations. To anticipate a person's current condition, the algorithm simply considers the last four days of their life at that moment. The first row of Table 4 illustrates this point. All of the machines were turned off at the same time, however, the machine was on for the fifth day at that point by pure coincidence. On the fifth day, the individual may have begun working a bit earlier. Algorithms designed for machine learning cannot represent these scenarios. The model that follows offers a solution to the problem. The next model provides a solution. C2: Note that in classification two consecutive rows in the dataset denotes the status between a ten-minute interval. For the learning algorithm to capture the situation shown in Table 5 it must have a way to model the data based on the sequential relationship between the rows of the dataset. This is done by stacking three successive rows on top of one another, as because it incorporates not just previous values from that minute's history but also values from the tenth and

Table 3 Representation of C1 data

Mday	Tday	Wday	Thday	Fday
000	010	010	010	010
000	010	010	010	010
000	010	010	010	010
000	010	010	010	010
000	010	010	010	010
000	010	010	010	010
000	010	010	010	010

twenty minutes after that minute, a model can accurately anticipate each output value. Table 6 shows an example of the dataset. We look at the nth, n+10th, and n+20th-min values, respectively. Then we make a prediction based on a 20-min forecast. The number 20 was chosen at random. Beyond the scope of this study, further research will be required to determine the best number of future tuples to employ.

Table 4 Model C1 fails to capture the following scenario

Mday	Tday	Wday	Thday	Fday
000	010	010	010	010
000	010	010	010	010
000	010	010	010	010
000	010	010	010	010
000	010	010	010	010
000	010	010	010	010
000	010	010	010	010

Table 5 The nth ticks values for the final four days are shown in columns 1–4. The values in columns 5–8 represent the (n + 10)th ticks. The (n + 20)th ticks data are displayed in columns 9–12. Column 13 displays the availability status on the nth minute

1	2	3	4	5	6	7	8	9	10	11	12	13
0	0	0	0	1	1	0	1	1	1	1	1	1
1	1	0	1	1	1	1	1	1	1	1	1	1

Table 6 The (n-20)th minute data are shown in columns 1–4. (n-10)th minute data are shown in columns 5–8 of the chart. Nine to twelve columns are devoted to the nth minute value. (n + 10)th min are shown in columns 13–16, and (n + 20)th min are shown in columns 16-20. As of the nth minute, the availability status is shown in column 21

1	2	3	4	5	6	7	8	9	10	11	12	13	14	15	#	17	18	19	20	21
		0	0	0	0	1	1	0	1	1	1	1	1	1	1	1	1	1	1

4.4 Non-dominated Sorting Genetic Algorithm II

Given that machines' uptime can be anticipated with reasonable precision, a method for assigning work to machines must be designed to make use of this predictive power. Problems of multi-objective optimization can be tackled in a variety of different methods. The optimization problem is solved using Deb and colleagues' evolutionary algorithm-based NSGA II [9]. This method ranks the population based on a simple and relatively quick approach that offers elitism and strong convergence near the real pareto optimal solution set. To retain the variety of the solution set, it utilizes a crowd distancing method. NSGA-I pseudocode is shown in Fig. 4.

An optimization problem with multiple goal functions is said to be non-dominated if there is no other solution in the solution set that is better or equal for all of the objectives. At first, create the starting population as your first task. The initial population is reached, and then, the population is ordered based on non-dominance, which is created. The first front is shown. comprises of all non-dominant solutions. In the second front, all solutions that are exclusively dominated by the first front's solution set are grouped, and so on. Based on which front an individual belongs, a rank (fitness) rating is awarded to that individual. If an individual belongs to the first front, they are awarded a fitness rating of 1. A value of 2 is assigned if they are allocated to the front of the second column, and so on.

4.4.1 Chromosome Operator

The chromosome must reflect a mapping from job to machine and no job must be mapped onto multiple machines. To represent a chromosome, a sequence of numbers is used in Fig. 5. The chromosome denotes the mapping that job 1 is mapped to machine 1, job 2 to machine 2, and so on. The chromosome in Fig. 6 shows the mapping that job 1 is allocated to machines.

```
1:   Initialize parameters
2:   Create random initial population P₀ and set generation number t = 0
3:   while t < max_gen (terminal condition) do
4:     –Generate a new offspring, Qₜ from Pₜ using recombination and selection operators
5:     –Combine parent and offspring population; Rₜ = Pₜ U Qₜ
6:     –Rank Rₜ to identify all non-dominated fronts F = fast-non-dominated-sort (Rₜ) where F = (F₁, F₂, ...)
7:     –Set Pₜ₊₁ = ø and i = 1
8:     while population size |Pₜ₊₁| + |Fᵢ| ≤ pop_size do
9:       Assign crowding-distance to individuals in Fᵢ
10:      Add front Fᵢ to the parent population Pₜ₊₁ = Pₜ₊₁ U Fᵢ
11:      Set i = i + 1
12:    end while
13:    if |Pₜ₊₁| ≤ pop_size then
14:      Sort Fᵢ in descending order according to the crowding distance of solutions
15:      Insert the first pop_size − |Pₜ₊₁|elements of Fᵢ to Pₜ₊₁, Pₜ₊₁ = Pₜ₊₁ U Fᵢ (i = 1 to pop_size − |Pₜ₊₁|)
16:      Set t = t + 1
17:    end while
18:    Return Pₘₐₓ_gen
```

Fig. 4 Pseudocode of the NSGA-II algorithm

Figs. 5 and 6 Flowlines

4.4.2 Mutation Operator

The swap mutation operator was employed for mutation. The swap mutation operator switches two chromosomal numbers. For chromosomes with non-repeating alleles, this method is better suited. This was achieved by using the 9 probability PMX crossover operator in conjunction with a swap mutation 75% chance of being utilized. Figure 7 denotes the mutation in the chromosomes.

5 Simulation Results

5.1 Results of Up-Time Prediction Model

Data were collected from twenty machines across four university labs. Two of the labs were used by research scholars and two by students for the practical classes. A good mix of machines with regular and irregular usage patterns was selected. The machines used for classes are expected to have more or less regular behavior and all machines belonging to a particular lab in which classes are held are expected to have somewhat similar usage patterns. However, on account of examination researchers, each machine is relied upon to have an exceptional utilization design. Information from fifteen of these machines was utilized to frame the preparation set and five of these machines were utilized to shape the test set. For the outcomes information from five machines was taken and the indicators were executed on this information. Out of the five machines some had regular usage patterns and others not so regular usage patterns. This was done to get pessimistic results on accuracy. Figure 8 shows the usage pattern of five of the five machines.

To evaluate the regression models, we use the metric R^2 value. Three learning algorithms (LA) were used for the regression model i.e., support vector regressor (SVR), random forest (RF), and NN. Table 7 summarizes the result of the training set. R2 is a measure of how well a model fit. There's a better match when it's closer to 1. As we can the model fits quite well on the training data. Table 8 shows the R^2 value on the test data.

As can be seen from Table 8 the results are very poor on the test set. That means even though the model fits well on the training set it fails to generalize well on

Fig. 7 Swap mutation

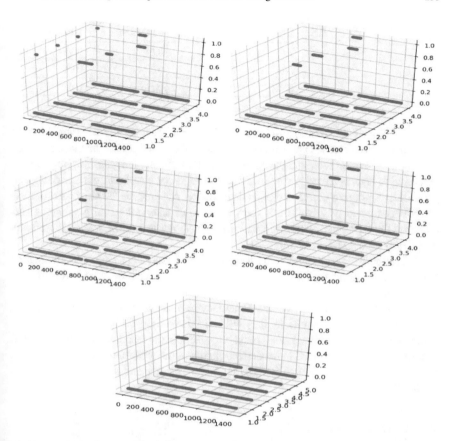

Fig. 8 Usage pattern for the five machines

Table 7 Performance on prctice set

L A	R^2
Support vector regressor	0.97
Random forest	0.96
Neural network	0.78

Table 8 Performance on the trial set

L A	R^2
Support vector regressor	−0.55
Random forest	0.057
Neural network	0.16

Table 9 Performance on a C1 practice set

LA	Recollet	Perfection	Exactnes
RF	0.78	0.83	0.95
NN	0.78	0.83	0.96
SVR	0.73	0.86	0.95

unknown data. For classification, there are four different results for four different classifiers. Three metrics are utilized to measure the execution of the classifiers i.e., recollect, perfection rate, and exactness. The ratio of instances that can be properly predicted to tests is called perfection. The review is the small part of tests delegated genuine encouraging points in the whole example. Accuracy is the small part of tests effectively named positive out of the multitude of tests delegated positive. After-effects of the C1 replica on the preparation set are shown in Table 9, while after-effects of the trial set are shown in Table 10.

Unlike the regression model, the score for the C1 model is acceptable. In the next two tables, the scores for the C2 model are shown. Table 11 is devoted to the practice set and Table 12 for demonstration of the trial set.

The above table shows a considerable increase in accuracy for the neural network model but taking the results as it is will be misleading. C1 will work better for some particular cases. These cases must be present in the test data. The difference of accuracy between C1 and C2 will depend on the number of these cases that exist on the test set. The results of the C3 classifier on the practice and trial sets are shown in Table 13.

Table 10 Performance on C1 the trial set

LA	Recollet	Perfection	Exactnes
RF	0.60	0.80	0.71
NN	0.60	0.80	0.71
SVR	0.6	0.80	0.78

Table 11 Performance on the C2 practice set

LA	Recollect	Perfection	Exactness
RF	0.85	0.86	0.97
NN	0.82	0.87	0.96
SVR	0.74	0.86	0.96

Table 12 Performance on a C2 trial set

LA	Recollect	Perfection	Exactness
RF	0.58	0.81	0.72
NN	0.70	0.80	0.76
SVR	0.42	0.80	0.65

Table 13 C2 performance on the practice set

LA	Recollect	Perfection	Exactness
RF	0.79	0.82	0.96
NN	0.79	0.80	0.95
SVR	0.78	0.81	0.96

From Table 14 we discovered that using regards before the estimated stretch prompts a lessen in precision from the C2 replica. In this way, we can assume that C2 is the best replica in the present circumstance. It ought to be referred to that all of the precision of the score, R squared characteristics, exactness, and audit are incredibly liable to the dataset used. In case there is inconsistency in the utilization plan, the scores will undoubtedly be high. In case no user configuration exists, the exactness scores will without a doubt be near to 0. We chose well-maintained machines, such as those whose usage configuration resembles that of a normal school playground machine. It was important to us to select labs where lessons are held with a high degree of consistency. In addition, we collected data from research professionals' standard machines. In contrast to courses, research experts do not utilize machinery following a predetermined schedule or schedule. Some models in their usage lead will be needed as time goes on, though. For the estimation, we purposefully chose to utilize data from the machine that did not have a predetermined usage plan. Follow-up execution was added in segment 2. We combined data for the following two game strategies. On one, a conventional execution is followed, whereas, on the other, an irregular execution is followed. To be fair, the pointer should be able to detect that the normal execution will be performed. Similarly, for non-standard or non-regular instructions, the pointer should return 0. For example, the pointer should indicate that the computer will be shut off. We trained our classifiers using a non-discontinuous, periodic dataset. Nine to eleven figures indicate the findings in six outlines. These three photos show the results of a nonperiodic dataset on C1, C2, C3, and they demonstrate the results of a periodic dataset on C1, C2, and C3. The outlines indicate the neural association model's presentation. We left out several models since the qualifying requirements are not very high (Figs 9, 10, and 11).

Table 14 C2 performance on a trial set

LA	Recollect	Perfection	Exactness
RF	0.57	0.77	0.70
NN	0.59	0.76	0.70
SVR	0.57	0.79	0.71

Fig. 9 Output of classifier C1 on non periodic execution trace

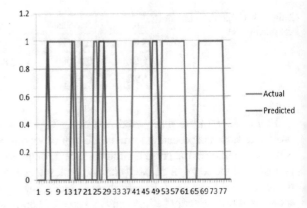

Fig. 10 Output of classifier C2 on non periodic execution trace

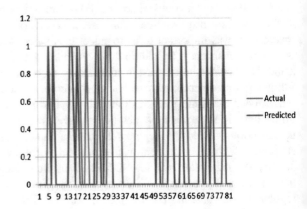

Fig. 11 Output of classifier C1 on non-periodic execution trace

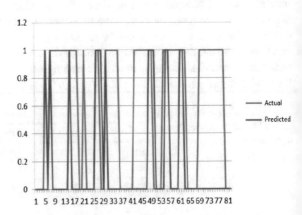

5.2 Results of the Job Allocation Model

To test the job allocation, algorithm an application has been developed. The application has two separate modules. One is called AdHoc Client and another is called AdHoc Server. AdHoc Server is to be executed on a single machine. After the AdHoc Client can be executed on multiple machines. Each AdHoc Client is designed to be connected to the AdHoc Server. On execution, the client asks for the IP address of the AdHoc Server. Then it registers itself with the AdHoc Server. To implement the genetic algorithm the JMetal framework has been used. As discussed, earlier, each job unzips a zipped file consisting of jpeg files and converts them from RGB to GrayScale. For the conversion, the OpenCV image editing library has been used.

5.3 Real-Life Scenario

So far it has been assumed that the tally of machines and quantity of jobs are equal but that is unlikely to be true in real-world scenarios. In the following sections, possible solutions to the scenarios when the number of machines and jobs are unequal are discussed.

5.3.1 Quantity of Jobs Are Less Than the Tally of Machines

If the quantity of jobs is less than the tally of machines then the solution is to create some dummy jobs. The dummy jobs will have a time requirement of 0 and they will have a size of 1. There is no unit associated with the dummy jobs size. To understand the rationale behind this, consider the following example: Table 15 shows three machines with the corresponding Dhrystone score and estimated uptime. In Table 16, D is the dummy job introduced. The dummy job has an estimated running time of 0 and size of 1. For every possible allocation the value will always be 1 for the dummy job and d will be equal to 0. So, objective 1 will heavily depend on the allocation of the other two actual jobs. For objective 2 the variable $gij = pi * X$ will be equal to the Dhrystone score of the machine it will be allocated to. So, this will discourage the allocation of the dummy job to machines with a high Dhrystone score (high processing power).

Table 15 Three machines, score of Dhrystone, uptime

Machine	C0	C1	C2
Dhrystone	57	73	36
Uptime	24	8	16

Table 16 .

Job	S3	S4	D
Size	293 MB	496 MB	1
Time	6	7	0

5.3.2 Quantity of Jobs Are More Than the Tally of Machines

A simpler solution is proposed in this scenario. On the off chance that the quantity of occupations is n and tally of machines is m and n is more than m them a first come first serve scheduling policy can be adopted. That is the first batch the first m jobs will be scheduled. After they have finished execution, the remaining jobs will be scheduled.

6 Conclusion

The main contribution of this thesis is a way to predict the uptime of machines used in environments like a university. A method is proposed that works satisfactorily. Then using the predicted uptime, a method for job allocation is proposed. The job allocation method is such that it can take into account multiple objectives while deciding on the allocation. Two objectives are proposed, but the model is easily extensible to more or different objectives. The major limitation of the uptime prediction method is that the number of minutes it looks before and after the prediction interval in the past data is fixed. It is hardcoded. There are deep learning-based techniques that can model sequential relations. An attempt was made to use such techniques in the hope that they will be able to automatically detect the sequential relationship between tuples of past data. Those attempts were unsuccessful. The algorithms used are very complicated and it is likely that with more extensive research with those algorithms better results will be possible. The synthetic job model developed is trivial due to lack of time. A more sophisticated job models whose runtime depends more strongly on the machine's processing capability needs to be developed for extensive testing.

Acknowledgements The authors would like to acknowledge the Computer Science Engineering Department, Jadavpur University, India.

Funding This research got no particular award from any financing office inside people in general, business, or not-revenue driven areas.

Declarations

Conflict of Interest The creators pronounce that there is no irreconcilable circumstance.
Code Availability Not Applicable.

References

1. Mell P, Grance T (2011) The NIST definition of cloud computing. Commun ACM 53. https://doi.org/10.6028/NIST.SP.800-145
2. McGilvary G (2014) Ad hoc cloud computing. PhD thesis, University of Edinburgh
3. Ramachandran K, Lutyya H, Perry M (2012) Decentralized approach to resource availability prediction using group availability in a P2P desktop grid. Futur Gener Comput Syst 28(6):854–860
4. Graham NC, Kirby G, Dearle A, Macdonald A, Fernandes A (2010) An approach to ad hoc cloud computing (2010). Corr arXiv: arxiv.org/abs/1002.4738
5. Dong B, Zheng Q, Qiao M, Shu J, Yang J (2009) BlueSky cloud framework: an elearning framework embracing cloud computing. In: Jaatun MG, Zhao G, Rong C (eds) CloudCom 2009, vol 5931. LNCS. Springer, Heidelberg, pp 577–582
6. Kirby G, Dearle A, Acdonald A, Fernandes A (2010) An approach to ad hoc cloud computing. In: DBLP: CoRR abs 1002.4738
7. Gan CT, Ooi BY, Liew SY (2014) Workstation's uptime analysis framework to identify opportunity for forming ad-hoc computer clusters. In: 2014 International conference on computer, communications, and control technology (I4CT), pp 234–238.https://doi.org/10.1109/I4CT.2014.6914181
8. Andrzejak A, Kondo D, Anderson DP (2010) Exploiting non-dedicated resources for cloud computing. In: 2010 IEEE network operations and management symposium—NOMS 2010, pp 341–348. https://doi.org/10.1109/NOMS.2010.5488488
9. Deb K, Pratap A, Agarwal S, Meyarivan T (2002) A fast and elitist multiobjective genetic algorithm: NSGA-II. IEEE Trans Evol Comput 6(2):182–197

Performance Improvement of SAC-OCDMA Network Utilizing an Identity Column Shifting Matrix (ICSM) Code

Mohanad Alayedi⬤, Abdelhamid Cherifi⬤, Abdelhak Ferhat Hamida⬤, Rima Matem⬤, and Somia A. Abd El-Mottaleb⬤

Abstract In general, SAC-OCDMA systems are confronted with many shortcomings that affect their performance, for instance, multiple access interference (MAI), phase-induced intensity noise (PIIN), complex structure, etc. For these purposes, this paper offers a novel technique to design a code based on an identity matrix and shifting feature in order to reduce the aforementioned challenges, namely Identity Column Shift Matrix (ICSM) code. Our proposed code features a zero cross-correlation property (ZCC) which restricts the effect of the MAI, cancels PIIN and thus improves the performance of the SAC-OCDMA system. Furthermore, it simplifies the SAC-OCDMA framework. Mathematical results appear the ability of ICSM code to improve the performance of SAC-OCDMA system as well outperform existing codes such as: Diagonal Permutation Shift (DPS), Dynamic Cyclic Shift (DCS) and Random Diagonal (RD) codes. Secondly, the design of those codes is more complicated than the design of our proposed code. Third, with respect to system capacity, the SAC-OCDMA system based on ICSM, RD, DCS, and DPS codes is 90, 58, 35, and 25 in a row. As a result, the greatest capacity of the system

M. Alayedi (✉)
Scientific Instrumentation Laboratory (LIS), Electronics Department, Faculty of Technology, Ferhat Abbas University of Setif 1, 19000 Setif, Algeria
e-mail: muhannadaydi@gmail.com

A. Cherifi
Technology of Communication Laboratory (LTC), Electronics Department, Faculty of Technology, Dr. Tahar Moulay University of Saida, 20000 Saida, Algeria

A. F. Hamida
Laboratory of Optoelectronics and Components (LOC), Electronics Department, Faculty of Technology, Ferhat Abbas University of Setif 1, 19000 Setif, Algeria

R. Matem
Advanced Communication Engineering Center of Excellence School of Computer and Communication Engineering (ACE CoE-SCCE), Universiti Malaysia Perlis (UniMAP), Perlis, Malaysia

S. A. A. El-Mottaleb
Department of Electronics and Communication, Alexandria Higher Institute of Engineering and Technology, Alexandria 21311, Egypt

© The Author(s), under exclusive license to Springer Nature Switzerland AG 2023
A. A. Abd El-Latif et al. (eds.), *Advances in Cybersecurity, Cybercrimes, and Smart Emerging Technologies*, Engineering Cyber-Physical Systems and Critical Infrastructures 4,
https://doi.org/10.1007/978-3-031-21101-0_21

263

was reached through the use of the ICSM code. On the other side, mimic result demonstrates that our proposed code is suitable for optical communications because meet their requirements through producing BER reaches 10^{-13} (<1e-9) and Q-factor reaches 7.2 dB (more than 6 dB).

Keywords SAC OCDMA · ICSM code · DPS code · PIIN · Quality factor

1 Introduction

Nowadays, optical networks have become important elements of telecommunications systems because of their greater speed of information transmission than switches and routers. They are featured by their transparence to data formats and protocols, which authorizes one to support the functionality of the network requirements, accomplish high quality of service (QoS) and cardinality, resist the electromagnetic effects, ... etc. [1, 2]. From time to time, changes have been oucc occurred for optical networks presented by inserting many technologies with a target of optimizing their efficiency like optical code division multiple access (OCDMA) technology which one of the internetworking and multiplexing techniques. It can encrypt and decrypt the signals utilizing simple and cost components, known as passive optical networks (PON) [3−5]. Furthermore, it has compulsory in communication networks since its ability to convey the massive quantity of information issued by high sub-carriers' number at the same time and common bandwidth between them over one fiber optic canal and where every sub-carrier owns a specific code [6−8]. The SAC-OCDMA (i.e. spectral amplitude coding-OCDMA) system offers a schema for OCDMA systems that was developed to upgrade the system performance by finally eradicating multiple access interference (MAI) by encrypting spectral components [9−12].

In SAC system, fiber Bragg grating (FBG) can be used as main part of encoder-decoder structure of each client. Nevertheless, when the number of active clients becomes large the sizes of FBG will become impractical and it is then recommended to be replaced with wavelength division multiplexing (WDM).

In an OCDMA system, physical effects, play an important role in arising noise, that caused by the system design itself; for example, relative intensity noise (RIN), shot noise, thermal noise and phase-induced intensity noise (PIIN) [13]. Accordingly, these noises contribute to impede their performance and limit the signal-to-noise ratio (SNR). This means that there is an overlap between the spectrum of users that produces PIIN and thus adversely affects system performance [14, 15].

In the state of the art, a lot of encrypting techniques in this regard have been reported to the SAC-OCDMA system, including Random Diagonal (RD) code [16], Dynamic Cyclic Shift (DCS) code [17], Diagonal Permutation Shift (DPS) code [18] etc. in literature investigation. Unfortunately, none of the aforementioned encryption techniques, is capable of restraining PIIN influence entirely because of the fixed in-phase cross correlation (IPCC) value amounts to one exactly, as the first obstacle.

Second, the extraction of the desired wavelengths needs to provide a great number of filters due to much long code length. Third and lastly, the process of design them is complicated.

In contrast, the RD code consists of two hybrid arrays: data segment and code segment [16]. As regards to DCS code, a mathematical rule is required to determine the position of ones and the application of shifting property to obtain the rest code sequences [17]. Finally, the DPS code matrix greatly relates to prime numbers where its parameters cannot be utilized for non-prime number [18]. Consequently, these issues have overall affected the performance of the SAC-OCDMA system and degraded its performance.

For these reasons, this paper is devoted to suggest a novel SAC-OCDMA encrypting technique termed as Identity Column Shifting Matrix (ICSM) code with zero cross correlation (ZCC) characteristic and with traits enable to select both desirable code weight and number of users facilely. In addition, our suggested ICSM code overcomes the complications aforementioned existing in each of them.

The organization of this paper is as follows.

Section 2 depicts the steps of designing our suggested code while Sect. 3 includes an explanation to the system performance. Concerning Sect. 4, it contains the numerical and emulation findings utilizing Matlab and Optisystem software, respectively, and at last, this study is terminated by a conclusion in Sect. 5.

2 ICSM Code Construction

At first, it should know the nature of characteristics of ICSM code before explaining how to design it. For that, two principal functions are responsible for them which are: auto correlation (AC) and cross correlation (CC). If we suppose "E" and "H" code sequences which equal to $E_j = \{E_1, E_2, E_3, \ldots \ldots, E_U\}$ and $H_j = \{H_1, H_2, H_3, \ldots \ldots, H_U\}$, successfully. CC and AC functions can be consecutively, defined as in Eqs. (1) and (2) below [19].

$$\left\{ \lambda_c = \sum_{u=1}^{U} E_u * H_{u+t} \right. \tag{1}$$

$$\left\{ \lambda_c = \sum_{u=1}^{U} E_u * E_{u+t} \right. \tag{2}$$

The ICSM code is distinctive by these parameters: $\{L, U, w\}$ where each symbol indicates code length, system capacity and code weight, respectively. In order to make the novel code more convenient for SAC-OCDMA systems overall, it's recommended to construct it with high AC as well null CC values. The design process of ICSM code can be dubbed according to the next five steps [20]:

A. Step 1

Arise an identity matrix (X_e) with order (e). In mathematics science, identity matrix is one of the distinctive matrices like zero matrix or magic matrix where it's deemed a square matrix constituted of binary numbers ("0" or "1") only. What distinct this matrix is that all elements are zeros unless the elements of the main diagonal in which equalizes one. Let's take an epitome which $(e = 3)$, it can be written in accordance to the following form:

$$X_3 = \begin{bmatrix} 1 & 0 & 0 \\ 0 & 1 & 0 \\ 0 & 0 & 1 \end{bmatrix}_{3\times3} \tag{3}$$

B. Step 2

Counting on shifting of columns characteristic and for an identity matrix that own (e) of order, therefor, X_3 requires two times of shifting based on $(e - 1)$ rule to the right. Similarly, the produced matrix is shifted by two times too. As an outcome, this allows us to acquire the next matrices:

$$X_3' = \begin{bmatrix} 0 & 0 & 1 \\ 1 & 0 & 0 \\ 0 & 1 & 0 \end{bmatrix} \& \ X_3'' = \begin{bmatrix} 0 & 1 & 0 \\ 0 & 0 & 1 \\ 1 & 0 & 0 \end{bmatrix} \tag{4}$$

C. Step 3

In order to more elucidate (X_3), write X_3 matrix in accordance with the form below

$$X_3 = \begin{bmatrix} L_1(X_3) \\ L_2(X_3) \\ L_3(X_3) \end{bmatrix} \tag{5}$$

where $L_1(X_3)$, $L_2(X_3)$ and $L_3(X_3)$: indicate successfully, the 1st row, 2nd row and 3rd row of X_3 matrix. Likewise, it is possible to also write $X_{,3}$ and $X"_3$ matrices as follows

$$X_3' = \begin{bmatrix} L_1(X_3') \\ L_2(X_3') \\ L_3(X_3') \end{bmatrix} \& X_3'' = \begin{bmatrix} L_1(X_3'') \\ L_2(X_3'') \\ L_3(X_3'') \end{bmatrix} \tag{6}$$

D. Step 4

Reform each of matrices X_3, $X'_3{}'$ and X''_3 into row vector as manifested below in Eq. (7).

$$\begin{cases} Re(X_3) = \left[L_1(X_3) \,,\; L_2(X_3) \,,\; L_3(X_3) \right] \\ Re(X_3') = \left[L_1\!\left(X_3'\right) \,,\; L_2\!\left(X_3'\right) \,,\; L_2\!\left(X_3'\right) \right] \\ Re\!\left(X_3''\right) = \left[L_1\!\left(X_3''\right) \,,\; L_2\!\left(X_3''\right) \,,\; L_3\!\left(X_3''\right) \right] \end{cases} \xrightarrow{leads} \begin{cases} Re(X_3) = \left[1\;0\;0\;0\;1\;0\;0\;0\;1 \right]_{1\times 9} \\ Re(X_3') = \left[0\;0\;1\;1\;0\;0\;0\;1\;0 \right]_{1\times 9} \\ Re\!\left(X_3''\right) = \left[0\;1\;0\;0\;0\;1\;1\;0\;0 \right]_{1\times 9} \end{cases}$$

$$(7)$$

E. Step 5

Eventually, by consolidating among these reshaped matrices or in other concept among row vectors in Eq. (7), this authorizes us to acquire the novel encoding technique: ZCC code as

$$ICSM = \begin{bmatrix} X_3 \\ X_3' \\ X_3'' \end{bmatrix} \xrightarrow{leads} ICSM = \begin{bmatrix} 1\;0\;0\;0\;1\;0\;0\;0\;1 \\ 0\;0\;1\;1\;0\;0\;0\;1\;0 \\ 0\;1\;0\;0\;0\;1\;1\;0\;0 \end{bmatrix}_{3\times 9} \tag{8}$$

Ultimately, to resume all five steps construction of our proposed code, a flowchart of ICSM code construction algorithm is added and presented in Fig. 1. According to above, the new ZCC code matrix that termed as ICSM code and owns a dimension of three rows and nine columns that express number of users (U) and code length (L), successfully. As a result, the relation between them can be deduced depending on the follows:

$$\begin{cases} U = e \\ L = e^2 \end{cases} \xrightarrow{leads} L = w^2 = U^2 \tag{9}$$

Additionally, it is observed that the code weight and number of users are equivalent. For this reason, Eq. (9) will be written in another formula as below

$$L_{ICSM} = w * U \tag{10}$$

Based on the ICSM code matrix, the code-word for each user would be

$$code - words = \begin{cases} 1^{st} \quad User \rightarrow \lambda_1, \lambda_5, \lambda_9 \\ 2^{nd} \quad User \rightarrow \lambda_3, \lambda_4, \lambda_8 \\ 3^{rd} \quad User \rightarrow \lambda_2, \lambda_6, \lambda_7 \end{cases} \tag{11}$$

To more elaborate the benefits obtained by ICSM code, Table 1 includes a simple comparison between DPS, DCS, RD and ICSM codes in term of code weight, code size, code length and cross correlation.

Fig. 1 Flowchart of ICSM
code development

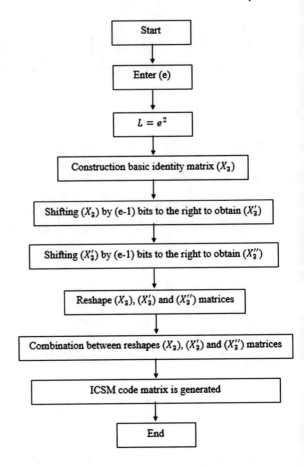

Table 1 Comparison between RD, DCS, DPS codes and ICSM code

Code	Code length	Cross-correlation	Code weight	Code size
RD code [16]	9	Variable	4	4
DCS code [17]	22	≥ 1	4	22
DPS code [18]	12	≥ 1	4	9
ICSM code	16	0	4	4

3 System Performance and Analysis

First, the SAC-OCDMA system can be analyzed simple without any complication
based on several suppositions should take account:

1. Un-polarized broadband light source (BBS) has a flat spectrum over the interval $[f_0 - \nabla f/2, f_0 + \nabla f/2]$ in the domain. (f_0) is the central optical frequency and (∇f) is the optical source bandwidth.
2. All transmitters are configured with equal power.
3. All components of the power spectral have an identical spectrum.
4. Achieving the synchronization property for each bit flux.

Second, to assess the SAC-OCDMA system based on ICSM code, two different noises are only taken account which are shot noise $\langle I_{sh}^2 \rangle$ and thermal noise $\langle I_{th}^2 \rangle$. in numerical analysis. Although ICSM code has a ZCC feature, therefore PIIN $\langle I_{PIIN}^2 \rangle$ is neglected in this case. Accordingly, the dark current noise $\langle I_{Total-noise}^2 \rangle$ can be given as shown in Eq. (12).

$$\langle I_{tot-noise}^2 \rangle = \langle I_{th}^2 \rangle + \langle I_{sh}^2 \rangle + \underbrace{\left\langle I_{PIIN}^2 \right\rangle}_{=0}$$

$$= \frac{4K_b B_e T}{R_l} + 2eB_e I_{DD} + 0 \tag{12}$$

where K_b, B_e, T, R_l, e and I_{SDD} mention to Boltzmann's constant, electrical bandwidth, absolute temperature, load resistor, electron charge and current issued by PD, consecutively. The output current of spectral direct detection (SDD) can be expressed by the following relation:

$$I_{DD} = \int_0^\infty S(f)df$$

$$= R \int_0^\infty \frac{P_r}{\nabla f} \sum_{u=1}^U d_u \sum_{j=1}^L [C_e(u) \times C_h(u)] \bigvee (f, n)df \tag{13}$$

where $S(f)$ mentions to the power spectral density (PSD), R mentions to photodiode (PD) responsivety, P_r mentions to the amount of power at receiver side, d_u indicates the data bit which is likely "0" or "1". The $\bigvee (f, j)$ function can be defined as:

$$\bigvee (f, n) = U\left\{ f - f_0 - \frac{\nabla f}{2L}(-L + 2n - 2) \right\} - U\left\{ f - f_0 - \frac{\nabla f}{2L}(-L + 2n) \right\}$$

$$\bigvee (f, n) = U\left\{ \frac{\nabla f}{L} \right\} \tag{14}$$

where $U(f)$ presents the unit step function defined as

$$U(f) = \begin{cases} 1 & f \geq 0 \\ 0 & f < 0 \end{cases} \tag{15}$$

Let $C_e(u)$ presents the uth element of Uth user and in accordance with ICSM code, the correlation properties of SDD technique can be written as

$$\sum_{j=1}^{L} C_e(u) \times C_h(u) = \begin{cases} 0 & for \quad e \neq h \\ w & for \quad e = h \end{cases} \tag{16}$$

where the upper term references to the CC function meanwhile the down term references to the AC function.

Accordingly, Eqs. (13), (15) and (16), permit to write the output photocurrent

$$I_{DD} = R\frac{P_r \times .w}{L} \tag{17}$$

Herein, the replace of Eq. (17) in Eq. (12) as well supposing the probability of "0" or "1" bit flux is equivalent and estimated by "0.5". As an outcome, the noise current expression will become:

$$\langle I_{tot-noise}^2 \rangle = \frac{4K_bB_eT}{R_l} + eB_eR\frac{P_r \times w}{L} \tag{18}$$

The average signal to noise ratio (SNR) can be given as

$$SNR = \frac{I_{DD}^2}{\langle I_{tot-noise}^2 \rangle} = \frac{\left[R\frac{P_{sr}w}{L}\right]^2}{\frac{4K_bB_eT}{R_l} + eB_eR\frac{P_r \times w}{L}} \tag{19}$$

The BER (i.e. bit error rate) can be estimated based on Gaussian approximation as follows [21]:

$$BER = \frac{1}{2} \times erfc\sqrt{SNR/8} \tag{20}$$

The QF (i.e. quality factor) can be derived from BER as follows [22]:

$$QF_{dB} = 20log_{10}\left[\frac{\sqrt{2}}{erfc(2 * BER)}\right] \tag{21}$$

4 Results and Debate

Utilizing Eq. (19), the QF of ICSM code is computationally examined and compared to other reporting SAC-OCDMA encoding techniques: RD, DCS and DPS for fixed parameter such as the code weight is chosen to be 4 of value. All values of parameters existent in Eq. (17) are accumulated in Table 2.

Figure 2 appears the QF alteration as a function of the number of concurrent users for configured parameters, such as an electrical bandwidth of 311 MHz and an optical bandwidth of 5 THz. As elaborated, our suggested code can accommodate a greater number of users, up to 90, whereas the other encoding techniques: DPS, DCS and RD can accommodate just 25, 35 and 58 users, respectively. As an outcome, the system capacity has been optimized by almost 3.6, 2.6 and 1.6 times from replacing one of these encoding techniques: DPS, DCS and RD used in the SAC-OCDMA system previously by our proposed encoding technique: ICSM, respectively. This means that the system can potentially meet the demands of optical networks.

Besides that, it can compute the increased percent resulting from the use of our suggested ICSM code as follows:

$$\frac{Suggested\ code - Compared\ code}{Compared\ code} \times 100$$

For that, we have

$$\begin{cases} \frac{90-25}{25} = 260\% \\ \frac{90-35}{35} = 160\% \\ \frac{90-58}{58} = 60\% \end{cases}$$

In consequence, the substitution of implemented codes in SAC-OCDMA system: DPS, DCS and RD by ICSM code allows to accomplish a percent increment whence system cardinality amounts to 260, 160 and 60%, respectively.

By transiting to Fig. 3, it appears the output PD current alteration as function of received power for 50 of users allowed them to be active. It is seemed that the ICSM code has minor output PD current which means that low signal information power can be produced. Therefore, it may produce a low SNR value and the system performance then could be impaired but results in Fig. 2 demonstrates the reversal of foregoing.

Table 2 Used parameters in numeral analysis

Parameter	Value	Parameter	Value
Number of active users (U)	50	Spectral width (Δf)	5 THz
Effective source power (P_r)	−10 dBm	Receiver load resistor (R_l)	1030 Ω
Boltzmann's constant (K_b)	1.38×10^{-23} j.s^{-1}	Electrical bandwidth (B_e)	0.311 GHz
Receiver noise temperature (T)	300 K	PD resposnsivity (R)	0.75 A/w

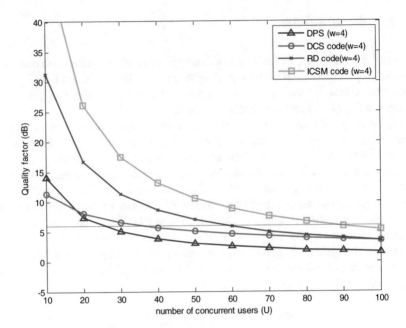

Fig. 2 QF versus number of concurrent users

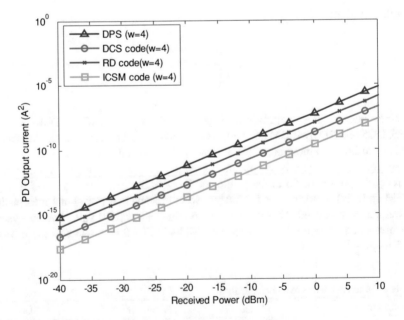

Fig. 3 Output PD current versus effective source power ($U = 50$)

For this reason, Fig. 4 is added that represent the dark current noise alteration as function of received power for the same activated users' number in the previous studied figure. First, it is remarked that all systems have the same noise alteration until −28 dBm value of received power. This signifies that all systems are affected by the same noise which is thermal noise because it is common between them. Second, the systems are affected by shot noise starting from −4 dBm. Third, although systems are affected by PIIN, so it can say that its effect starts from −28 dBm meanwhile its effect is disappeared in our proposed system relying on Eq. (12). Fourth and finally, it is noted that our proposed system is less vulnerable to noise effects than other systems. So that, this leads to increase the SNR value making very high although high number of active users and data rate.

On the other hand, with target of validate the numerical results and clear the motivation of ICSM code, a mimic examination was also proceeded utilizing Optisystem software. It was used wavelength division multiplexing (WDM) and fiber Bragg grating (FBG) as encoder and decoder, respectively. The performance examination of SAC-OCDMA system employing ICSM code can be performed by depending on one of two main criteria in optical communications: QF and BER in which should be more than 6 dB and less than 1e-9, consecutively.

Figure 5 appears a plain graphic composed of 3 users. This examination was done with assist of the employed parameters tabulated in Table 3 according to ITU-T G.652 standards as well taking into account the effects of non-linear in SMF (i.e. single mode fiber). Table 4 appears the ICSM code sequences for three users with code weight 3 and the selected conforming wavelengths. The transmission part

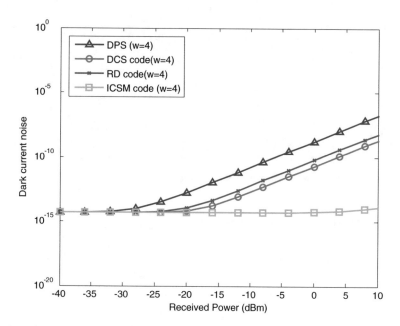

Fig. 4 Dark current noise versus effective source power ($U = 50$)

shows the encoder of each customer; the code is generated through the choose of wavelengths from the optical banner of the broadband. The wavelengths are selected by the code to pass through the channel. The signals are controlled by means of the modulator in accordance with the data given. The codes of different clients are attached before they are delivered onto the SMF. With regard to the receiving part, the received signal passes through the encoder which performs a comparative wavelength structure corresponding to that of the encoder.

Relying on Fig. 6, it reveals the eye diagram of ICSM code at SMF length amounts to 40 km. Although the generated QF and BER values are: 7.2 dB and 2.5×10^{-13}, successfully so they are agreeable in accordance with the conditions of optical communication networks mentioned above.

Here, it can say that the SAC-OCDMA system is granted a good performance thanks to our suggested ICSM code in addition to its ability of meeting the demands of optical communications.

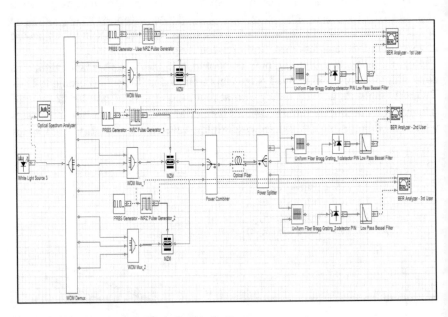

Fig. 5 ICSM code schematic block diagram for 3 users

Table 3 Utilized parameters in mimic examination

Parameter	Value	Parameter	Value
Data rate	622 Gbps	SMF length	40 km
Bandwidth of FBG	0.625 nm	Reference wavelength	1550 nm
Thermal noise	1.8×10^{-23} W/Hz	Dark current	5×10^{-9} A
Dispersion	18 ps/nm/km	Attenuation	0.25 dB.km^{-1}
Transmitted power	-115 dBm	Cutoff frequency	466.5 MHz

Table 4 IRSM code with K = w = 4, L = 16 and the conforming wavelengths

L		1	2	3	4	5	6	7	8	9
Wavelength (nm)		1548	1548.625	1549.25	1549.875	1550.5	1551.125	15,511.75	1552.375	1553
Code sequences	1st user	1	0	0	0	1	0	0	0	1
	2nd user	0	0	1	1	0	0	0	1	0
	3rd user	0	1	0	0	0	1	1	0	0

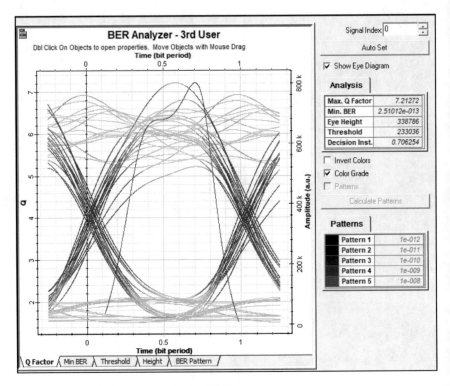

Fig. 6 Eye diagram of three users utilizing ICSM code

5 Conclusion

A novel encrypting technique termed as ICSM code has been suggested for SAC-OCDMA systems in this research. The distinctive feature of this technique is that it belongs to the ZCC code families, the easy construction steps depend on the identity matrix and shifting property. In addition, the QF, output PD current and dark current noise performances in different terms, of the ICSM code have been assessed at high transmission rate SAC-OCDMA system basing both the numerical computational utilizing Matlab software and the mimic test utilizing Optisystem software that aims more validate of viability of ICSM code. Thanks to SDD technique applied in the part of the receiver, the MAI phenomena has been canceled considerably other encrypting techniques suggested previously and this matter reflected positively in the system performance presented by an increment has been fulfilled and amounted to 260, 160 and 60% compared with DPS, DCS and RD codes, successfully. In the future, it can develop this work in carrying out, for instance, an amendment for ICSM code matrix or employing it as encrypting-decrypting scheme in various multi-dimensional-OCDMA systems, in striving towards obtaining a performance more responding to than exhibited in this paper.

References

1. Alayedi M, Cherifi A, Ferhat Hamida A, Mrabet H (2022) A fair comparison of sac-ocdma system configurations based on two dimensional cyclic shift code and spectral direct detection. Telecommun Syst 79(1):193–212. https://doi.org/10.1007/s11235-021-00840-8

2. El-Mottaleb SAA, Fayed HA, Ismail NE, Aly MH, Rizk MRM (2020) MDW and EDW/DDW codes with AND subtraction/single photodiode detection for high performance hybrid SAC-OCDMA/OFDM system. Opt Quantum Electron 52(2)

3. Matem R, Aljunid SA, Junita MN, Rashidi CBM, Shihab Ahmed I (2019) Photodetector effects on the performance of 2D Spectral/Spatial code in OCDMA system—ScienceDirect. Optik 178. https://doi.org/10.1016/j.ijleo.2018.10.068

4. Ghoumid K, Ghadban A, Boukricha S, Ar-reyouchi EM, Yahiaoui R, Mekaoui S, Lepers C (2020) Spectral coded phase bipolar OCDMA technological implementation thanks to low index modulation filters. Telecommun Syst 73(3):433–441. https://doi.org/10.1007/s11235-019-00610-7

5. Mrabet H, Mthali S (2016) Performance enhancement of OCDMA systems for LAN consideration. IET Optoelectron 10(6):199–204. https://doi.org/10.1049/iet-opt.2016.0044

6. Alayedi M, Cherifi A, Hamida AF, Bouazza BS, Aljunid SA (2021) Performance improvement of optical multiple access cdma network using a new three - dimensional (spectral/time/spatial) code. Wireless Pers Commun 118(4):2675–2698. https://doi.org/10.1007/s11277-021-08149-0

7. Alayedi M, Cherifi A, Hamida AF (2019) Performance enhancement of sac-ocdma system using a new optical code. In: Proceedings—2019 6th international conference on image and signal processing and their applications, ISPA 2019, pp 1–4. https://doi.org/10.1109/ISPA48434.2019.8966912

8. Raad R, Inaty E, Maier M (2019) Dynamic bandwidth allocation algorithms for improved throughput and latency performance in CDMA-based next-generation ethernet passive optical networks. OSA Continuum 2(11):3107. https://doi.org/10.1364/osac.2.003107

9. Alayedi M, Cherifi A, Hamida AF, Rahmani M, Attalah Y, Bouazza BS (2020) Design improvement to reduce noise effect in CDMA multiple access optical systems based on new (2-D) code using spectral/spatial half-matrix technique. J Opt Commun. https://doi.org/10.1515/joc-2020-0069

10. Tseng SP (2019) A new polarization-sac scheme suitable for compact ocdma-fso networks. IEEE Syst J 13(2):1333–1335. https://doi.org/10.1109/JSYST.2018.2875042

11. Zaeer Dhaam HS, Ali FM, Wadday AG (2020) Evaluation performance of two-dimensional multi-diagonal code using polarization and wavelength of ocdma system. Al-Furat J Innov Electron Comput Eng 1(3):22. https://doi.org/10.46649/150920-04

12. Cherifi A, Bouazza BS, Al-ayedi M, Aljunid SA, Rashidi CBM (2021) Development and performance improvement of a new two-dimensional spectral/spatial code using the pascal triangle rule for ocdma system. J Opt Commun 42(1):149–158. https://doi.org/10.1515/joc-2018-0052

13. Ahmed G, Djebbari A (2012) New technique for construction of a zero cross correlation code. Opt-Int J Light Electron Opt 123(15):1382–1384. https://doi.org/10.1016/j.ijleo.2011.08.017

14. Chaudhary S, Tang X, Sharma A, Lin B, Wei X, Parmar A (2019) A cost-effective 100 Gbps SAC-OCDMA–PDM based inter-satellite communication link. Opt Quant Electron 51(5):1–10. https://doi.org/10.1007/s11082-019-1864-2

15. Alayedi M, Cherifi A, Ferhat Hamida A, Bouazza BS, Rashidi CBM (2021) Performance enhancement of sac-ocdma system using an identity row shifting matrix code. In: Proceedings of international conference on information technology and applications (ICITA). https://doi.org/10.1007/978-981-16-7618-5_48

16. Fadhil HA, Aljunid SA, Ahmad RB (2010) Design considerations of high performance optical code division multiple access: a new spectral amplitude code based on laser and light emitting diode light source. IET Optoelectron 4(1):29–34. https://doi.org/10.1049/iet-opt.2009.0010

17. Abd TH, Aljunid SA, Fadhil HA, Ahmad RB (2012) A new algorithm for development of dynamic cyclic shift code for spectral amplitude coding optical code division multiple access systems. Fiber Integr Opt 31(6):397–416. https://doi.org/10.1080/01468030.2012.733905
18. Ahmed HY, Gharsseldien ZM, Aljunid SA (2016) An efficient method to construct diagonal permutation shift (DPS) codes for SAC OCDMA systems. J Theor Appl Inf Technol 94(2):475–484
19. Alayedi M, Cherifi A, Ferhat Hamida A, Matem R, Abd El-Mottaleb AS (2021) Improvement of SAC-OCDMA system performance based on a novel zero cross correlation code design. In: Proceedings of international conference on advances in communication technology, computing and engineering. RGN Publications. https://doi.org/10.26713/978-81-954166-0-8
20. Alayedi M, Cherifi A, Hamida AF, Rashidi CBM, Bouazza BS (2020) Performance improvement of multi access OCDMA system based on a new zero cross correlation code. In: IOP conference series: materials science and engineering, vol 767. https://doi.org/10.1088/1757-899X/767/1/012042
21. Alayedi M, Cherifi, A, Ferhat Hamida A, Bouazza BS, Aljunid SA (2021) Improvement of multi access ocdma system performance based on three dimensional-single weight zero cross correlation code. In: Proceedings—2021 3rd international conference on computer and information sciences ICCIS 2021
22. Meftah K, Cherifi A, Dahani A, Alayedi M, Mrabet H (2021) A performance investigation of SAC - OCDMA system based on a spectral efficient 2D cyclic shift code for next generation passive optical network. Opt Quant Electron 53(10):1–28. https://doi.org/10.1007/s11082-021-03073-w

Localization of Pashto Text in the Video Frames Using Deep Learning

Syeda Freiha Tanveer, Sajid Shah, Ahmad Khan, Mohammed ELAffendi, and Gauhar Ali

Abstract Object detection has remained an attractive and challenging task for the computer vision research community. A along with other objects, researchers tried to detect the texts in the images and videos as well. Earlier, the handcrafted features were used to detect text in the images and videos. These features have low discriminative power, leading to poor performance of the underlying machine learning model. Furthermore, more features are added to boost the discriminative ability of features, resulting in large data dimensionality. When dimensionality is increased, the performance of conventional machine learning usually falls. Deep learning can learn a feature by itself, which is known as representation learning, but it also performs better on high-dimensional data due to its data-hungry nature. Deep Neural network is an end-to-end system that is fully automated and does not require any handcrafting. Earlier, Arabic and Urdu and few other languages were detected in videos, but they mostly used handcrafted features to localize text in videos which shows the low performance on high dimensional data. Pashto language being the superset of Arabic, Urdu, and Persian, was remained unattended by the researchers. The contribution of this work is two folded: (i) dataset generation and annotation (ii) using a deep learning model for the Pashto text localization. Since it is pioneering

S. F. Tanveer (✉) · S. Shah · A. Khan
COMSATS University, Abbottabad Campus, Islamabad, Pakistan
e-mail: syedakazmi94@outlook.com

S. Shah
e-mail: sajidshah@cuiatd.edu.pk; sshah@psu.edu.sa

A. Khan
e-mail: ahmadkhan@cuiatd.edu.pk

S. Shah · M. ELAffendi · G. Ali
EIAS Lab, CCIS, Prince Sultan University, Riyadh, KSA, Saudi Arabia
e-mail: affendi@psu.edu.sa

G. Ali
e-mail: gali@psu.edu.sa

A. A. Abd El-Latif et al. (eds.), *Advances in Cybersecurity, Cybercrimes, and Smart Emerging Technologies*, Engineering Cyber-Physical Systems and Critical Infrastructures 4, https://doi.org/10.1007/978-3-031-21101-0_22

work on Pashto text location, that is why comparison with the state of the art is not conducted. We obtained good results with IOU more than 80% and recall is 0.98.

Keywords Localization · Pashto · Deep learning · YOLO · Darknet

1 Introduction

Object detection is the most challenging and fundamental task in the field of computer vision. To gain a comprehensive understanding of images, one should go beyond classification [1, 2] and try to discover the locations of things in images [3], which is known as object detection. An object detection algorithm generates the coordinates of an object contained in the image. In the field of computer vision, objects are localized with the help of bounding boxes. With the advancement of object detection, researchers turned their attention to text detection. Traditional methods require handcrafted features and preprocessing steps like Histogram of gradient (HOG), Scale-Invariant Feature Transform (SIFT) [4] Speeded Up Robust Feature (SURF) [5], etc. Moreover, feature engineering is carried out separately. While in the end to end system, it is done automatically. In deep learning systems, there is a glut of pre-trained models for object localization in videos and images like YOLO, RCNN, Fast RCNN, Mask RCNN, etc. https://towardsdatascience.com/evolution-of-object-detection-and-localization-algorithms-e241021d8bad [Accessed: 21-APR-2022].

Since the development of deep learning, many academics have considered this challenging task for indexing and retrieving information from videos. Because of the rapid rise of big data which includes photographs and videos. The data is diverse and needs to retrieve useful information from it. In the recent era, information acquisition from videos and images turn out to be the most significant task because of the rapid growth of machine learning and data mining techniques. Text in videos plays an important role as a source of information. Unar et al. [7] categorize texts in the videos i.e. Artificial text and scene text. Artificial text includes captions, subtitles, annotations, and is entrenched during editing of the video. However, scene text naturally appears in various objects e.g. billboards, walls, images of buildings, and vehicles. Artificial texts in the videos always been a challenging task, particularly for large scale. It becomes a more complex and open problem when there is a variety of sizes and multifaceted background. Text localization inspired many fields. In computer vision and machine learning, many algorithms, and techniques are used to localize texts in video frames. With the advent of deep learning, many of these methodologies have become obsolete. A detailed survey article is presented in [6] about text detection and localization.

Cursive languages, such as Arabic, Urdu, and Persian, have attracted the attention of the machine learning community [7–9]. Mansouri et al. [10] proposed a technique to classify Urdu text and non-Urdu text. They have used classical machine learning technique which requires hand crafted feature extraction. Moreover, Jamil et al. [11] also used preprocessing techniques such as Vertical Gradient, Mean Gradient,

RLSA, Edge Density based Noise Detection to detect Arabic text in videos and images. Yousfi et al. [12] used deep auto encoder to develop deep learning system to detect text in the images and videos. The performance of their system is very low. The authors in [13] have worked on Urdu text detection in signboards. Mosannaf et al. [14] have worked on farsi text location and detection in videos and images. Arabic text localization in natural scene is performed in [15]. Up to the best of our knowledge no has worked on Pashto text localization that is why there is no related literature.

Researchers have previously worked on detecting the above languages mostly using conventional machine learning approaches, but these models are poor in performances to detect the text of corresponding languages. Pashto being the super set of its siste languages These models are unable to detect and localize Pashto text. So, the motivation behind this research is to pioneer the work of Pashto text location. Furthermore, this work has aimed to build and annotate a dataset for the Pashto language and to design an end-to-end system that will be able to detect the Pashto language.

The output of text localization will be feed to Optical Character Recognition (OCR), content based retrieval system. Text detection and recognition in images and video frames, which points at integrating innovative optical character recognition (OCR) and text-based exploring technologies, is now accepted as a key element in the growth of advanced image and video annotation and retrieval systems.

The data for this work is acquired through various Pashto Tv channels and programs and YouTube Pashto channels. Pashto Television channels includes Khyber TV, BBC Pashto, Voice of America (Pashto), Hewed TV , Pashto TV, and YouTube Channels includes Bahij Virtual University, AVT Khyber Channels and various other TV channels. Video frames were obtained from the videos and were annotated manually by human being. YOLO (You Only Look Once) was implemented for text localization. The obtained results are quite encouraging. We got more than 80% of IoU (Intersection over Union) and 0.98 as recall.

2 Related Work

More research is conducted on the text extraction either captured by cameras or just embedded in the videos and images. Formerly, the text extraction was carried out by means of various filters and transformation algorithms. These algorithms required some preprocessing and handcrafted features in order to retrieve information from the frames. Khare [16, 17] exploits handcrafted features like Sobel, Canny edge detector, Gabor filter approaches to get the textual regions. Similarly, Raza et al. [18] presented a method to perceive multilingual artificial text that computes the textual areas in frames using the wavelet transform. Mansouri et al. [19] uses segmentation techniques to detect Arabic text in videos. Machine learning approaches involve learning techniques that are split into two broad categories i-e supervised and unsupervised learning. Unsupervised techniques required the extensive image evaluation and use segmentation methods in order to distinguish text from the various other parts in frame. Likewise, supervised learning techniques simply apply machine learning techniques to identify the text regions in the images. Thus, the supervised method involves classification and training while unsupervised learning involves gradient, connected component (CC), texture and color clustering-based methods. Salahuddin et al. [19] proposed a novel approach to detect artificial Urdu text in video frames by applying Sobel and canny edge detector and stroke width to validate text regions and finally SVM is used to classify the text and non-text regions. Many authors have worked on the scene and artificial text detection using deep learning techniques. Earlier a deep neural network ICPT-CNN proposed to detect the text that was captured by cameras. Similarly, [20] used VGA-16 architecture to detect scene text and used text boxes to localize the texts. Likewise, [21] proposed a method to localize the English text in frames. By the Connected components and text, region indicator candidate text lines are obtained and then fed into ANN(Artificial Neural Network) for the classification of textual and non-textual regions. Existing literature employed in localization of different languages like English, Urdu and Arabic and Persian. Various other native and non-native languages are also considered by the researchers to localize them in videos. But localization is carried out mostly using the classical machine learning using handcrafted features and a classifier.

3   Methodology

This section is dedicated to explain data acquisition and annotation (labelling of text as Pashto and non-Pashto). Moreover, the YOLO network design is discussed, as well as the loss function (IoU).

3.1   Dataset

The dearth of data is more remarkable in some languages like Pashto. To the best of our knowledge there is no such dataset which can be used for text localization. Ahmed et al. [22] proposed the dataset for Pashto text which contains only cropped images which are insufficient for localization of the text in the videos and images. Cropped images do not show the context of text in the original images. The proposed research prepared a Pashto dataset by annotating the text information from the videos and labeled them. The text was annotated according to the architecture used for the localization of the images. Those Videos were specifically picked for the present study which has embedded texts in them specially Pashto and its sister languages like Urdu and Arabic just to ensure the accurate classification among them.

3.2 Data Acquisition and Annotation

Talk shows, interviews, documentaries, weather reports, and sports are among the many TV programs that are broadcasted on television networks. Data is acquired through various Pashto Tv channels and programs and Pashto YouTube channels. Pashto television channels include Khyber TV, BBC Pashto, Voice of America (Pashto), Hewed TV, Pashto TV, and YouTube Channels includes Bahij Virtual University, AVT Khyber and various other TV channels.

All the videos were downloaded using a library called youtube-dl. After downloading 160 videos, a python script was used to extract all the frames from these videos. To remove redundancy from the dataset, similar frames were removed by using a technique called Structured Similarity Index Matrix (SSIM) [23]. It is the method which is used to calculate the similarity between two frames. The comparison of the images is carried out pixel by pixel. In SSIM the threshold was set to 95. All those frames were removed whose similarity was found greater or equal to 95%. We labeled the frames with a YOLO Annotation Master Tool https://github.com/ManivannanMurugavel/Yolo-Annotation-Tool-New-/issues. This technique perfectly labels and annotate texts in the images. Pashto text was labeled with class 1 and non-Pashto text was labeled as class 0. A bounding box around the text was represented by the following attributes: $< Classid > < classid > < X_o/X > < Y_o/Y > < W/X > < H/Y >$

- Class id: Label index of the class to be annotated.
- X_o: X coordinate of the bounding box's center
- Y_o: Y coordinate of the bounding box's center
- W: Width of the bounding box
- H: Height of the bounding box
- X: Width of the image
- Y: Height of the image.

3.3 Implementation

We have trained our model using incremental learning approach. There were 4000 total examples out of which 3000 were Pashto and the rest were non-Pashto. A deep learning based model called YOLO (You Only Look Once) was used. The flow diagram of the proposed work is shown in Fig. 1.

4 YOLO

You only look once (YOLO) is the object detection deep neural network. It is an algorithm that uses convolutional neural networks for object detection. YOLO is one of the faster object detection algorithms among all other state of the art algorithms.

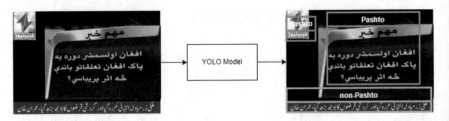

Fig. 1 YOLO architecture

When we need real-time localization and detection, it outperforms better without sacrificing accuracy. It is single convolutional neural network concurrently predicts numerous bounding boxes as well as class probabilities. YOLO trains on full images and directly boosts detection performance [24].

YOLO has versions with better performance improvement and detection accuracy. Speed also improved in the latest versions. YOLO introduced YOLOv1, YOLOv2, YOLOv3 and recently developed YOLO9000 [25]. Darknet19 developed based on YOLOv2. As its name depicts it has 19 convolutional layers. Further Darknet53 also introduced with the better capabilities and strengths based on YOLOv3 [26]. The implementation in this research is conducted using Darknet19 based on YOLOv2.

4.1 YOLO Architecture

YOLO consists of 19 convolutional layers followed by 4 max pool layers and finally two fully connected layers. Convolutional layers create the feature maps by convolving with the filters. Convolutional layer predicts which feature belong to which class. Relu (Rectified Linear Units) activation [27] function is used in the hidden layers.

Mathematically ReLU can be expressed as:

$$y = max(0, x) \tag{1}$$

The purpose of pooling layer is down sampling. It reduces the amount of information in features maps returned by convolutional layer. Pooling layer retains only important information and discards unnecessary ones. In order to get the important information, it takes several rounds by convolution and pooling layer. Finally, fully connected layers flatten the matrix into vector obtained from the previous layer. First fully connected layer takes the input from the feature map and applies weights to predict the correct labels. The second layer predicts the class for each object by calculating probabilities. The activation function used here is softmax [28]. Softmax is best suited for the multi-class predictions.

Mathematically softmax can be expressed as:

$$Softmax(x_i) = \frac{\exp(x_i)}{\sum_j \exp(x_j)} \qquad (2)$$

The distribution of filters on each layer can be studied in [24].

5   Loss Function

YOLO foretells several bounding boxes per grid cell. To calculate the loss for the true positives, only one bounding box with highest IoU (Intersection over Union) is considered. YOLO uses sum-squared error between the predictions and the ground truth to calculate loss.

6   Results and Discussion

The dataset was splitted into testing and training data with the ratio of 20–80. The neural network was trained over the dataset using incremental learning approach. The learning rate was set to 0.001 with the momentum 0.9 and decay of 0.005. Batch size was set to 64 with the subdivisions of 8. Weights obtained at every 100th iterations. Image size is 416 × 416. Firstly, 1000 images were subjected to neural network and calculates the recall and intersection over union. The values obtained for these metrics are 0.75 and 0.61 respectively. In the next increment 2000 samples were used. An improvement in the performance was seen. Finally, 1000 examples were fed to neural network. A further boost in performance was observed. The obtained recall is 0.98 and the area over union (IoU) is 0.8.

6.1   Recall

Recall also called sensitivity is the fraction obtained when the correctly detected instances are divided by the total number of instances in the dataset. As shown in Fig. 2, there is gradual increase in recall up to 400 iterations.

6.2   Intersection over Union (IoU)

Intersection over union is defined as the intersection of area occupied by the overlapping of predictor box and the ground truth and area of union of both boxes. The IoU obtained with the passage of training is shown in Fig. 3.

Fig. 2 Recall

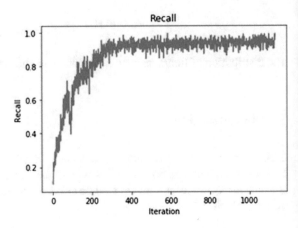

Fig. 3 Intersection over union

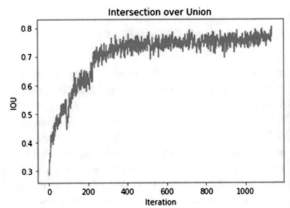

$$IOU = \frac{Area\ of\ intersection\ (overlap)}{Area\ of\ union} \tag{3}$$

Our model shows the balanced state for the training and validation loss. Initially it was 19.5 in the first iteration but gradually by updating weights it reduces to 0.22 which is highest fall of loss. The validation and training loss throughout the process is shown in Fig. 4.

7   Comparison with Existing Predictor

Since it is pioneering work on Pashto text location and no existing predictor is found that is why its performance cannot be compared with previous ones. But still the obtained results are very encouraging and acceptable for problem of text localization.

Fig. 4 Training and validation loss

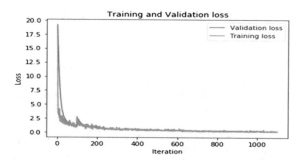

8 Conclusion and Future Work

In this research work an end to end system was proposed which was able to localize the Pashto text in videos. Earlier, the machine learning model was proposed for the text localization of Arabic and Urdu and various other languages as well, but those models now have been obsolete, and does not perform well on high dimensional data. So, a fully automated system proposed in this research that has a capability to learn feature itself and predicts the location of the Pashto text in the videos. A dataset was built and annotated to train YOLO neural network. The model shows the highest IOU of 0.8 and Recall of 98%. In future, this research will be extended to recognition stage.

References

1. Felzenszwalb PF, Girshick RB, McAllester D, Ramanan D (2009) Object detection with discriminatively trained part-based models. IEEE Trans Pattern Anal Mach Intell 32(9):1627–1645
2. Zhang D, Han J, Cheng G, Yang M-H (2021) Weakly supervised object localization and detection: a survey. IEEE Trans Pattern Anal Mach Intell
3. Zhao Z-Q, Zheng P, Xu S-T, Wu X (2019) Object detection with deep learning: a review. IEEE Trans Neural Netw Learn Syst 30(11):3212–3232
4. Karami E, Shehata M, Smith A (2017) Image identification using sift algorithm: performance analysis against different image deformations. arXiv preprint arXiv:1710.02728
5. Bay H, Tuytelaars T, Gool LV (2006) Surf: speeded up robust features. In: European conference on computer vision. Springer, pp 404–417
6. Shilpa M, Rajneesh R (2021) Text detection and localization in scene images: a broad review. Artif Intell Rev 54(6):4317–4377
7. Unar S, Jalbani AH, Jawaid MM, Shaikh M, Chandio AA (2018) Artificial urdu text detection and localization from individual video frames. Mehran Univ Res J Eng Technol 37(2):429–438
8. Moradi M, Mozaffari S, Orouji AA (2010) Farsi/arabic text extraction from video images by corner detection. In: 2010 6th Iranian conference on machine vision and image processing. IEEE, pp 1–6
9. Moradi M, Mozaffari S (2013) Hybrid approach for farsi/arabic text detection and localisation in video frames. IET Image Process 7(2):154–164
10. Mansouri S, Charhad M, Zrigui M (2017) Arabic text detection in news video based on line segment detector. Res Comput Sci 132:97–106

11. Jamil A, Siddiqi I, Arif F, Raza A (2011) Edge-based features for localization of artificial urdu text in video images. In: 2011 international conference on document analysis and recognition. IEEE, pp 1120–1124
12. Yousfi S, Berrani S-A, Garcia C (2015) Deep learning and recurrent connectionist-based approaches for arabic text recognition in videos. In: 2015 13th international conference on document analysis and recognition (ICDAR), 2015. IEEE, pp 1026–1030
13. Arafat SY, Ashraf N, Iqbal MJ, Ahmad I, Khan S, Rodrigues JJPC (2022) Urdu signboard detection and recognition using deep learning. Multimedia Tools Appl 81(9):11965–11987
14. Mosannafat M, Taherinezhad F, Khotanlou H, Alighardash E (2022) Farsi text detection and localization in videos and images. In: 2022 9th Iranian joint congress on fuzzy and intelligent systems (CFIS). IEEE, pp 1–6
15. Khalil B, Qahtani Abdulrahman M, Omar A, Habib D, Alimi Adel M (2022) Reduced annotation based on deep active learning for arabic text detection in natural scene images. Pattern Recognit Lett 157:42–48
16. Khare V, Shivakumara P, Paramesran R, Blumenstein M (2017) Arbitrarily-oriented multilingual text detection in video. Multimedia Tools Appl 76(15):16625–16655
17. Jamil A, Abidi A, Siddiqi I, Arif F (2012) A hybrid approach for artificial urdu text detection in video images. In: Proceedings of the ICPR, pp 1944–1947
18. Raza A, Siddiqi I, Djeddi C, Ennaji A (2013) Multilingual artificial text detection using a cascade of transforms. In: 2013 12th international conference on document analysis and recognition. IEEE, pp 309–313
19. Mansouri S, Charhad M, Zrigui M (2017) Arabic text detection in news video based on line segment detector. Res Comput Sci 132:97–106
20. Liao M, Shi B, Bai X, Wang X, Liu W (2017) Textboxes: a fast text detector with a single deep neural network. In: Thirty-first AAAI conference on artificial intelligence
21. Thilagavathy A, Aarthi K, Chilambuchelvan A (2012) Text detection and extraction from videos using ann based network. Int J Soft Comput, Artif Intell Appl (IJSCAI) 1(1)
22. Ahmad R, Afzal MZ, Rashid SF, Liwicki M, Breuel T, Dengel A (2016) Kpti: Katib's pashto text imagebase and deep learning benchmark. In: 2016 15th international conference on frontiers in handwriting recognition (ICFHR). IEEE, pp 453–458
23. Wang Z, Bovik AC, Sheikh HR, Simoncelli EP (2004) Image quality assessment: from error visibility to structural similarity. IEEE Trans Image Process 13(4):600–612
24. Redmon J, Divvala S, Girshick R, Farhadi A (2016) You only look once: unified, real-time object detection. In: Proceedings of the IEEE conference on computer vision and pattern recognition, pp 779–788
25. Redmon J, Farhadi A (2017) Yolo9000: better, faster, stronger. In: Proceedings of the IEEE conference on computer vision and pattern recognition, pp 7263–7271
26. Redmon J, Farhadi A (2018) Yolov3: an incremental improvement. arXiv preprint arXiv:1804.02767
27. Agarap AF (2018) Deep learning using rectified linear units (relu). arXiv preprint arXiv:1803.08375
28. Kanai S, Fujiwara Y, Yamanaka Y, Adachi S (2018) Sigsoftmax: reanalysis of the softmax bottleneck. In: Advances in neural information processing systems, vol 31

Exploration of Epidemic Outbreaks Using Machine and Deep Learning Techniques

Farah Jabeen, Fiaz Gul Khan, Sajid Shah, Bilal Ahmad, and Saima Jabeen

Abstract Contagious diseases and their intensity have shattered human beings all over the world. Infections due to viruses spread so rapidly that it becomes so hard for nations to deal with them timely and effectively with traditional approaches. There are different sources of these epidemic outbreaks and some regions of the world got more affected against certain types of the epidemic for several reasons. Researchers have investigated different automated techniques for effective and, timely handling of these epidemics. This study summarizes some of the latest approaches adopted in this direction for four major epidemics i.e., Zika, Congo, Dengue, and Corona virus, in recent times. A detailed comparison and discussion of these techniques have been presented along with the future directions.

Keywords Epidemics · Outbreaks · Corona virus · Artificial intelligence · Survey · Machine learning · Deep learning · CNN and LSTM

F. Jabeen (✉) · F. G. Khan · S. Shah · B. Ahmad
COMSATS University, Abbottabad Campus, Islamabad, Pakistan
e-mail: farah.jabeen.ahmad@gmail.com

F. G. Khan
e-mail: fiazkhan@cuiatd.edu.pk

S. Shah
e-mail: sshah@psu.edu.sa

B. Ahmad
e-mail: bilalcuiatd@gmail.com

S. Shah
EIAS Lab, CCIS, Prince Sultan University, Riyadh, KSA, Saudi Arabia

S. Jabeen
Department of IT and Computer Science, Pak-Austria Fachhochschule: Institute of Applied Sciences and Technology, Mang, Haripur, Pakistan
e-mail: saima.jabeen@fecid.paf-iast.edu.pk

1 Introduction

Epidemic means a common amount of a transferable disease in a public at a specific time, while an outbreak is the unexpected occurrence of something unwanted, such as war or disease. An epidemic happens when an infectious disease spread quickly in many peoples and an unexpected rise in the number of cases of a disease. The epidemic outbreak may affect numerous countries. An epidemic outbreak may last for days or weeks, even for many years. Artificial Intelligence (AI) enabled us to design intelligent models that can predict future values based on past data from which it learns like human beings. Machine and deep learning are the subfields of AI that are being widely used for predictive modeling in different fields. The most important use of these AI-based predictive modeling techniques is in the field of disease predictions. Predicting the disease can be utilized to lessen the mortality rate [1]. Several different domains made use of ML (Machine Learning) such as self-driving cars, discourse identification, effective network search, and an enhanced impression of the social group. Today, AI is available all over the place so that without knowing it, one can utilize it quite often. AI systems find automated fitness examples or records that have high-dimensional examples and various data sets. Pattern recognition is the subject of MLT (multiple-level tracking) that suggests backing to foresee and make choices for analysis and to design a cure. MLA (Machine Learning Adaptation) is skilled to oversee a large amount of data, joining data from divergent properties, and adding the related material in the study [2].

Nowadays, epidemic outbreaks caused by different kinds of viruses have attracted the focus of researchers. Many approaches have been employed to investigate various machine learning algorithms for virus analysis. In this study, a detailed survey is presented about several ML and deep learning techniques used in recent studies to analyzing four major types of epidemics in recent history. Figure 1 depicts different global epidemics in the recent past.

The rest of the paper is organized as followed, in Sect. 2 we presented the contribution of the machine and deep learning techniques to epidemic outbreaks caused by different viruses in the recent past. Section 3 presents a comparison of reviewed approaches, along with their limitation and strengths. Finally, Sect. 5 ends with a conclusion along with the future directions.

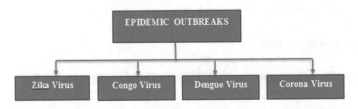

Fig. 1 Different global epidemics in the recent past

2 Machine and Deep Learning Uses in Epidemic Outbreaks

Several approaches adopted for the prediction of different kinds of diseases caused by viruses have been discussed in this section. The traditional steps in the computer-aided diagnosis of viruses are image acquisition, preprocessing, feature extraction, and classification. This work aims to discuss the different machine and deep learning based classification models in the recent past for the prediction of different epidemics shown in Fig. 1. Figure 2 shows the different traditional steps in the computer-aided models for the diagnosis of different epidemics. The literature is presented in such a way that each sub section combines machine learning/deep learning techniques used for a specific virus type. As mentioned before that we have discussed only four virus types in this work.

2.1 Dengue Virus

Dengue is a bone-breaking illness that belongs to the virus family, is spread by female mosquitoes. The mosquito's family which causes the dengue virus also becomes a source of some other diseases such as Zika, yellow fever, and chikungunya [3, 4]. Different machine learning methods to identify this disease, have been devised.

Dave Kaveri Atulbhai presented a survey to predict and detect Dengue for which DNA microarray dataset was used having material of the gene's expression responsible for dengue virus [5]. Different classification techniques were investigated for this purpose such as Naive bayes. Naive Bayes performed better than others by achieving 92% accuracy. A framework was devised to predict and forecast Dengue disease using a decision tree and SVM by researchers in [1]. The dataset consisting of

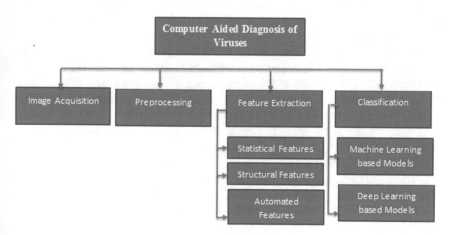

Fig. 2 Traditional steps in computer aided diagnosis of viruses

ninety-five entries used for classification was collected from the District Headquarter Hospital Jhelum. Fisher Filtering was utilized to create a decision tree while SVM was used to forecast. Authors in [6] worked on a medical profile about multiple serological diagnostic markers of dengue disease. The data of 180 dengue fever patients, showing all conducted tests named NS1, IgM, IgG (dengue virus types) as positive was taken as the input. Patients with positive results of all tests were considered for a cross- sectional study. To detect early Dengue severity, SVM, Naive Bayes were proposed [7].The obtained results show that SVM performed better. In another study, data of Hospitals' patients was gathered to validate various machine learning techniques. SVM, L2 Regularized Logistic Regression, and Naive Bayes Classifier were investigated on the collected data to predict the disease. A 5-Fold Stratified Cross-Validation was used to train all the models which showed good forecasting skills and Support Vector Machine (SVM) obtained good results in terms of Receiver operating characteristic area. To predict Dengue fever, a data mining approach was proposed in [8]. Fever, Tiredness, Blood, Metallic Flavor, and more features were included in the dengue dataset which was considered as input. REP Tree, J48, SVM, and Random Tree classifiers were used by making use of the WEKA tool. The comparison of obtained results indicated that SVM and J48 showed better accuracy i.e., 84% and 76%. A similar approach was use by authors in [9]. A data mining tool named WEKA worked on a chunk of the dataset which was gained from District Headquarter Hospital, Jhelum. REP Tree (an extension of C45), Naive Bayes, Regression Tree (RT), J48, and SMO (Sequential Minimal Optimization) were classification methods used to analyze the test performance. NB and J48 showed good performance by providing 92% and 88% accuracy respectively. A decision trees based work is presented in [10]. The dataset was gathered from the hospital, Ho Chi Minh City, Vietnam. Around 8000 medical records were collected. For the creation of a decision tree, C4.5 algorithm was used which achieved high accuracy of 98%. The work of K.A. Shakil et al., investigated denque using Naive Bayes, J48 in [11]. They have used WEKA for their experiments. In the conducted experiments, Naive Bayes and J48 techniques were employed and these techniques showed good results. NB (Naive Bayes) and J48 showed 100% and 99.70% accuracies respectively. An ANN (Artificial Neural Networks) was used to predict Dengue outbreaks in Sri Lanka [12]. Each separate variable with dengue correlation analysis was applied. They performed correlation analysis for feature selection. They got 68.5% accuracy. Classification Rules using Decision Trees have been investigated for predicting Dengue disease [13]. Different sources were used to gather 2424 records to forecast the probabilities of dengue disease in the area. To predict dengue cases deep learning method was used by authors in [14]. LSTM and RNN were used for the forecast of dengue. Data was taken from 20 selected cities in mainland China. Basic demographic information was contained in the data (for example, sex, age, country, and residence address). The training set data was used from 2005 to 2016 and test set data was used from 2017 to 2018. LSTM Shows very good results with 96% accuracy as compared to GAM (generalized additive model), and SVR (support vector regression) model. An SVM based work in presented in [15]. Average temperatures of previous years used as a dataset. The model successfully executes dengue epidemic classification based on different temperatures from the data. Temperatures between 27 and 29 °C were the typical range in which an outbreak or illness transmission can occur. The SVM model showed a good result with 97% accuracy.

2.2   Congo Virus

Congo fever is a common disease between humans and animals. To predict and identify people at risk for this disease by the use of essential tests, there are numerous data mining and machine learning methods that were used in literature.

A C4.5 decision trees based work is discussed in [16]. Data with 965 records with 29 features were collected from the health center of Iran. Quadratic programming high- light choice technique was utilized for choosing the factors which were viable on the model. Afterward, the C4.5 decision tree model was implemented by utilizing input factors. They obtained 95% accuracy. A computational model based on Gaussian processes was used to work on the spatiotemporal data for the prediction of infectious diseases [17]. They investigated the task of modeling Crimean-Congo hemorrhagic fever cases in Turkey between 2004 and 2015. BRT (boosted regression trees), RF (random forests), and Gaussian processes were used. It was noted that Gaussian process formulation obtained good results which were 85% than the other two. The global distribution of Crimean-Congo hemorrhagic fever was investigated in [18]. Boosted regression trees (BRT) was used for this purpose and an extensive database of human CCHF records was used. The resulting dataset contained 1721 occurrence records in total Boosted regression trees (BRT) showed 92% accuracy. Multiple classification techniques such as (k-NN) k-nearest neighbors, (ANN) artificial neural network, random forest (RF), and (SVM) support vector machine with Chou's pseudo amino acid composition were used in [19]. The model as trained on the IEDB database. ANN algorithm performed the most accurate prediction with an accuracy of 90%. To confront different viruses included Congo, a next generation sequencing technique is presented in [20].

2.3   Zika Virus

The Zika virus is a flavivirus spread by mosquitos that originated in monkeys. It was later discovered in humans in Uganda and the United Republic of Tanzania in 1952. Zika virus epidemics diseases have been recorded in Asia, the America, and Africa.

MLT (Medical Laboratory Technology) was used to record the transmission risk of the Zika virus [21]. To predict the outbreaks of Zika virus, three popular machine learning models were used named backward propagation neural network (BPNN), gradient boosting machine (GBM), and random forest (RF). Datasets were used to get the results to mimic the possibility of ZIKV infection. The BPNN model achieved the highest predictive accuracy based on the training dataset. By facing data sparsity, possible origins of Zika virus spillover infection among primates were discovered in [22]. In the recent era, the American Zika virus epidemic was ranked the largest outbreak. The Bayesian multi-label machine learning method was combined with a multiple imputation method to confront the sparsity of the dataset. The dataset contained 50 features for 376 species of primates. With 82% accuracy, subsequent models (BMLPL) distinguished flavivirus positive primates. For Tracking Health Misinformation on Twitter, work on Crowd-sourcing and machine learning based applications were carried on [23]. Machine learning, health informatics, and social network analysis were used to identify the health- related gossips on social media sites. On social media, the daily thoughts and actions of millions of users have been extracted for this purpose. For real-time forecasting of the Zika epidemic, a dynamic neural network model was proposed to forecast the earthly spread of epidemics [24]. Regions at high risk during the out break have been identified by the proposed model. Above 85% accuracy was shown by the developed model for predicting the expansion of Zika geographically in America. Using Deep Learning Techniques, a work on real time mosquito species identification has been proposed in [25]. To identify mosquito species, several techniques showed encouraging results. For this purpose, the WINGBEATS dataset was used. From 6 mosquito species, 279,566 mosquito wingbeat records were obtained. Convolutional Neural Networks being the favorite techniques for image classification was used to minimize dimensionality. Multi-Layer CNN achieved 86% accuracy while XGBoost achieved 85.8% accuracy. The hospitalization of suspects having an arboviral infection is predicted by a model based on machine learning with high accuracy [26]. The signs, symptoms, and laboratory data were considered to predict hospitalization status by testing seven algorithms using a machine learning approach. In the dataset, SISA and SISAL were capable to forecast arbovirus hospitalization with good accuracy of 94% by considering only symptoms and signs while 92% accuracy was obtained by considering laboratory data as well. A technique was proposed to analyze social network images with deep learning models to combat Zika Virus [27]. Two customized deep neural networks were planned to analyze Instagram images. Only supervised learning was used on the challenging dataset. Deep Convolutional Neural Network was capable of attaining decent training results (82.5%). Few images which were collected from Instagram were tested and 73.33% classification accuracy was achieved. An ML application, based on RF for Adding Mass Spectrometry-Based Metabolomic data, which was a method for patients with Zika Virus, was proposed [28]. An improved technology based on high-resolution mass spectrometry (HRMS) and the RF algorithm was designed to get boosted accuracy. The RF algorithm performed well on selected best features and also on the original complete feature space. For discovering latent transferable diseases using social media data, an unsupervised machine learning model was planned [29]. A latent infectious disease in a given location was identified by using an unsupervised machine learning model. Google Trends was used to forecast Zika Epidemics dynamically [30]. Correlation data was used from Zika related online searches. ARIMA model was selected for the prediction of the Zika virus. ML and NLP methods were combined to identify what is being discussed by people about Zika [31]. It was predicted by using the machine learning techniques that the number of tweets by females and males was similar. Performance of Tree-Based Classifiers was evaluating by using Zika Virus Dataset [32]. Experiments are performed by using filters and data mining algorithms. J48, decision tree, SVM, and RF algorithms were shown well results. A very recent work presents Zika virus prediction using an AI-driven system based on hybrid optimization algorithms [33].

2.4 Coronavirus

The simple acute respiratory illness coronavirus 2 (SARS-COV-2, also known as 2019-nCov) has created a worldwide outbreak of a typical and person-to-person contagious pneumonia. The outbreak was first detected in Wuhan, China, but has since spread throughout the world. Researchers in machine learning and computer scientists have worked for the development of systems that recognize and categorize COVID-19. As a result, developing a precise computer-aided strategy to assist clinicians in identifying COVID-19-infected patients is critical. A huge amount of research work is published until now, but we have selected few papers for review in this section. By using CT images, Deep learning algorithms were planned to screen Corona Virus Disease (COVID-19) [34]. 453 CT images were collected to perform the test. Internal validation yielded an overall accuracy of 82.9%. The total accuracy of the external test set was 73.1%. ML was used to figure out the main issues which affect recovery in MERS Co-V patients [35]. Data were gathered for analysis from the ministry of health's website. SVM, a conditional inference tree, NB, and J48 are examples of machine learning algorithms. The key factors were identified by using the machine learning algorithms. ML analysis of genomic signatures offers evidence of associations between Wuhan 2019-nCoV [36]. In Wuhan, many deaths occurred due to the coronavirus. 100% accuracy achieved for the classification of 2019-nCoV. To Screen Coronavirus Disease 2019 Pneumonia, the Deep Learning System was offered [14]. Using deep learning techniques, the total of 618 CT samples were gathered. Data set, containing a total of 618 CT records: 219 from 110 patients with COVID-19, 224 CT records from 224 patients with Influenza-A viral 5/29, pneumonia, and 175 CT records from healthy people, showed 86.7% accuracy. COVID-19 was detected by identifying the viral nucleic acid [37]. They were comparable to RT-PCR and non-contrast chest CT in terms of compassion. They looked at 51 patients who had severe respiratory symptoms from an unknown source. For identifying COVID-19 infection, RT-PCR had a sensitivity of 71%, whereas non-contrast chest CT had compassion of 98%. Developed an automated deep learning model for the detection of coronavirus infected patients and non-infected patients through analysis of CT images [38]. The model was trained by two datasets, one dataset has 56 coronavirus infected patients and the other one has 51 Chinese non-infected patients and 49 more non-infected U.S patients, which was collected from different institutes of China. The model enables the categorization of patients with an accuracy of 95%. For the accurate detection of COVID-19 pneumonia, a novel deep learning model UNet++ was developed by [39]. For the training and testing of the model, they collect 46,096 anonymous records, of which 51 were COVID-19 infected patients and 55 controlled patients, from 106 admitted patients in Renmin Hospital of Wuhan University. The model achieved 95.24% accuracy for twenty-seven consecutive patients CT scans. Proposed a deep-learning based automated segmentation model name as VB-Net for the detection of infected areas in CT scans [40]. The model trained on 249 COVID-19 patients and validated on 300 COVID-19 patients. For the training samples generation, they proposed human involved model iterations. The proposed model automatically counters the infected regions from CT scans and also calculate the volumes and percentage of infection with an accuracy of 91.6%. Features relation extraction, neural network (DRE-Net) was a deep neural network that extracts deep details from CT scans to obtain picture level predictions [41]. They used CT images from 88 COVID-19 patients, 101 bacterial pneumonia patients, and 86 healthy people. They both recorded a 94% accuracy rate. A most recent work is presented in [42] about COVID-19 diagnostic system using deep learning for noisy and corrupted images. A combination of CNN and LSTM is used to develop a COVID-19 detecting system in [43]. In some cases, they have achieved 100% accuracy. They have used X-ray and CT scan image in their experiments. The major difficulty in detecting COVID-19 accurately is the availability of reliable and sufficient amount of data. Researchers in [44] have presented two novel data augmentation techniques to increase the size of dataset. The reported results show an improvement in the performance after data augmentation.

3   Comparison of Reviewed Approaches

A comparison of reviewed approaches, adopted for analysis of infectious diseases, has been made in different four Tables (see Tables 1, 2, 3 and 4) which show the used methodology, kind of input data, and achieved accuracy of the models. By looking into these tables, the accuracy of Congo virus detection is relatively less while that Corona virus is the maximum. The accuracy of dengue is stable.

Table 1 Comparison of adopted approaches to analyze dengue virus

References, year	Method and algorithm	Dataset	Findings
[1], 2017	Decision tree, SVM	Dengue Dataset	SVM: 96% accuracy
[5], 2017	SVM, Naive Bayes, RF	DNA dataset	Naive Bayes: 92% accuracy
[10], 2016	Decision tree	8000 medical records	Accuracy: 98%
[14], 2020	Random forest	Demographic features	Accuracy: 97%
[15], 2020	SVM	Temperature of cities	Accuracy: 97%

Table 2 Comparison of adopted approaches to analyze congo virus

References, Year	Method and algorithm	Dataset	Findings
[16], 2017	C4.5	965 records	Accuracy: 95%
[17], 2018	Gaussian processes	Data of 2004 and 2015	Accuracy: 92%
[18], 2015	Boosted regression tree	1721 records	Accuracy: 85%
[19], 2020	KNN, ANN, SVM, RF	IEDB Database	ANN: 90% accuracy

Table 3 Comparison of adopted approaches to analyze zika virus

References, year	Method and algorithm	Dataset	Findings
[21], 2018	BPNN,GBM and RF	Multidisciplinary	BPNN: 96% accuracy
[22], 2019	Bayesian multi-label learning	33 features for 364 primate species	82% accuracy achieved
[23], 2017	Machine learning	Dataset of 13 million tweets	User actions are extracted
[25], 2017	Multi-layer CNN and XGBoost	WINGBEATS	Multi-Layer CNN: 86% accuracy XGBoost: 85.8% accuracy
[26], 2020	LCCA models	1160 patients, 412 males, 748 females	Accuracy: 92%
[27], 2018	Deep neural network	1Google images	Accuracy: 73.33%
[28], 2018	Random forest	Features	Accuracy: 95%
[29], 2017	Unsupervised learning	Social media data	Accuracy: 80.20%
[30], 2017	ARIMA	Correlation data	Accuracy: 95%
[31], 2017	NLP	Tweets data	Accuracy: 99%
[32], 2017	J48, decision tree SVM and RF	Zika Virus data	Classification results are similar to experimental

Table 4 Comparison of adopted approaches to analyze corona virus

References, year	Method and algorithm	Dataset	Findings
[34], 2020	Transfer learning	453 CT images	82.9% and 73.1% accuracy
[35], 2019	SVM, NB and J48	33 Health website	ML identifies important variables
[36], 2020	Decision tree	Gnomic dataset	Accuracy:100%
[14], 2020	CNN	618 CT images	Accuracy: 86.7% accuracy
[38], 2020	2D, 3D Deep learning	157 patients	Sensitivity: 98.2%
[39], 2020	UNet++	46,096 CT images	Accuracy: 95.25%

4 Challenges and Future Work

In this study, machine learning-based approaches are studied and presented in a summarized way by highlighting their significant contributions. It is observed that for more than one disease, SVM and Naive Bayes gave promising results. Deep learning methods are also being investigated in recently conducted work for analysis of these diseases and showing very encouraging results. In most of the adopted approaches, researchers have made use of the WEKA data mining tool, MATLAB as well as Python. There are a few weaknesses to ZIKV diagnosis because of the absence of accessibility of diagnostic tools. There is as yet a need to look for an enhance, more sensitive, and reliable detection techniques ZIKV. Data mining and machine learning techniques have proven very useful after their extensive use for the prediction of diseases. However, it is found that approaches strive for a good dataset with useful features. There is a need to record patient's related data according to such a format which could help the machine learning techniques to get applied effectively. There are some limitations in the Congo virus as well. There is a need to collect useful data for the diagnosis of the Congo virus. To achieve a milestone for efficient diagnosis of viruses, novel techniques for data sampling and data processing is needed to be introduced. NB and J48 are the top performance classifier techniques they achieved an accuracy of 92 and 88% for the diagnosis of dengue virus. However, there are some limitations that if more datasets are available, than these techniques can show good promising results. There is a need to record the patient's connected information as indicated by such an organization which could help AI strategies to get applied effectively. There is a need to work on investigating different machine learning methods with a combination of various preprocessing techniques. Deep learning methods are being used to analyze outbreaks but there is still scope to fully exercise their strengths. There is also a need to work on collecting data of COVID-19 as a bench-mark dataset so that comprehensive research could be conducted.

5 Conclusion

An important dilemma of epidemic outbreaks has prompted the need for research by exploiting several techniques. This paper summarizes different approaches based on the machine learning methods such as J48, decision tree, SVM, Random forest algorithms, deep neural networks, and unsupervised learning. Different publically available datasets were used in the conducted studies but there is still a need to have a bench- mark collection for some of the diseases such as COVID-19.

References

1. Kumar DA, Devi C, Karthick G (2017) Dengue disease prediction using decision tree and support vector machine
2. Rambhajani M, Deepanker W, Pathak N (2015) A survey on implementation of machine learning techniques for dermatology diseases classification. Int J Adv Eng Technol 8(2):194
3. Baheerati M (2014) Natural therapy for dengue fever. Res J Pharmacy Technol 7(2):269–271
4. Bimal MK, Kaur L, Kaur M (2016) Assessment of knowledge and practices of people regarding dengue fever. Int J Nurs Educ Res 4(2):174–178
5. Dave KA, Serasiya S (2017) A survey: predection and detection of dengue mining methods and techniques
6. Morlawar R, Kothiwale V et al (2017) A study of clinical profile in different serological diagnostic parameters of dengue fever. Indian J Health Sci Biomed Res (KLEU) 10(2):178
7. Caicedo-Torres W, Paternina A, Pinzon H (2016) Machine learning models for early dengue severity prediction. In: Ibero-American conference on artificial intelligence. Springer, pp 247–258
8. Bhavani M, Vinod Kumar S (2016) A data mining approach for precise diagnosis of dengue fever. Int J Latest Trends Eng Technol 7(4)
9. Kumar P, Ankushe RT, Kuril BM, Doibale MK, Hashmi SJ, Pund SB (2016) An epidemiological study of fever outbreak in aurangabad, maharashtra, India. Kumar P et al. Int J Community Med Public Health 1107–1111
10. Duc Tong P, Le Duc V, Hieu D, Nguyen H, Snasel V (2016) Decision trees for diagnosis of dengue fever
11. Shakil KA, Anis S, Alam M (2015) Dengue disease prediction using weka data mining tool. arXiv preprint arXiv:1502.05167
12. Nishanthi P, Perera A, Wijekoon H (2014) Prediction of dengue outbreaks in srilanka using artificial neural networks. Int J Comput Appl 101(15)
13. Rao NK, Varma GS, Rao D, Cse P (2014) Classification rules using decision tree for dengue disease. Int J Res Comput Commun Technol 3(3):340–343
14. Xu X, Jiang X, Ma C, Du P, Li X, Lv S, Yu L, Chen Y, Su J, Lang G et al. (2002) Deep learning system to screen coronavirus disease 2019 pneumonia. arXiv preprint arXiv:2002.09334
15. Pandey H, Sharma S (2020) Dengue outbreak prediction model using machine learning algorithm 3(5)
16. Esmaeeli GR, Esmaeeli GE, Shafiei M (2017) Detection of crimean congo fever using c4. Decis Tree 5:108–121
17. Ak C, Ergonul O, Sencan I, Torunoglu MA, Gonen M (2018) Spatiotemporal prediction of infectious diseases using structured Gaussian processes with application to Crimean-Congo hemorrhagic fever. PLoS Neglect Trop Diseas 12(8):e0006737
18. Messina JP, Pigott DM, Golding N, Duda KA, Brownstein JS, Weiss DJ, Gibson H, Robinson TP, Gilbert M, William Wint G et al (2015) The global distribution of crimean-congo hemorrhagic fever. Trans R Soc Trop Med Hygiene 109(8):503–513
19. Nosrati M, Mohabatkar H, Behbahani M (2020) Introducing of an integrated artificial neural network and Chou's pseudo amino acid composition approach for computational epitope-mapping of crimean-congo haem- orrhagic fever virus antigens. Int Immunopharmacol 78:106020
20. Quer J, Colomer-Castell S, Campos C, Andres C, Pinana M, Cortese MF, Gonzalez-Sanchez A et al (2022) Next-generation sequencing for confronting virus pandemics. Viruses 14(3):600
21. Jiang D, Hao M, Ding F, Fu J, Li M (2018) Mapping the transmission risk of zika virus using machine learning models. Acta Trop 185:391–399
22. Han BA, Majumdar S, Calmon FP, Glicksberg BS, Horesh R, Kumar A, Perer A, von Marschall EB, Wei D, Mojsilovic A et al (2019) Confronting data sparsity to identify potential sources of zika virus spillover infection among primates. Epidemics 27:59–65
23. Ghenai A, Mejova Y (2017) Catching Zika fever: application of crowdsourcing and machine learning for tracking health misinformation on Twitter. arXiv preprint arXiv:1707.03778

24. Akhtar M, Kraemer MU, Gardner L (2018) A dynamic neural network model for real-time prediction of the zika epidemic in the americas. bioRxiv 466581
25. Kiskin I, Orozco BP, Windebank T, Zilli D, Sinka M, Willis K, Roberts S (2017) Mosquito detection with neural networks: the buzz of deep learning. arXiv preprint arXiv:1705.05180
26. Sippy R, Farrell DF, Lichtenstein DA, Nightingale R, Harris MA, Toth J, Hantztidiamantis P, Usher N, Aponte CC, Aguilar JB et al (2020) Severity index for suspected arbovirus (sisa): machine learning for accurate prediction of hospitalization in subjects suspected of arboviral infection. PLOS Neglect Trop Diseas 14(2):e0007969
27. Barros PH, Lima BG, Crispim FC, Vieira T, Missier P, Fonseca B (2018) Analyzing social network images with deep learning models to fight zika virus. In: International conference image analysis and recognition. Springer, pp 605–610
28. Melo CFOR, Navarro LC, De Oliveira DN, Guerreiro TM, Lima EDO, Ori JD, Dabaja MZ, Ribeiro MDS, de Menezes M, Rodrigues RGM et al (2018) A machine learning application based in random forest for inte- grating mass spectrometry-based metabolomic data: a simple screening method for patients with zika virus. Front Bioeng Biotechnol 6:31
29. Lim S, Tucker CS, Kumara S (2017) An unsupervised machine learning model for discovering latent infectious diseases using social media data. J Biomed Inf 66:82–94
30. Lim S, Tucker CS, Kumara S (2017) An unsupervised machine learning model for discovering latent infectious diseases using social media data. J Biomed Inf 66:82–94
31. Miller M, Banerjee T, Muppalla R, Romine W, Sheth A (2017) What are people tweeting about zika? an exploratory study concerning its symptoms, treatment, transmission, and prevention. JMIR Publ Health Surveill 3(2):e38
32. Mahesh JU, Reddy PS, Sainath N, Kumar GV (2017) Evaluating the performance of tree based classifiers using zika virus dataset. In: Innovations in computer science and engineering. Springer, pp 63–72
33. Dadheech P, Mehbodniya A, Tiwari S, Kumar S, Singh P, Gupta S (2022) Zika virus prediction using AI-driven technology and hybrid optimization algorithm in healthcare. J Healthcare Eng 2022
34. Wang S, Kang B, Ma J, Zeng X, Xiao M, Guo J, Cai M et al (2021) A deep learning algorithm using CT images to screen for Corona Virus Disease (COVID-19). Europ Radiol 31(8):6096–6104
35. John M, Shaiba H (2019) Main factors in uencing recovery in mers co-v patients using machine learning. J Infect Publ Health 12(5):700–704
36. Randhawa GS, Soltysiak MPM, El Roz H, de Souza CPE, Hill KA, Kari L (2020) Machine learning analysis of genomic signatures provides evidence of associations between Wuhan 2019-nCoV and bat betacoronaviruses. BioRxiv 2020–02
37. Fang Y, Zhang H, Xie J, Lin M, Ying L, Pang P, Ji W (2020) Sensitivity of chest ct for covid-19: comparison to rt-pcr. Radiology 200432
38. Gozes O, Frid-Adar M, Greenspan H, Browning PD, Zhang H, Ji W, Bernheim A, Siegel E (2020) Rapid ai development cycle for the coronavirus (covid-19) pandemic: initial results for automated detection and patient monitoring using deep learning ct image analysis. arXiv preprint arXiv:2003.05037
39. Chen J, Wu L, Zhang J, Zhang L, Gong D, Zhao Y, Chen Q et al (2020) Deep learning-based model for detecting 2019 novel coronavirus pneumonia on high-resolution computed tomography. Sci Rep 10(1):1–11
40. Shan F, Gao Y, Wang J, Shi W, Shi N, Han M, Xue Z, Shen D, Shi Y (2020) Lung infection quantification of COVID-19 in CT images with deep learning (2020). arXiv preprint arXiv:2003.04655
41. Song F, Shi N, Shan F, Zhang Z, Shen J, Lu H, Ling Y, Jiang Y, Shi Y (2020) Emerging 2019 novel coronavirus (2019-ncov) pneumonia. Radiology 295(1):210–217
42. Mohamed H, Tawalbeh L, Iliyasu AM, Sedik A, Abd El-Samie FE, Alkinani MH, Abd El-Latif AA (2022) Efficient multimodal deep-learning-based COVID-19 diagnostic system for noisy and corrupted images. J King Saud Univ-Sci 34(3):101898

43. Sedik A, Hammad M, El-Samie A, Fathi E, Gupta BB, El-Latif A, Ahmed A (2021) Efficient deep learning approach for augmented detection of Coronavirus disease. Neural Comput Appl 1–18

44. Sedik A, Iliyasu AM, El-Rahiem A, Abdel Samea ME, Abdel-Raheem A, Hammad M, Peng J et al (2020) Deploying machine and deep learning models for efficient data-augmented detection of COVID-19 infections. Viruses 12(7):769

Propaganda Identification on Twitter Platform During COVID-19 Pandemic Using LSTM

Akib Mohi Ud Din Khanday, Qamar Rayees Khan, Syed Tanzeel Rabani, Mudasir Ahmad Wani, and Mohammed ELAffendi

Abstract Online social media has been evolved as a universal platform for sharing information. Termination being shared on these platforms can be dubious or filthy. Propaganda is one of the systematic methods by which behavior of user can be manipulated. In this work, various machine learning methods are used for detecting such types of information on online social media. Data is collected d from Twitter using its API with the help of various ambiguous hashtags. The results showed that proposed Long Short Term Memory (LSTM) based propaganda identification showed better results than other machine learning techniques. An accuracy of 77.15% is achieved using the proposed approach. In the future BERT model can be used for achieving better Accuracy.

Keywords Propaganda · Machine learning · Online social media · Twitter

This research work is supported by the EIAS Laboratory, Prince Sultan University, Riyadh, Saudi Arabia

A. M. U. D. Khanday (✉) · Q. R. Khan · S. T. Rabani
Baba Ghulam Shah Badshah University, Rajouri, Jammu and Kashmir, India
e-mail: akibkhanday@bgsbu.ac.in

S. T. Rabani
e-mail: syedtanzeel@bgsbu.ac.in

M. A. Wani · M. ELAffendi
College of Computer and Information Sciences (CCIS), Prince Sultan University (PSU), Riyadh 11586, Saudi Arabia
e-mail: mwani@psu.edu.sa

M. ELAffendi
e-mail: afffendi@psu.edu.sa

1 Introduction

Online Social Networks (OSNs) play an important role of the Internet in this technological age. Online Social Networks arose with the introduction of Web 2.0 in the early 2000s. The social influence of OSN services drew immediate interest from internet users all over the world in a short period of time. OSNs use computer-mediated tools to help create and distribute content, concepts, business interests, and other modes of communication through virtual links [1]. In modern age OSNs are used for advertisement, political campaigns, image management, viral marketing, product launches and updates, customer reaction, and other purposes. With the eruption of OSNs the classical way of communicating with people have been changed. Social media has become part and parcel of life, everyone wants to make himself/herself aware of things that are occurring around him. But the adverse use of these platforms has created mass panic. Malicious use of social media is popular nowadays, with the platform being used to spread deceptive or false information that poses a cultural, economic, and politicalinanger [2]. Fearmongers are using social media sites to effectively spread terror and fear [3]. The majority of news organizations are investing in digital journalism. They post content online and use social media and the internet to extend their scope. It has its drawbacks, despite the advantages of greater expansion across emerging technologies. Significant expenditures have been made in creating and improving digital journalism, producing material for their online environments, and extending their networks via social media and the Internet by major news organizations. According to labiki [4], the three types of assaults in cyber networks are physical, syntactic, and semantic assaults. Physical assaults are cyber-attacks that target a computer's hardware. Syntactic attacks are caused by technology and political danger there is no human involvement in these assaults. The most dangerous attacks are semantic attacks, which alter the content or context of information [4]. This type of attack is becoming more popular in online social network organizations sites. Semantic cyber-attacks are distinct from the other two types of attacks. Human-computer interactions are targeted in semantic assaults. Semantic assaults do not have the same apparent impact as other kinds of attacks (Physical, Syntactic) [5]. Overt assaults (such as phishing, spam, and other forms of spam) and covert attacks are two types of semantic attacks. Covert assaults, such as misinformation, deception, and propaganda, would be our focus [6, 7]. Researchers are critically studying this area, which is considered an emerging field of inquiry. Many authors have proposed that the trustworthiness of information may be determined by the social computing characteristics of users on OSNs.

Social media sites are often utilized to disseminate false information. Various fearmongers share various misleading information [8, 9]. Common users easily fall into the trap of these accounts [10, 11]. They are active during some particular time like at times of election, festivals, etc. with the spread of Corona Virus around the globe these types of users use this topic for creating panic and community gag between these assaults. Various misleading information was shared during the peak time of COVID-19. Propaganda plays a vital role in this type of event that have gained very much interest around the world. Propagandists use COVID-19 as one of the important event for sharing there views. On social media various rumors were spread regarding the deadly virus. As in India various hashtags were used for sharing the information regarding coronavirus. Some of the ambiguous hashtags that were used were #CORONAJIHAD, #CORONAMUSLIM, #CORONA-TERRORISM #BioWeapon etc. Using these types of hashtags the propagandists used the Virus for spreading a bad image of a community. Which when validated was a false propaganda regarding such community.

2 Related Work

The ever-increasing allure and beauty of utilizing social media have an impact on our everyday lives, whether directly or indirectly. By instilling confidence in other people's thoughts and recommendations while making little or large choices, such as purchasing low-cost goods & voting in elections to establish a new government. Unsurprisingly, social media has become a tool for influencing public opinion by disseminating misinformation. On social media, false material and propaganda are widely utilized and need to be recognized and resisted. Gao et al. [12] investigated features of fraudulent accounts, and focused on messages including URLs in textual form.

With the increase in social media usage, political events and government programs are more debated, & social media abuse has become widespread. Ratkiewicz [13] developed a technique to detect & track political abuse through social media. The network structure exhibits slight deviation from the usual structure at the time of the election. During the election campaign, Halu et al. [14] proposed a framework for opinion dynamics, demonstrating that major parties survive by accumulating a limited number of votes at election time. Ramakrishnan et al. [15] examined and mined data from online social networks on civil unrest. They extracted data from the particular event, identified contributing variables, and analyzed the event's evolution. Underneath traditional work connections, Liu et al. [16] identified inner circles of political power holders and studied how the chosen political groupings develop and evolve. For this, they used three different methods. Community identification, network construction, and community track. Bhat et al. [17] developed a new method for identifying Stealthy accounts in online social networks. Detecting Communities at Node Level, Features, Classification, and Identifying Stealthy Sybils were all part of the process. Wong et al. [18] used tweets, retweets, and re-tweeters to calculate political leaning. The calculated behavior of Twitter users is a methodical process. Various occurrences occur simultaneously due to the enormous use of social media. Zarrinkalam et al. [19] detected social network events. Three models are utilized to identify undefined events on OSNs: LDA for topic modeling, TF/IDF for clustering documents, and Wavelet Analysis for clustering features. Lightfoot [20] investigated the political impact of social bots. Do they believe that social bots solely had a negative impact on politics? They discovered that social bots play a critical role in the propagation of false news and that accounts that consistently disseminate disinformation are much more likely to be bots. People who hold extreme views solely use social media to promote hatred and terror. Ashcroft et al. [21] developed a semantic graph-based method to detect radicalizations via social media. The results showed that Pro-ISIS users prefer to talk about religions, historical events, and ethnicity, while anti-ISIS users talk about politics, physical areas, and counter-ISIS actions. They also posted shady information on social media. Kumar et al. [22] Using cognitive psychology, we were able to detect disinformation on online social networks. "Consistency of message, coherency of message, trustworthiness of source, and general acceptance of message" are all part of the cognitive process. Using Twitter API, utilizing were collected using terms such as Syria, Egypt, and others. The socially mediated form of populist communication described by Mazzoleni et al. [23] is significantly influenced by social media. They based their assessment of populist communicative ideology on appeals to public, attacks on the elite, and exclusion of others. The following conclusion can be drawn from the above literature reviews.

- Clustering algorithms are used for identifying propaganda, classification algorithms may improve the accuracy and it needs further investigations.
- Majority of work is being done on Newslet datasets, work can be explored on Online Social Media Platforms.
- Semantic Nature of the propagandistic posts, due to the semantic nature of the posts it is a challenge to detect propaganda.
- Feature engineering can be explored by choosing various Emphatic features.

3 Methodology

Propaganda detection can be viewed as a classification or clustering problem in machine learning: unknown propaganda text should be clustered into different clusters depending on certain attributes detected by the algorithm. On the other hand, we can reduce this problem to classification after training a model on a large dataset of propaganda and non-propaganda files.

This challenge can be simplified down to classification only for known propaganda—with a small set of classes, to one of which the propaganda sample undoubtedly belongs, it is easier to identify the right class, and the result is more accurate than with clustering methods.

An explained investigative and experimental research is carried out based on a wide written evaluation and related discussions with field experts on the subject. In light of the many factors that influence the model's presentation, data is collected from Twitter, which was supplied to the model, which provided definitive results. There are five major phases in the whole system. Figure 1 depicts the overall picture of the suggested system that we utilized in our research.

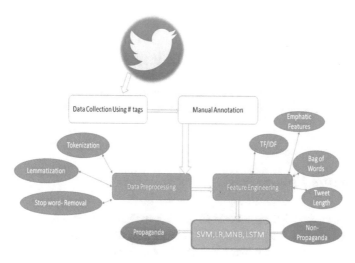

Fig. 1 Proposed methodology

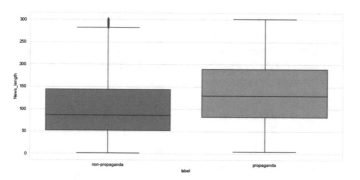

Fig. 2 Labeled classes (Propaganda and non-propaganda with their length in characters)

3.1 Data

Data is extracted from one of the online social media platforms 'Twitter' using its API [27]. Various ambiguous hashtags were used for the collection of data. Hashtags like #Coronamuslim, #CoronaJihad, #CoronaTerrorism, #Propaganda, #Hoaxes, #Bioweapon, etc. were used. About 1.5 million tweets were extracted using 25 keywords out of which only six keywords were identified as ambiguous. The data was mostly extracted during the peak time of COVID-19 that is in April-May 2020. The extracted data were annotated into binary classes "propaganda' and 'non-propaganda'. Manual annotators were hired for annotating the tweets into binary classes using various techniques of propaganda. About 30k tweets were annotated manually. The labeled dataset with their character length is shown in Fig. 2.

3.2 Preprocessing

After building the corpus of tweets regarding propaganda and non-propaganda, data was refined using preprocessing techniques. The tweets were cleaned by removing stop words, links, digits etc. Text is tokenized by eliminating accents from words, and each token is labelled with grammatical features by assigning them to morphosyntactic classification.

3.3 Feature Engineering

Feature selection plays a vital role in training the classifier, if relevant features are not selected the algorithm cannot show good results. While selecting features various techniques were used such that the machine learning algorithm can perform better. We used hybrid features for classifying the text into binary classes. The features were chosen based on the following techniques.

TF/IDF Term frequency/Inverse document frequency technique is used for extracting the features that are frequently used in the propagandistic and non-prlabeledstic text. The most frequent words were selected as the morphosyntactic technique selects the most frequent words. The following Equation is used to compute the TF/IDF about our corpus.

$$TFIDF(t, w, D) = TF(t, w) * IDF(t, D) \qquad (1)$$

$$IDF(t, D) = \log \frac{|D|}{1 + |(w \in D : t \in w)|} \qquad (2)$$

where t stands for the term feature, w stands for each tweet in the corpus, and D stands for the total number of tweets/news in the dataset (Document space).

Bag of Words Using this technique various words are selected on the basis of Bigrams, and Trigrams such that we can get the semantics/context of the tweet/message. In our work we used Bigrams and Trigrams.

Emphatic Features Various Emphatic features were extracted using this technique such that our model can perform better.

Tweet Length Since the length of tweets is limited to 280 characters we used the length of a tweet as a feature such that we can get more insight into the data.

In our work, we merged all these feature extraction techniques for selecting the best features. We choose 100 relevant features for performing this binary classification task. Since 100 features need more computations we reduced the features into the most relevant 40 features using dimensionality reduction techniques.

3.4 Classification

For performing classification, we used various Machine Learning algorithms. Various machine learning algorithms have been fine-tuned to provide better outcomes. The algorithms that were used for performing binary classification were Multinomial Naïve Bayes, Random Forest, and Support vector Machine. After training and testing these models the support vector machine overperformed other algorithms. Since deep learning algorithms are also used for performing classification tasks we trained and tested one of the deep learning approaches Long Short Term Memory (LSTM). The classification process is carried out to classify the given substance into two distinct types of classes. The two classes are Propaganda (a tweet that is propagandistic), Non-Propaganda (a normal tweet). LSTM algorithm is finetuned and is used to classify the tweet into these classes. The LSTM is finetuned and hybrid features that are selected in feature engineering are supplied to it. Algorithm 1 shows the overall steps of the proposed approach for detecting propaganda from social media, that is Twitter.

Algorithm 1 LSTM based Propaganda Identification(LSTMBPI)

Require: Filtered Tweets (T_{input})
Ensure: Propaganda Tweet (T_{Pr}) and Non-Propaganda Tweet (T_{Np})
1: Filtered Tweets $\rightarrow T_{input}$
2: Propaganda Tweets $\rightarrow T_{Pr}$
3: Non-Propaganda Tweets $\rightarrow T_{Np}$
4: Tokenization $\rightarrow$ T
5: StopWordRemoval $\rightarrow$ SW
6: Stemming $\rightarrow$ S
7: Total Number of Tweets $\rightarrow$ n
8: Term Frequency/Inverse Document Frequency $\rightarrow$ TF/IDF.
9: Bag of Words $\rightarrow$ B
10: **START**
11: **for** i $from$ 1 to n **do**
12:    $C[i] = T_{input}[i] + Label$ //Manual Annotation
13:    $T[i] = Tweetlength(C[i])$
14: **end for**
15: **for** i $from$ 1 to n **do**
16:    $Pro[i] = T(C[i])$
17:    $Pro[i] = SW(Pro[i])$
18:    $Pro[i] = S(Pro[i])$
19: **end for**
20: **for** i $from$ 1 to n **do**
21:    $F[i] = B(TF/IDF(Pr[i]))$
22:    $F[i] = F[i] + T[i]$
23: **end for**
24: $LSTM(F[i])$(CLASSIFIER)
25: **END**

A. M. U. D. Khanday et al.

	Features	Standard Deviation	Mean	Minimum	Maximum
2	cardinal	0.029656995	0.14257	0	1
3	ford	0.018193369	0.1146	0	1
4	continue	0.013808926	0.09728	0	1
5	come	0.024063467	0.13523	0	1
6	say	0.016012491	0.10581	0	1
7	vote	0.01594771	0.10996	0	1
8	now	0.015949827	0.10744	0	1
9	president	0.017218344	0.11264	0	1
10	pope	0.020469633	0.13311	0	1
11	take	0.019260909	0.1179	0	1
12	state	0.015109876	0.10569	0	1
13	time	0.015201134	0.10522	0	1
14	ur	0.022776423	0.13501	0	1
15	trump	0.023046996	0.12755	0	1
16	go	0.016218233	0.10721	0	1
17	judge	0.016106411	0.10986	0	1
18	many	0.028394832	0.13741	0	1
19	first	0.024561689	0.13314	0	1
20	kavanaugh	0.019281869	0.12065	0	1
21	wro	0.026997508	0.13874	0	1
22	one	0.015090731	0.10773	0	1
23	report	0.018210036	0.12399	0	1
24	also	0.015907013	0.1068	0	1
25	see	0.020471401	0.1201	0	1
26	get	0.026714662	0.14038	0	1
27	mccarrick	0.020174947	0.12202	0	1
28	court	0.016533577	0.10887	0	1
29	know	0.027349941	0.13791	0	1
30	people	0.016872763	0.11161	0	1
31	may	0.054810034	0.19477	0	1
32	would	0.016422881	0.11489	0	1
33	email	0.031358579	0.14968	0	1
34	like	0.019312089	0.11526	0	1
35	even	0.026195422	0.13305	0	1
36	make	0.024135267	0.13266	0	1
37	church	0.043605471	0.17499	0	1
38	claim	0.033382729	0.15819	0	1
39	arrange	0.017434389	0.11173	0	1
40	story	0.013268811	0.09848	0	1
41	guardian	0.026272882	0.13926	0	1
42					

Fig. 3 Statistical derivations

4   Experimental Results and Discussion

The experimentation is performed on a Workstation having a configuration of 16 GB RAM and GPU. The data set is split into many ratios. First, then data was split into 50:50 that is 50% of the data was used for training and 50% for testing. It was found that the accuracy was less as compared to when data was split into 60:40. After performing classification on this dataset it was concluded that the dataset, when split into 80:20 that is 80% of data, is used for training and 20% for testing should the best accuracy. While training and testing the models the LSTM overperformed other algorithms by showing an accuracy of 77.15%. Factual circumstances are used to conduct quantifiable calculations such as mean and standard deviation. Figure 3 depicts the statistical results.

The objective of the factual examination is to gather as much information as possible about the data that will be used in the classification task. The preceding

Table 1 Classification report of algorithms

Algorithm	Precision (%)	Recall (%)	F1-score (%)	Accuracy (%)
LR	65	68	60	68.2
MNB	69	69	59	68.6
SVM	78	77	74	76.8
DT	68	70	63	69.65
LSTMBPI	79	77	74	77.15

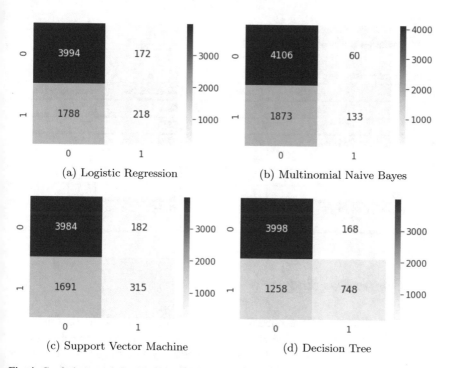

(a) Logistic Regression   (b) Multinomial Naive Bayes

(c) Support Vector Machine   (d) Decision Tree

Fig. 4 Confusion matrices of traditional machine learning algorithms

work was completed using the Python programming language and the Spyder IDE. The data is presented in the form of a table with different quality for the 40 attributes that were selected. The provided task is being carried out using several libraries. Pandas, Sckitlearn, and NLTK, for example, are among the pre-owned libraries. The comparison of different methods for conducting binary classification is shown in Table 1.

Figure 4 illustrate the confusion matrices of the conventional Machine Learning methods, whereas Fig. 5 shows the confusion matrix of LSTMBPI.

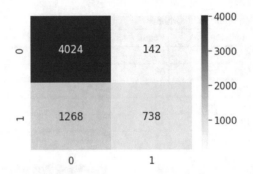

Fig. 5 Confusion matrix of LSTMBPI

Table 2 Comparison with existing work

Algorithm	Precision (%)	Recall (%)	F1-score (%)	Accuracy (%)
LR	64	67	59	67.8
MNB	67	67	58	68.2
SVM	76	75	72	74.8
DT	67	69	62	68.65
LSTMBPI	78	76	75	76.15
Mario et al. [28]	56.54	47.37	51.55	NA
Jurkiewicz et al. [29]	59.95	41.65	49.15	NA

4.1 Comparative Study

For validating the results generated using proposed approach with the existing work. The similar dataset as used by other researchers was experimented based on proposed methodology. The results showed that our methodology performed better than other previous studies. Table 2 shows the comparison of the proposed work with existing approaches in this domain.

5 Conclusion

As the number of social networking sites grew, so did the amount of dubious information that is shared. The filthy data, a disseminator text that may be genuine or fraudulent, has been the subject of this study. Propaganda is often used to acquire influence among ordinary people in order to influence their choices. For classification, we used LR, MNB,SVM, DT and LSTM based on our proposed approach. The results showed that SVM has about 76.8% accuracy, with F1-Scores of 0.84 for Non-Propagandists and 0.64 for Propagandists, respectively. Logistic Regression, on the other hand, showed 68.2% accuracy with F1-Scores of 0.78 for Non-Propagandists and 0.58 for Propaganda. It was also discovered that propagandistic sentences are longer than non-propagandistic ones. SVM also outperforms all other conventional machine learning methods, with Recalls of 0.99 and 0.57 and Precision of 0.80 and 0.76 for two classes, respectively.

The existing machine learning methods demonstrated less accuracy when compared to the suggested LSTMBPI approach. Precision was 0.79, recall was 0.77, F1-Score was 0.74, and accuracy was 77.15% on the LSTMBPI. In the future, Ensemble Machine Learning Algorithms may be used to achieve higher accuracy, as well as a higher F1-Score, Precision, and Recall. To extract relevant characteristics, further feature extraction methods would be used.

Acknowledgements This research work is supported by the EIAS Laboratory, Prince Sultan University, Riyadh, Saudi Arabia.

References

1. Kietzmann JH, Hermkens K, McCarthy IP, Silvestre BS (2011) Social media? Get serious! Understanding the functional building blocks of social media. Bus Horiz 54(3):241–251
2. World Economic Forum (2017) The global risks report 2017, 12th edn. Global Compet Risks Team 103
3. Gupta A, Lamba H, Kumaraguru P (2013) $1.00 per RT #BostonMarathon #PrayForBoston: analyzing fake content on twitter. eCrime Res Summit eCrime
4. Arts THE, Policy C, Justice C, Security N, Safety P, Security H, RandMg877
5. Arpinar IB, Kursuncu U, Achilov D (2016) Social media analytics to identify and counter islamist extremism: systematic detection, evaluation, and challenging of extremist narratives online. In: Proceedings of the - 2016 international conference collabration technology system CTS 2016, pp 611–612
6. Seyednezhad SMM, Cozart KN, Bowllan JA, Smith AO (2018) A review on recommendation systems: context-aware to social-based. arXiv:1811.11866
7. Ashcroft M, Fisher A, Kaati L, Omer E, Prucha N (2016) Detecting Jihadist messages on twitter. In: Proceedings of the - 2015 European intelligence and security informatics conference EISIC 2015, pp 161–164
8. Mehta B, Hofmann T, Fankhauser P (2007) Lies and propaganda: Detecting spam users in collaborative filtering. In: International conference on intelligent user interfaces, proceedings of the IUI, pp 14–21
9. Theory the of Political Propaganda Author (s), (2012) Harold D. Lasswell Reviewed work Published by American Political Science Association 21(3):627631
10. Orlov M, Litvak M (2019) Using behavior and text analysis to detect propagandists and misinformers on twitter. Commun Comput Inf Sci 898:67–74
11. Arth SJ, O'Donnell V (1986) Propaganda and persuasion
12. Gao H, Hu J, Zhao BY, Barbara S, Barbara USC, Chen Y (2010) Detecting and characterizing social spam campaigns categories and subject descriptors. In: Proceedings of the internet measurement conference (IMC)
13. Ratkiewicz J, Conover M, Meiss M, Alves BG, Patil S, Flammini A, Menczer F (2011) Truthy: mapping the spread of astroturf in microblog streams. In: Proceedings of the 20th international conference companion on world wide web. ACM, pp 249–252
14. Halu A, Zhao K, Baronchelli A, Bianconi G (2013) Connect and win: the role of social networks in political elections. Epl 102(1)

15. Ramakrishnan N, Tech V, Chen F, Arredondo J, Mares D, Summers K (2013) Analyzing civil unrest through social media. Computer (Long Beach Calif), pp 80–84
16. Liu JS, Ning KC, Chuang WC (2013) Discovering and characterizing political elite cliques with evolutionary community detection. Soc Netw Anal Min 3(3):761–783
17. Bhat SY (2014) Abulaish M (2014) Communities a gainst deception in online social networks 1 the platform 2 the mischef. IEEE 2:8–16
18. Wong FMF, Tan CW, Sen S, Chiang M (2016) Quantifying political leaning from tweets, retweets, and retweeters. IEEE Trans Knowl Data Eng 28(8):2158–2172
19. Zarrinkalam F, Bagheri E (2017) Event identification in social networks. Encycl Semant Comput Robot Intell 01(01):1630002
20. Lightfoot S (2018) Political propaganda spread through social bots 0–22
21. Esposito A (2006) The semantic web. Adv Inf Technol Electromagn 29–44
22. Kumar KK, Geethakumari G (2014) Detecting misinformation in online social networks using cognitive psychology. Human-Centric Comput Inf Sci 4(1):1–22
23. Mazzoleni G, Bracciale R (2018) Socially mediated populism: the communicative strategies of political leaders on facebook. SSRN
24. Description of logistic regression algorithm. https://machinelearningmastery.com/logistic-regression-for-machinelearning/ (Accessed 15th May 2021)
25. Description of multinomial Naive Bayes algorithm. https://www.3pillarglobal.com/insights/document-classification-usingmultinomial-naive-bayes-classifier (Accessed 15th May 2021)
26. Khanday AMUD, Khan QR, Rabani ST (2020) SVMBPI: support vector machine-based propaganda identification. Proc CISC Cognit Inf Soft Comput 445
27. Verma P, Khanday AMUD, Rabani ST, Mir MH, Jamwal S (2019) Twitter sentiment analysis on Indian government project using R. Int J Recent Technol Eng 8(3):8338–41
28. Morio G, Morishita T, Ozaki H, Miyoshi T (2020) Hitachi at semeval-2020 task 11: an empirical study of pre-trained transformer family for propaganda detection. In: Proceedings of the fourteenth workshop on semantic evaluation, pp 1739–1748
29. Jurkiewicz D, Borchmann L, Kosmala I, Gralinski F (2020) Applicaai at semeval-2020 task 11: on roberta-crf, span cls and whether self-training helps them. arXiv preprint arXiv:2005.07934

Model of the Internet of Things Access Network Based on a Lattice Structure

A. Paramonov, S. Bushelenkov, Alexey Tselykh, Ammar Muthanna, and Andrey Koucheryavy

Abstract The article discusses the difficulties inherent in constructing access networks in high-density communication networks and proposes a method for estimating the maximum data transfer rate that can be achieved, considering the mutual influence of network nodes and building a logical network structure. The study uses analytical methods to estimate the data transfer rate, considering the mutual influences of network nodes and methods of percolation theory to estimate the probability of network connectivity. The paper shows that the logical structure of a high-density network can be described in the form of a lattice with the required properties. The work results are a method for estimating the rate of data transmission, taking into account interchannel interference, and a method for assessing connectivity built based on the provisions of percolation theory for a network with a lattice structure. The practical significance of the results obtained consists in proposing a method for selecting the logical structure of a high-density access network with specified properties based on the provisions of percolation theory.

Keywords Access network · Internet of Things · Connectivity · Percolation theory · Lattice structure

1 Introduction

The evolution of information communication systems has resulted in the introduction of new domains, such as the Internet of things and ultra-low latency networks [1–5], which enhance the system's capabilities in terms of information availability and communication service quality. The number of gadgets connected to communication

A. Paramonov · S. Bushelenkov · A. Muthanna (✉) · A. Koucheryavy
The Bonch-Bruevich Saint-Petersburg State University of Telecommunications, St. Petersburg 193232, Russian Federation
e-mail: ammarexpress@gmail.com; muthanna.asa@spbgut.ru

A. Tselykh
Department of Information and Analytical Security Systems, Southern Federal University, Rostov-On-Don, Russia

networks has already surpassed the planet's population and is continuing to grow. The density of devices in communication networks is increasing, and projections indicate that it will eventually approach, and possibly exceed, one device per square meter [6].

The access networks undergo the greatest changes in such conditions. At this level of the network, devices are connected and interact with each other. As the network's density increases, the complexity of creating the access level increases proportionately. The evolution of mobile networks affects the user access layer to the greatest extent.

This level has always been the most complex from the technological point of view. Therefore it has always been the "bottleneck" of the network and determined the possibilities in terms of quality of communication services.

In networks with a high device density and networks with high latency requirements (in networks with ultra-low latency), special requirements are imposed on access networks. When using wireless technology, they are associated with the need to meet conflicting requirements, on the one hand, to ensure high data rates and low latency, and on the other hand, a high user density and, as a consequence, to work in conditions of strong intra-channel interference.

These issues are addressed in 5G networks through a complex combination of various methods [7]: expanding the band used in the radio frequency spectrum, using space–time separation of channels, using narrow antennas with a controlled radiation pattern, utilizing various technologies (standards) for access network organization, and distributing traffic between different networks.

5G networks and future communications networks involve extensive interaction with networks built using different technologies to offload traffic. This makes it possible to offload the network's base stations and manage the quality of traffic service.

All these processes must be performed at the subscriber access level, which requires multiple traffic management methods and network structures to ensure efficient use of resources and provide the required quality of service provision. In order to apply such methods, it is necessary to be able to describe the management objects.

The purpose of this work is to develop a model of the access network to describe its property management tasks.

2   Problem Statement

The Internet of Things network includes several layers, which can be conditionally divided into the access layer, transport layer, and service layer, Fig. 1.

This division is very arbitrary because different purpose TDF networks may not include all of these levels or combine some of them within a single technology. However, this representation of the IS structure is the most general, and we will adhere to it.

The most specific in this structure is the level of the access network. It is the implementation of this level of the TR network that most works are devoted to.

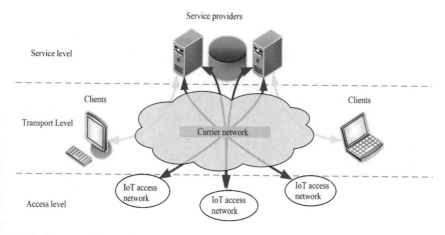

Fig. 1 Structure of the IoT network

As a rule, the access network of TR is a network built using wireless technology, often referred to as a wireless sensor network. Its construction can be used in various wireless technologies, such as WiFi, Bluetooth, ZigBee, LoRaWan, and others. They allow you to build networks of sensors (activators) distributed over a given service territory. As a rule, such networks have a star, tree, or mesh structure.

The choice of the structure is dictated by the purpose of the network. If units and tens define the number of network nodes, its structure is determined by the location of these nodes in space. However, in high-density networks, where the node density is equal to or more than the units per square meter, the development of the logical structure already allows for the formation of a set of arbitrary structures at the developer's request. Figure 2 illustrates a structure's organization near a square (Fig. 2a) and a triangular lattice in a high-density network (Fig. 3b).

If the node density is high, there is a high probability that a node in the network will be near the target point, for example, the node position in the square grid. Thus, the higher the density of nodes, the structure obtained in the network will be closer to the realized model. This trait also applies to networks containing movable nodes. The high probability of a new node moving into the area where the structural element was placed compensates for migrating an existing node. Thus, by modifying the functionality of the nodes in the logical structure, it is possible to maintain a stable network topology.

Indeed, most wireless communication technologies used for these purposes provide a range (radius) of communication nodes from 50 to 100 m (possibly more). This is the case when the communication zone of a single node varies between around 300 and more than 600 nodes in the network. Naturally, when such a dense network of nodes exists, their mutual influence–interchannel interference–is exceedingly strong. This results in a fall in the signal-to-noise ratio and, as a result, a decrease in the radio channel's quality under certain situations.

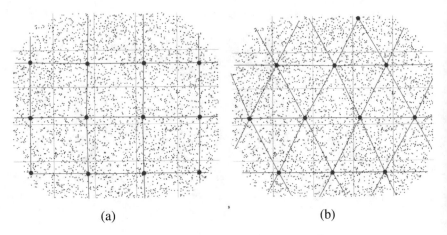

(a)   (b)

Fig. 2 Example of constructing a square grid in a high-density network

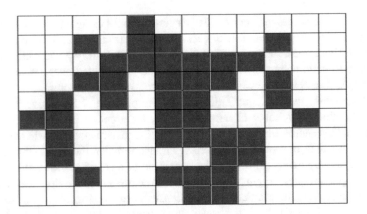

Fig. 3 Model of percolation lattice

The problem of structure selection, in particular, can be reduced to the task of constructing a lattice structure, the properties of which are well enough known and can be used to assess the performance of the resulting network.

3 Problem Solution

Let us assume that the density of the network is so high that a lattice-shaped structure can be built in this network. In this case, we will assume that the lattice is rectangular. In the general case, the lattice can be not only rectangular, for example, triangular, hexagonal, or some other. A particular kind of lattice can be chosen according to the peculiarities of the problem to be solved.

We will also assume that the main indicators of the quality of the network are:

- Data transmission speed,
- The probability of establishing a connection between two arbitrary nodes.

1. To estimate the rate of data transmission, we can apply the model of estimation of the signal to noise ratio in which the noise power is determined by int-channel interference from neighboring nodes of the network. The power of these interferences at point O can be defined as [6]

$$p_{OI} = \sum_{j=1}^{n} p_{oj} I_j, \text{ W} \tag{1}$$

p_{oj}–signal power of node j at point O (Bт); n–total number of nodes; I–indicator function.

$$I_j = \begin{cases} 1 & p_{oj} \leq p_m \\ 0 & p_{oj} > p_m \end{cases} \tag{2}$$

where pm is the signal power threshold at point O, the channel state is considered occupied (Bт).

The power level of interference from the j-th node at point O can be estimated by selecting the signal attenuation model a (d_{oj})

$$\tilde{p}_{oj} = \tilde{p}_j - a(d_{Oj}) \text{ dBm} \tag{3}$$

When $\tilde{p}_j$–power level at the output of the j-th node (dBm);
d_{Oj}–is the distance between the point O and the j-th node of the network (m);
$a(d_{Oj})$–the model of signal attenuation versus distance between nodes (dB).

The squiggle marks are power levels in dBm, and without the squiggle marks are power levels in watts.

Then the signal to noise ratio can be estimated as

$$SNR = \tilde{p}_S - \tilde{p}_{OI}, \text{ dB} \tag{4}$$

When $\tilde{p}_S$–is the power level of the useful signal at point O.

Different models can be chosen as the $a(d_{Oj})$ attenuation model in (3), depending on the conditions of the access network (open space, industrial premises, residential buildings, etc.), for example [8, 9].

Based on expression (4), we can get an estimate of the data transfer rate between nodes, which is determined by the specific standard used to organize the radio channel

$$br = f(SNR), \text{ bit/c} \tag{5}$$

When $f(SNR)$–the functional dependence of the data rate on the value of the signal to noise ratio, according to the radio channel standard used.

2. The probability of establishing a connection between two arbitrary network nodes. In this paper we propose to use the theory of percolation (percolation) [8] to estimate the possibility of establishing a channel between two arbitrary network nodes. When describing the network structure by means of a lattice, the application of this theory makes it possible to estimate the possibility of establishing a channel.

Percolation theory considers different types of lattices, both planar and three- and more-dimensional. A lattice describes the structure of a medium. Lattice cells can be of two types: conducting and non-conducting. A lattice contains some (random) number of cells of both types. One of the main tasks of this theory is to describe the state of the medium, characterized by the ratio of the number of conducting and non-conducting cells, which causes "flow" through the medium. This condition is also called the formation of a percolation cluster, i.e. a cluster of adjacent conducting cells, which ensures flow through the medium. An example for a flat square lattice is shown in Fig. 3.

In the figure above, conductive cells are colored red, and non-conductive cells are left white. The set of adjacent red cells is a percolation cluster that conducts, for example, liquid from the upper boundary of the lattice to the lower boundary.

Percolation theory is widely used in physics to describe processes such as diffusion in metals, whose structure is described by various types of three-dimensional lattice structures, such as those shown in Fig. 4.

Figure 4 shows a simple cubic lattice (Fig. 4a), a volumetrically centered cubic lattice (Fig. 4b), and a face-centered cubic lattice (Fig. 4c).

The percolation probability is equivalent to the probability of establishing a connection between two arbitrary network nodes whose structure is described by the corresponding type of lattice in this theory's terminology.

The probability of percolation, estimated for an infinitely large lattice with a finite number of cells, depends on the lattice's features [8, 9].

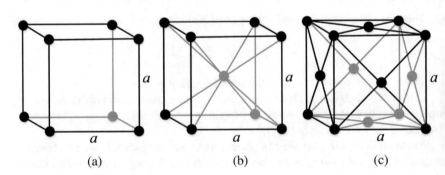

(a) (b) (c)

Fig. 4 Examples of cubic lattices

Table 1 Percolation thresholds for some grid types

N	Lattice type	Percolation threshold
1	Simple cubic	0,310
2	Volumetrically centered cubic	0,243
3	Faceted cubic	0,195
4	Hexagonal	0,200

The notion of flow threshold also applies to an infinite lattice. With finite lattice dimensions, the threshold varies from case to case, i.e., it is a random value. However, the values it takes to fall into some regions of width δ [8].

$$\delta(N) \approx \frac{c}{\sqrt[v]{N}} \tag{6}$$

When N is the number of cells (nodes) of the lattice, c is a constant ($c \approx 0,5$), and v–is the critical index (index of the correlation radius).

As the number of cells increases to infinity, the interval (6) narrows to a point and the probability of percolation is defined by the following limit

$$p_c = \lim_{N \to \infty} (p_c(N)) \tag{7}$$

For each type of lattice, there is a different threshold of flow. The percolation problems are considered either relative to the nodes or the lattice bonds. In some cases, the percolation threshold can be obtained analytically. For example, for a square two-dimensional lattice, for percolation of nodes $p_c = 0,59$, for a triangular lattice $p_c = 1/2$ (for the same problem).

Not all cases are amenable to analytical solutions. In such cases, the values are obtained by numerical simulation methods. For example, for the three-dimensional case, the values of percolation thresholds for several lattice types are given in Table 1.

As can be seen from Table 1, the pc values vary from 0.195 for a face-centered cubic lattice to 0.310 for a simple cubic lattice. With some approximation, these values can be considered the fraction of conducting lattice nodes at reaching (and exceeding) which the structure transits to the coupled state. In terms of a communication network, this ratio tells how many operable nodes a given structure must have for it to be able to construct a route between two arbitrary network nodes.

4 Conclusions

Access networks are a complex element of a modern communications network and generally determine the capabilities of the entire network. Increasing device density and service quality requirements create new conditions in which it is possible to use methods that were previously inapplicable. It has been shown in this work that increasing network node density creates the potential for increasing the logical structure of links. And the link structure itself, due to the high density of nodes, can be chosen from considerations dictated by the solution of specific problems.

If the achievable data transmission rate and the probability of establishing a connection between two arbitrary network nodes are used as primary indicators of network operator quality, then selecting the appropriate structure can achieve the required data transmission rate while taking into account the influence of neighboring nodes and the traffic they generate.

The logical structure of connections in an IoT wireless access network can be built in the form of a lattice. With a high density of nodes, the network structure can be quite close to the structure of the model, and known mathematical methods can be used for it. In particular, the lattice structure can be used methods of percolation theory, which allows for estimating the probability of connectivity between two arbitrary network nodes.

Aknowldgment This research is based on the Applied Scientific Researchunder the SPbSUT state assignment 2022.

References

1. Muthanna MSA, Wang P, Wei M, Rafiq A, Josbert NN (2021) Clustering optimization of LoRa networks for perturbed ultra-dense IoT networks. Information 12:76
2. Muthanna MSA, Muthanna A, Rafiq A, Hammoudeh M, Alkanhel R, Lynch S, Abd El-Latif AA (2021) Deep reinforcement learning based transmission policy enforcement and multi-hop routing in QoS aware LoRa IoT networks. Comput Commun 183:33–50
3. Muthanna MSA, Wang P, Wei M, Abuarqoub A, Alzu'bi A, Gull H (2021) Cognitive control models of multiple access IoT networks using LoRa technology. Cogn Syst Res 62–73. ISSN 1389-0417
4. Chaaf A, Saleh Ali Muthanna M, Muthanna A, Alhelaly S, Elgendy IA, Iliyasu AM, Ahmed A (2021) Energy-efficient relay-based void hole prevention and repair in clustered multi-AUV underwater wireless sensor network. Secur Commun Netw
5. Rafiq A, Ping W, Min W, Muthanna MSA (2021) Fog assisted 6TiSCH tri-layer network architecture for adaptive scheduling and energy-efficient offloading using rank-based Q-learning in smart industries. IEEE Sens J 21(22):25489–25507
6. Muthanna MSA, Alkanhel R, Muthanna A, Rafiq A, Abdullah WAM (2022) Towards SDN-enabled, intelligent intrusion detection system for internet of things (IoT). IEEE Access 10:22756–22768. https://doi.org/10.1109/ACCESS.2022.3153716
7. Rafiq A, Wang P, Wei M, Ali Muthanna MS, Josbert NN (2022) Mitigation impact of energy and time delay for computation offloading in an industrial IoT environment using levenshtein distance algorithm. Secur Commun Netw 1939–0114
8. Aboulola O, Khayyat M, Al-Harbi B, Muthanna MSA, Muthanna A, Fasihuddin H, Alsulami MH (2021) Multimodal feature-assisted continuous driver behavior analysis and solving for edge-enabled internet of connected vehicles using deep learning. Appl Sci 11:10462
9. Chiu SN, Stoyan D, Kendall WS, Mecke J (2013) Stochastic geometry and its applications. John Wiley & Sons, p 583

Study of Methods for Remote Interception of Traffic in Computer Networks

Maxim Kovtsur ⓘ**, Ammar Muthanna** ⓘ**, Victoria Konovalova** ⓘ**,
Abramenko Georgii** ⓘ**, and Shterenberg Olga**

Abstract The development of technology and the growth of cyber-physical systems lead to an increase in attacks on critical infrastructure, which creates the need for its monitoring to prevent an attack. Network monitoring is based on monitoring the information contained in the data transmitted over it. There is a need to analyze information passing through network devices where there are vulnerabilities, which means that there is a need to get this data for your own analysis. Currently, there are several ways to monitor the network, some of which are already actively used in critical infrastructures, and some have great potential, but are not fully used or have not yet been fully studied. The article discusses a method of remote traffic interception aimed at monitoring traffic and analyzing information transmitted over the network.

Keywords Critical infrastructure · Network administration · RPCAP · SPAN · RSPAN · ERSPAN · Cyber-physical systems · Research of data interception methods · Remote traffic interception

1 Introduction

Every day technologies are developing and becoming an integral part of our life, and it is difficult to imagine modern life without the latest means or computer systems. All the above is used by us in various spheres of life, such as medicine, industry, transport, nuclear energy, etc. These areas define critical processes that need to be protected and constantly monitor the processed information or traffic that is transmitted inside the critical infrastructure.

M. Kovtsur (✉) · A. Muthanna · V. Konovalova · A. Georgii · S. Olga
M. A. Bonch-Bruevich Saint-Petersburg State University of Telecommunications, Saint-Petersburg, Russia
e-mail: maxkovzur@gmail.com

A. Muthanna
University of Russia, RUDN University, 6 Miklukho-Maklaya St, 117198 Moscow, Russia

© The Author(s), under exclusive license to Springer Nature Switzerland AG 2023
A. A. Abd El-Latif et al. (eds.), *Advances in Cybersecurity, Cybercrimes, and Smart Emerging Technologies*, Engineering Cyber-Physical Systems and Critical Infrastructures 4,
https://doi.org/10.1007/978-3-031-21101-0_26

Individual systems in critical infrastructure can be automated, for example, collecting Internet of Things data, monitoring a patient's condition remotely, or managing processes at a nuclear power plant. Systems that implement such processes can absolutely be called cyberphysical, since they are an intermediate link between man and the world of machines.

In some cases, disruption of critical processes because of a failure or a centralized attack on cyber-physical systems in critical infrastructure can harm the health of others or damage an entire city or even a country. To avoid this, it is necessary to use a reliable infocommunication system with suitable protection for a particular area of critical infrastructure. An important and key factor in the protection of such infrastructure is the need for constant monitoring of information or traffic within the network or system.

This article focuses on the effectiveness of traffic monitoring in an infrastructure that uses protocols that are part of the popular TCP/IP protocol stack. Such protocols allow transmitting information over wired and wireless communication channels. In addition to protocols for data transmission, there are various techniques for intercepting transmitted data.

Traffic interception, or traffic collection, is listening to a network interface using special libraries. The ability to intercept information can be used for both good and bad purposes. For example, it can be used by hackers for malicious purposes, as well as by network administrators for monitoring, detecting looped traffic, detecting viruses and malware, etc.

2 Problematics

Analysis and systematization of information about researches makes it possible to identify existing methods of intercepting traffic, determine their difference, and most importantly confirm the relevance of the topic and the degree of its development. This requires familiarization with existing research in this area [1–9]. It is worth paying attention to the fact that when studying the information, none of the cited articles on the research topic contain fully descriptive methods for remote traffic interception. In particular, the method of remote traffic interception for the administration of network devices, the method of remote traffic interception for a remote training laboratory. Moreover, in the considered works, there was no evaluation of the effectiveness of the selected interception methods. This means that the need for a detailed description of little-studied methods of remote traffic interception in corporate networks is confirmed.

There are several ways to intercept traffic. The first is collected through specific hosts of the network infrastructure. Usually these are L3 devices, for example, firewalls, switches, computers. The advantages of this method are that it is quick and easy to implement, but method restricted by one device and low performance.

The second way is traffic or port mirroring. In this case, traffic is collected from physical or logical interfaces of network devices. The source port copies the traffic flowing through it. Then the copied data stream is sent to the monitoring device. Port mirroring copies each packet and sends it through the interface to the monitoring device. This method has a minus–a large amount of traffic transmitted. At the same time, the advantage is that you do not need to change the network topology.

The last method discovered is based on the use of system resident processes or daemons. These methods are used, for example, in a SIEM system based on the SYSLOG process. Another resident process is *rpcapd*, created using the libpcap library.

The process of interacting with network traffic is handled by a software tool that runs in the OS environment and allows you to interact with the network inter-face drivers for the device. This requires a remote RPCAPD daemon application that captures and sends the data back, and a local client that sends the appropriate commands and receives the captured data.

The RPCAPD software is based on the RPCAP (Remote Packet Capture) protocol, which is designed to monitor network traffic and capture packets arriving at a remote device on the network to control and analyze the transmitted data streams. Traffic interception is carried out by copying the transmitted packets, after which, using the RPCAP transmission protocol, they are transmitted to the desired remote host.

3 Materials and Methods

The RPCAP protocol implies the interaction of a remote network device and a network data analysis program (packet analyzer) according to the server-client scheme (Fig. 1). The RPCAP daemon is launched on the remote device, which accepts connection requests from client applications, performs authentication and starts servicing authorized clients: "listens" on the network and sends the requested packets to the client for processing and analysis [10, 11].

Thus, the RPCAPD system consists of two separate processes: a server (or agent) that captures network traffic on a remote system, and a client that, upon request, receives and processes these packets.

During the establishment of the connection, information was received containing the exchange of packets between the server and the client at the time of connection (Fig. 1).

Let is describe each stage in more detail:

– Authentication request. The Authentication type header field ranges from 0 to 1, (0)–no authentication, and (1)–authentication used. The RPCAP header does not encrypt its data, the username and password of the client can be found when intercepting the session using a regular sniffer.
– Authentication reply—response to a server request from the client side. In the case when the client sets up a method using authentication on its side, and the server sends an incorrect login or password when registering a session, the client sends an authentication error response.

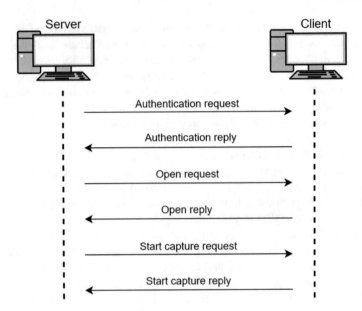

Fig. 1 An example of establishing an RPCAPD session

- Find all interfaces request—server request to get a list of all available client interfaces.
- Find all interface reply—reply with information about available interfaces.
- Open request—start of traffic interception at the specific interface.
- Start capture request—a request with connection settings is sent to a remote user.
- Start capture reply—response to connection settings request. Data is sent to create a new TCP session with information about which port to use.

During transmission over a communication channel, a packet from a remote machine is encapsulated in an RPCAP protocol header, after which a transport layer header, such as the TCP protocol, is superimposed on top. Then the headers of the network, link and physical layers are added.

The Remote Packet Capture header can be a maximum of 28 bytes. To study it, tests were conducted on remote packet capture with various options installed when starting the daemon on a remote client. The result was a list of RPCAP protocol header fields. Table 1 shows the characteristics of the RPCAP protocol.

Table 1 Characteristics of the RPCAP Protocol

Name	RPCAP
Layer according to the TCP/IP model	Application layer
Port	2002/TCP, 2003/TCP
Purpose of the protocol	Transferring Captured Packets

To study the protocol, perform experimental assessments of the protocol's impact on the performance of low-power devices, such as Raspberry Pi. An evaluation of the effectiveness of the RPCAPD protocol has also been established.

4 Methodology for Analyzing Efficiency Evaluation

To study the efficiency of data transmission over a communication channel, it is necessary to study the transmitted packets in the network. To do this, at the time of the research, files of the ".pcap" format were created using Wireshark and Tcpdump sniffers. These files contain all captured packets. Figure 2 shows a captured burst load created example generated by the RPCAPD daemon [13–15] (Figs. 3 and 4).

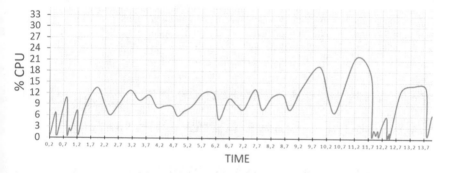

Fig. 2 Changing the percentage of CPU load when playing a low-quality image

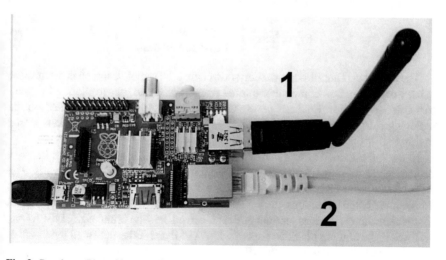

Fig. 3 Raspberry Pi working sample

Efficiency

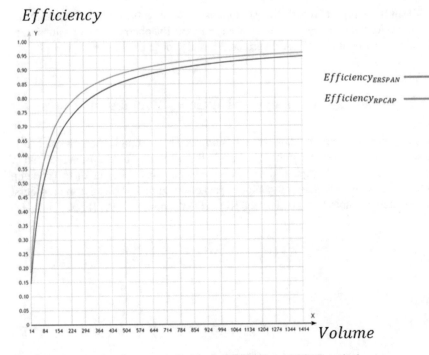

Volume

Fig. 4 Evaluation of performance indicators for ERSPAN and RPCAP methods

To evaluate the effectiveness of using the RPCAPD method, an efficiency evaluation formula was defined, consisting of the ratio of the volume of useful traffic to the volume of intercepted traffic (1).

$$Efficiency = \frac{V \text{useful traffic}}{V \text{intercepted traffic}} \tag{1}$$

The packet capture files in question showed that the remote daemon is performing batch operations, i.e., combines captured packets into one, encapsulating them for transmission over an Ethernet network. These headers are deencapsulated on the receiving side, thereby splitting the received packet into a few other packets stored in it.

In the file obtained after the study of traffic interception using the Raspberry Pi device, while watching a low-quality video, Frame No. 175 contains 10 parts. The RPCAP header is 28 bits per frame. Thus, we get that in the case when the demon does not use batch operations, the load on the network increases according to formula (2). Where Data is the new amount of data received, x is the initial size of the frame, 82 bytes is the amount of added data of which (14 bytes–Ethernet II, 20 bytes–IPv4, 20 bytes–TCP and 28 bytes–RPCAP).

$$Data = x + 82\,(byte) \tag{2}$$

In the case when a batch operation is used, for example, from 10 frames of different lengths, the load on the network increases according to formula (3), where x_i is the initial volume of the frame i.

$$Data = 14 + 20 + 20 + 28 + x_1 + 28 + x_2 + \cdots + 28 + x_{10}\,(byte) \tag{3}$$

Reducing formula (3), we obtain the network load formula using batch operations (4).

$$Data = 54 + 28 * n + \sum_{i=1}^{n} x_i\,(byte) \tag{4}$$

where n is the number of frames in the batch operation and i is the sequence number.

From formulas (2) and (4) we get that when sending a batch operation, the amount of data is approximately 0.4 times less than when sending the same packets in separate frames.

At the same time, during encapsulation, the useful volume increases by about 82 bytes, which means that the efficiency of adding additional headers when using RPCAPD can be calculated using formula (5).

$$Efficiency_{RPCAP} = \frac{x}{x + 82} \tag{5}$$

Research is conducted in different experimental setups. One consists of a Windows 10 PC, a low-powered Raspberry Pi, and a wireless LAN router. The Raspberry Pi has an additional physical wireless adapter installed through which the device communicates with the router. The Raspberry Pi (Fig. 3) is also wired to the PC. The PC uses a wireless communication channel to exchange data with the router and access the Internet [12]. On Fig. 3, number 1 is the wireless network adapter, number 2 is the Ethernet adapter.

For comparison, a frame containing the ERSPAN header was used. The TCP packet is transported encapsulated in a GRE header; it is of the ERSPAN protocol type. The analyzer at the end of the ERSPAN tunnel will see the already de encapsulated traffic, starting with the Ethernet II header. The GRE header can vary from 4 to 16 bytes depending on which options are enabled. From the example, you can calculate the length of the ERSPAN header, calculate its efficiency and compare with the RPCAPD method. So, we get that the length of the header is 24 bytes. Calculations were made based on the smallest possible length of each example header and the difference between the total size (116 bytes) and these values.

With these values, the payload increases by about 62 bytes, which means that the efficiency of adding additional headers when using RPCAPD can be calculated using formula (6).

$$Efficiency_{ERSPAN} = \frac{x}{x + 62} \qquad (6)$$

Let's take an existing packet with the ERSPAN header and a packet with the RPCAP header as an example and plot the effectiveness of using remote interception methods. Let's compare the results.

Figure 4 shows the efficiencies ESPAN and RPCAP built according to formulas (5) and (6). The X-axis shows the change in the useful volume of traffic. The Y-axis shows the change in efficiency relative to useful traffic.

Thus, we got the result when the use of the RPCAP protocol shows excellent results of influencing the network. It should be noted that the change in the efficiency indicators using the ERSPAN method is not much higher than the efficiency of using RPCAP. At the same time, the RPCAP protocol has many advantages, for example, the ability to use it on any device without being tied to vendors.

5 Practical Performance Analysis

The Remote interception was carried out on the wireless network adapter on the Raspberry side, and the captured packets were sent over a wired communication channel on the PC side. PC from the address of the availability of a request on the route, access to the Internet, the possibility of watching videos, medium and high quality on the YouTube service.

The wireless network was scanned using the wireless network adapter installed on the Raspberry Pi, which was switched to "monitoring" mode before the interception began. This is a data capture mode that allows you to use a Wi-Fi adapter in a format or «promiscuous» protocol. In this case, the adapter intercepts any type of Wi-Fi packets: Management (including Beacon packets), Data and Control [16].

The tools provided on the Raspberry Pi receive a table output showing the CPU load of the low power device while the RPCAPD process is running. After processing the result of the output to the screen, a graph of the change in the percentage of CPU load was obtained relative to the time the RPCAPD process was running and the load on its load (Fig. 2).

The graph clearly shows how the load on the processor has changed in relation to the amount of data transferred and the time of the process. At the time of peaks, video playback data were transmitted wirelessly between the home router and the PC.

During the test, the rate of reception of wireless network packets was approximately 450 Kbps on average (Fig. 5). The figure also shows the dependence of the speed of the transmitted data and the change in the indicator of the percentage of CPU utilization relative to the time of the RPCAPD process. Speed peaks correspond to network load peaks. The outline of the graphs is identical.

A similar study was conducted while watching medium quality (720p) and high quality (2160p) videos. Figure 6 shows a graph of the percentage of CPU utilization

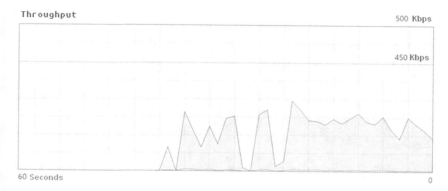

Fig. 5 Bit rate during low-quality video playback

versus the time the RPCAPD process was running at the time of playing medium quality video.

The graph clearly shows how the load on the processor has changed in relation to the amount of data transferred and the time of the process. At the time of peaks, video playback data were transmitted wirelessly between the home router and the PC. During the test, the wireless network packet reception rate averaged approximately 7.7 Mbps, and the data transmission over the wired communication channel increased dramatically during the launch of the medium quality video (Fig. 7). Also, based on the graphs, you can see the dependence of the speed of the transmitted data and the change in the indicator of the percentage of CPU utilization relative to the time of the RPCAPD process. Speed peaks correspond to network load peaks. The outline of the graphs is identical. It is worth noting that with an increase in image quality, the speed of transmitted data also increased.

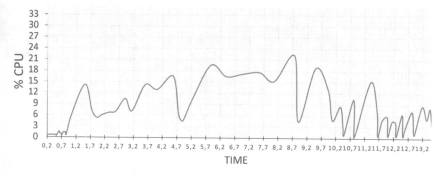

Fig. 6 Changing the percentage of CPU utilization relative to RPCAPD running time

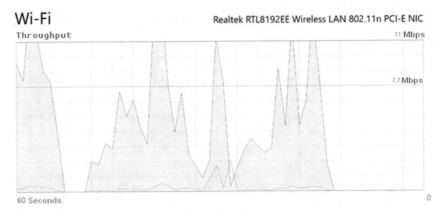

Fig. 7 Activity of sending and receiving data at the time video playback of medium quality

Figure 8 shows a graph of the change in the percentage of CPU utilization on the Raspberry Pi from the time the RPCAPD process was running on the device while watching high quality video (2160p).

During the test, the wireless network packet reception rate averaged approximately 32 Mbps (Fig. 9). Also, based on the graphs, you can see the dependence of the speed of the transmitted data and the change in the indicator of the percentage of CPU utilization relative to the time of the RPCAPD process. Speed peaks correspond to network load peaks. The outline of the graphs is identical. It is worth noting that with an increase in image quality, the speed of transmitting data also increased.

As a result of practical research, there is a low CPU load. RPCAP also allows you to solve all the necessary tasks for intercepting and monitoring traffic. In addition, its advantages include the fact that this protocol is an opensource solution, unlike ERSPAN.

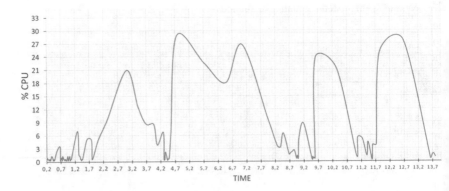

Fig. 8 Changing the percentage of CPU utilization relative to the time the RPCAPD process was running when playing high-quality video

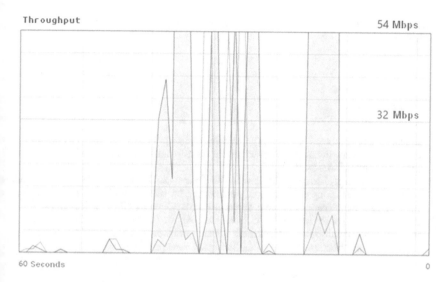

Fig. 9 Activity of sending and receiving data at the time of high-quality video playback

6 Conclusion

Low-power devices were used for the proof. One of them was the Raspberry Pi. Based on his work, it was possible to compare the impact on the CPU of a weak network load device. So, by overlaying graphs of changes in CPU load indicators, you can see that the greater the load on the network, the higher the busyness of the process on the CPU. At the same time, for the entire time of research, the load value was not higher than 30%, which is a good indicator for a low-power device. The RPCAPD daemon itself is an open solution, which allows it to be used on any device and not be tied to vendors.

Having studied this direction, we can say that the choice of a transmission method depends on several factors; several options can be used in the network at once for convenience and saving both finances and power resources. Further work involves the study of the use of traffic filtering when using the RPCAPD daemon, the creation of methods [17] for remote traffic interception and a more in-depth study of remote traffic interception methods.

The development of the high-tech industry makes it possible to integrate computing and physical components combined into unified systems designed to solve an increasingly wide range of social and technical tasks efficiently and quickly, making the human environment easier, safer, and more comfortable. Smart city, smart buildings, digitalization of the manufacturing sector–all these are examples of the introduction of cyber-physical systems, which are becoming more complex and universal every day.

Acknowledgements This paper has been supported by the RUDN University Strategic Academic Leader-ship Program.

References

1. Stellios I, Kotzanikolaou P, Psarakis M, Alcaraz C, Lopez J (2018) A survey of iot-enabled cyberattacks: assessing attack paths to critical infrastructures and services. IEEE Commun Surv Tutor 20(4):3453–3495
2. Maxim K, Anton K, Anastasiya M, Pavel P, Vladimir V (2011) Research of wireless network traffic analysis using big data processing technology. In: 13th international congress on ultra modern telecommunications and control systems and workshops (ICUMT). IEEE, pp 115–121
3. Kushnir D, Kovtsur M, Muthanna A, Kistruga A, Akilov M, Batalov A (2022) Developing instrument for investigation of blockchain technology. In: Robotics and AI for cybersecurity and critical infrastructure in smart cities. Springer, Cham, pp 123–141
4. Branitsky AA, Kotenko IV (2016) Analysis and classification of methods for detecting network attacks. Inform Automat 2(45):207–244
5. Krasov AV, Yagudin IR (2018) Analysis of active network attacks: arp-spoofing and dns-spoofing. In: Actual problems of infotelecommunications in science and education (APINO 2018). pp 520–526
6. Bernacki J, Grzegorz K (2015) Anomaly detection in network traffic using selected methods of time series analysis. Int J Comp Netw Inform Sec 7(9):10–18
7. Shahid MR, et al (2018) IoT devices recognition through network traffic analysis. In: IEEE international conference on big data (big data). IEEE, pp 5187 5192
8. Perminov GV, Zarubin SV (2019) Analysis of the relevance of the problem of passive traffic interception in modern computer networks. Publ Saf Legal Law Order III Millennium 5(2):176–179
9. Gladkikh AM (2020) Basic methods of network traffic analysis. Issues Sci Educ 19(103):23–28
10. Krasov AV, Sakharov DV, Stasyuk AA (2020) Designing an intrusion detection system for an information network using big data. High-tech Technol Earth Space Res 12(1):70–76
11. Yurkin DV, Nikitin VN (2014) Intrusion detection systems in IEEE 802.11 broadband radio access networks. Inform Control Syst 2(69):44–49
12. Targonskaya AI, Tsvetkov AY (2019) Development of a secure web interface for managing devices on the network. In: Actual problems of infotelecommunications in science and education (APINO 2019). pp 734–739
13. Krasov AV, Sakharov DV, Ushakov IA, Losin EP (2017) Ensuring the security of multicast traffic transmission in IP networks. Inform Protect Insider 3:34–42
14. Krasov AV, Levin MV, Shterenberg SI, Isachenkov PA (2016) Methodology of traffic flow management in a software-defined adaptive network. In: Bulletin of the St. Petersburg State University of Technology and Design. Series 1: Natural and Technical Sciences 4. pp 3–8
15. Savinov NV, Tokareva KA, Ushakov IA, Krasnov AV, Sakharov DV (2019) Research of a data center network model based on Cisco ACI policies. Inform Protect Insider 4:32–43
16. Grigoriev VA, Nikitin VN, Kuznetsov VI, Tarakanov SA, Kovtsur MM (2014) Analysis of the bandwidth of the IEEE 1609 radio communication network. Telecommunication 1:13–15
17. Sakharov DV, Shterenberg SI, Levin MV, Kolesnikova YA (2016) Development of a model for ensuring fault tolerance of a data transmission network. In: News of higher educational institutions, vol 34.4. Light Industry Technology, pp 14–20

Printed in the United States
by Baker & Taylor Publisher Services